D0143388

INTRODUCTION TO FRACTURE MECHANICS

Kåre Hellan

University of Trondheim
Norway

McGraw-Hill Book Company

New York St. Louis San Francisco Auckland Bogotá Hamburg
London Madrid Mexico Montreal New Delhi
Panama Paris São Paulo Singapore Sydney Tokyo Toronto

INTRODUCTION TO FRACTURE MECHANICS
INTERNATIONAL EDITION

Copyright © 1985
Exclusive rights by McGraw-Hill Book Co. — Singapore for manufacture and export. This book cannot be re-exported from the country to which it is consigned by McGraw-Hill.

1 2 3 4 5 6 7 8 9 FSP 8 9 4 3 2 1 0 9

Copyright © 1984 by McGraw-Hill, Inc. All rights reserved. Except as permitted under the United States Copyright Act of 1976, no part of this publication may be reproduced or distributed in any form or by any means, or stored in a data base or retrieval system, without the prior written permission of the publisher.

This book was set in Times Roman.
The editor was Madelaine Eichberg.
The production supervisor was Leroy A. Young.

Library of Congress Cataloging in Publication Data

Hellan, Kare, date.
Introduction to fracture mechanics.

Bibliography: p.
Includes index.
1. Fracture mechanics. I. Title.
TA409.H44 1984 620.1'126 83-12022
ISBN 0-07-028048-7

When ordering this title use ISBN 0-07-Y66350-5

Printed in Singapore

CONTENTS

PREFACE

Two modest ambitions and one idiosyncrasy have shaped this book. Its scope represents a double interpretation of the word "Introduction." The presentation of a particular field of applied mechanics to a reader with no previous experience in it (though preferably some in elasticity and plasticity) is coupled with a brief review of research activities through bibliographical references and notes. This second purpose is at odds with the author's opinion that a textbook should appear self-contained and not as a catalogue of references to outside performances. The form chosen has been to keep the introductory text and the introductory review reasonably far apart by placing the bibliographic material in an appendix. An effort has been made to keep mathematical manipulation to a minimum.

Fracture of metals is the physical basis of this text as of most literature, largely as a reflection of economical and technical realities. It must be conceded, however, that metals do normally fracture with some consideration for the analyst in the sense that pertinent properties are *relatively* easily systematized. Contrary cases exist such as composite materials, often having polymeric constituents with time-dependent properties, which are manifestly unruly both individually and as a class. Only a very few references linked to Chaps. 3 and 8 apply to these and other important materials, like ceramics, polymers, rocks, and concrete. It is still worthy of note that the macromechanical treatment, which is the substance of the present text, may be the same even if micromechanisms are totally different. For example, linear fracture mechanics finds direct application among ceramics over a wide range of temperatures, while a concept like the yield zone in metals may have its gross analogs in yield or craze zones in polymers and microcrack zones in rocks and concrete. Thus, even with more exotic fields of application in view, one would profit by going through metals as an introductory matter of reference.

The material is ordered by subject rather than level of complexity. *For rapid insight with minimum distraction by theory or special adaptation the following shorter path may be followed*: *Chap*. 1, *Secs*. 2.1 (*summarily*) *to* 2.6, *Chap*. 3 (*excluding Secs*. 3.3.4 *and* 3.3.5), *Secs*. 4.1 *and* 4.3 *to* 4.6, *Chaps*. 6 *to* 8 (*excluding Sec*. 8.3.2), *and Sec*. 9.1.

Some of the material in this book appeared in a Norwegian text by the author published in 1979. The process of extending, revising, and updating was initiated during the author's stay at Ohio University, College of Engineering and Technology, during the spring of 1981. This opportunity offered by my hosts and the C. Paul and Beth K. Stocker Endowment is gratefully acknowledged, as is the competent assistance in preparing the manuscript on both sides of the ocean by Mrs. Linda Stroh and Miss Gry-Tove Nilsen. I am also indebted to Dr. E. Folmo and Professor N. Ryum for providing the electronmicrographs used in Chap. 8.

Kåre Hellan

INTRODUCTION TO FRACTURE MECHANICS

CHAPTER

ONE

INTRODUCTION

A solid body responds to extreme loading by undergoing large deformation and/or fracture. The second phenomenon, i.e., loss of contact between parts of the body, is the topic of primary interest in fracture mechanics. The concern here is partly with the microscopic mechanisms which govern separation and partly with establishment of fracture criteria and predictions from a macroscopic point of view. In this last respect one of the most interesting parameters is of course the fracture load itself.

As implied above, fracture may occur without warning through large deformation. Such a reaction is possible even with material having a rather high ductility. Obviously, the phenomenon is undesirable—and sometimes catastrophic—supplying strong motivation for the great effort devoted to the study of fracture mechanics during the last few decades.

A total analysis with regard to carrying capacity should therefore include the investigation of a possible fracture before the occurrence of large deformations as well as in conjunction with them. In the last case the critical load may be the limit load or instability load, as determined by a strain analysis of the intact body. Even if a final separation follows in the wake of this critical state of deformation, explicit study of the fracture event can be avoided for such ductile behavior.

The discussion here will be centered on primary fracture, as may occur prior to any state of global collapse. The starting assumption is that in a body there may be sites for crack initiation, introduced by design, e.g., notches, or accident, e.g., flaws due to manufacture of handling. The analyst seeks to investigate whether crack growth can be initiated under a given load at a given site and to predict the further development of the crack.

Some basic concepts referring to such cracking will be defined as follows. The *fracture process zone* is a small region surrounding the crack where fracture develops through the successive stages of inhomogeneous slip, void growth and coalescence, and bond breaking on the atomic scale. Obviously, a macroscopic theory (classical continuum mechanics) cannot describe what happens in this zone. The inverse statement can be used to define the extent of the process zone, and the surface separating this zone from the continuum region will be viewed as the crack surface in discussions to follow. This anticipates an assumption which is essential for the analysis to be limited to a continuum, i.e., that the size of the process zone is *negligible* compared with the macroscopic dimensions of the body, including the crack length. Some implications of this assumption will be touched on later. With these preliminaries, the concepts defined below are of macroscopic content.

The *crack front* is the line connecting all adjacent sites where separation may occur subsequently. During continued separation, this line will move along a geometric surface termed the *fracture surface*. Obviously, the area of this surface, i.e., the developed crack area, will increase as the crack grows. Possible decrease ("crack healing") will not be considered here.

The *fracture mode* designates the separation geometrically. In Irwin's notation, mode I denotes a symmetric opening, the relative displacements between corresponding pairs being normal to the fracture surface, while modes II and III denote antisymmetric separation through relative tangential displacements, normal and parallel to the crack front, respectively. The modes are illustrated in Fig. 1.1. In general, for a crack moving in a homogeneous body the separation can be described as some combination of these modes. The importance of distinguishing between the three will become apparent as we proceed, but it can be noted here that crack growth usually takes place in mode I or close to it. This is probably also the type of separation which is easiest to conceive intuitively.

Less stringent notation is generally applied to the development of fracture by such adjectives as *ductile* on the one hand and *brittle* on the other. These may refer to events on the micro or on the macro scale, and the concepts in the two cases are not uniquely connected. Within the process zone of a crystalline material a brittle fracture is normally understood to be the cleavage of grains (transcrystalline fracture) and/or a separation along grain boundaries (intercrystalline fracture), such that the grain structure will appear in a microscopic view of the

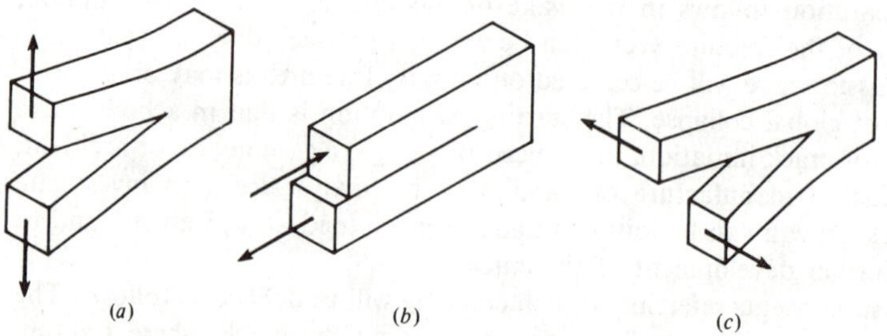

Figure 1.1 The three modes of cracking: (*a*) mode I, (*b*) mode II, and (*c*) mode III.

fractured faces. An expression like *granular fracture* might therefore be more suggestive. A locally ductile fracture is normally understood to be the growth and coalescence of voids formed around inclusions and can be characterized as giving *fibrous* faces. In general a locally ductile, normally fibrous fracture consumes more energy than a locally brittle, normally granular one. In a *global* characterization, a ductile fracture is understood to proceed upon evident permanent deformation (such that the two parts cannot be fitted together after complete fracture), whereas a brittle fracture has the element of surprise mentioned above. A globally ductile fracture clearly depends on the local process being ductile, whereas the analogy is not necessarily true. Globally brittle fracture may evolve even from a locally ductile process if the geometry has served to concentrate sufficient energy at the separation site (deep cracklike notches). One should be aware of this duplicate notation (Table 1.1) to avoid confusion when reading the literature. It must also be recognized that intermediate or atypical forms of fracture may occur. A special type of crack growth associated with fatigue will be discussed later.

Table 1.1 Notation

Development	Ductile	Brittle
Local	High-energy Fibrous ↓	Low-energy Granular →↓
Global	Large deformation	Small deformation

Later in the book we first investigate stresses and displacements near the edge of a cracklike notch or flaw, also termed a stationary crack, in order to gain some insight into quantities which may be assumed to contribute to a critical state. Actions will then be assumed to be quasi-static and monotonically increasing, whereas some effects of dynamic loading will be mentioned later. A critical state is understood to be one which governs the onset or continuation of crack growth. Criteria for determining whether or not a state is critical may have different forms, depending on the conditions (material, type of action, stage of growth, environment, etc.) and certain underlying assumptions. Fracture may evolve quasi-statically (stably, slowly, controlled) or dynamically; in the latter case the crack front can attain large velocities, comparable to that of sound, with total failure as a near consequence.

Whereas a possible slow fracture is linked to a monotonic action in the first discussion, we shall proceed later to that type of crack propagation which takes place under a pulsating or cyclic load. This phenomenon of fatigue is not easily accessible through basic principles, but there is a large body of experimental data upon which semiempirical predictions can be based.

A further discussion should pertain in some detail to experimental fracture mechanics, both on the macro and the micro scale. Such experience is, of course, a necessary basis for all analysis and should be referred to whenever needed. Due attention should also be paid to the general background provided by the theories of elasticity and plasticity; some of this background is relegated to the appendixes to avoid interrupting the main exposition. Some guidelines and experiences related to applied fracture mechanics form a natural continuation. A detailed bibliography is provided in Appendix H for reference and further study.

To illustrate the potential of simple analysis and to raise some questions which must be dealt with later, we conclude this introduction by looking into an elementary exercise in concepts of fracture mechanics.

Example 1.1 An elastic plate of unit thickness is attached along its upper and lower edge to horizontal guides (Fig. 1.2). A horizontal crack extends from one side through the thickness of the plate. With the notation of Fig. 1.2, $a \gg H$ and $b \gg H$ are assumed. We want to find how far apart the guides must be moved, as expressed by a critical value of their relative displacement u, for the existing crack to propagate.

With the assumptions made, we roughly consider the material above and below the crack to be free of stress, while the plate to the right of the crack has a constant stress corresponding to the vertical strain $\varepsilon = u/H$. (Conditions near such a crack tip will be studied extensively later.) From Hooke's law this stress is found to be

$$\sigma = \eta E \varepsilon = \frac{\eta E u}{H} \tag{1.1}$$

where E is Young's modulus; η may be regarded as a known multiplier depending on Poisson's ratio ν and the geometry

$$\eta = \frac{1}{1-\nu^2} \qquad \text{or} \qquad \eta = \frac{1-\nu}{(1+\nu)(1-2\nu)} \tag{1.2}$$

under vanishing stress or strain normal to the plane of the plate. These are the limiting cases, plane stress or plane strain, respectively, when the plate being considered is very thin or just a slice of a cylindrical body having large

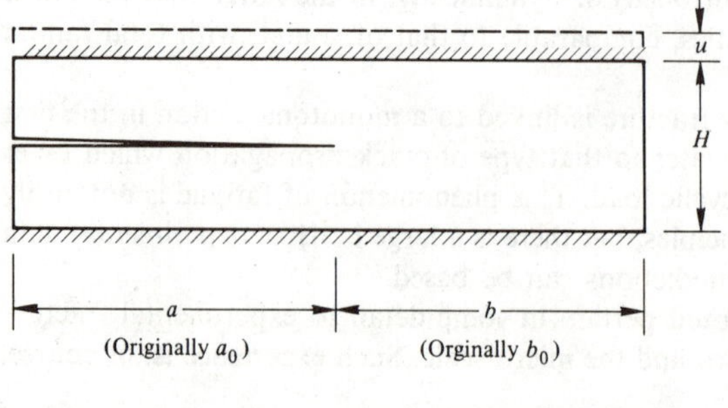

Figure 1.2 A cracked striplike specimen under translational action.

extension normal to the paper.† By the displacement, the plate with its crack will have the following supply of potential elastic-strain energy:

$$\Phi = \tfrac{1}{2}(\text{force} \times \text{relative displacement})$$
$$= \frac{1}{2}(\sigma b)u = \frac{1}{2}\left(\eta E \frac{u}{H} b\right)u = \frac{\eta E b}{2H}u^2 \tag{1.3}$$

Assume now that the crack has started to grow from its initial length a_0 through the whole remaining ligament so that fracture is complete. If this happens without kinetic-energy production, i.e., quasi-statically, the potential energy Φ will gradually approach zero since the fractured pieces obviously are free of stress. At each stage of the process (when $b_0 > b \gg H$) Φ is given by expression (1.3).

We are then left with the question: What has happened to the original supply of potential energy Φ? The answer can only be that it has been consumed during crack propagation. Specifically, this means that work has been done on the process zone with the result that new crack surfaces have been generated. In 1920 Griffith [2] also considered this work from a microscopic viewpoint, as that needed to take neighboring atoms out of their mutual fields of attraction. From this point of view the work of separation, as referred to each unit of new crack surface, should appear as a material constant. Denoting it by γ, we complete the analysis on the basis suggested by Griffith as follows.

With Φ given by (1.3) we see that per unit of generated fracture surface (negative unit of b) there is a loss of potential energy equal to

$$\frac{\eta E}{2H}u^2 \equiv \mathcal{G} \tag{1.4}$$

In a general context this is often termed the *crack driving force*. The symbol $\mathcal{G}$ relates to Griffith, who claimed that this quantity must also equal the work 2γ necessary to move the crack front through a unit area when the state is critical. The factor of 2 arises because the crack surface doubles the fracture surface. By the equality

$$\frac{\eta E}{2H}u^2 = 2\gamma$$

we then obtain the critical displacement

$$u = 2\sqrt{\frac{\gamma H}{\eta E}} \tag{1.5}$$

as the solution to the problem posed.

†A constraint which defines the constant η is the neglect of all longitudinal strain. This is dictated by the wall contact. The reader may wish to verify the detailed expressions.

The procedure in such analyses will be discussed in more general terms later, where it will appear that (1.5) is a rather special result in that the critical value turns out to be independent of the crack length.

With regard to the behavior of real materials, however, not all problems have been cleared away by the simple reasoning above. It might well be that an increase of u past the critical value would be possible without causing propagation because it would take more to initiate growth than to provide the potential for further motion (initiation barrier). It is well established that most materials do not fracture through the ideal cleavage assumed by Griffith, and that much of the potential energy will be consumed in plastic dissipation outside the process zone. These facts may suggest some modification of the Griffith hypothesis or the use of quite different criteria if fracture is to be realistically predicted. The simple example and the comments upon it are intended to illustrate some of the problems and uncertainties which remain to be discussed.

CHAPTER

TWO

THE STATIONARY CRACK UNDER STATIC LOADING

Our first aim is to study local fields near the sharp edge of a cracklike notch or flaw in the body. We assume for now some quasi-static action due to the monotonic increase of a characteristic load or displacement, so that dynamic effects and (for plastic states) local unloading can be disregarded. After a few typically elastic solutions have been reviewed, attention will be focused on plastic fields, which must appear in these cases of intense strain. Our description will be based on the usual first-order theory, whereby all displacements and their gradients are formally treated as small. Isothermal conditions are assumed, in bodies which are homogeneous and isotropic in the domain considered.

2.1 TWO-DIMENSIONAL ELASTIC FIELDS AS A POWER-SERIES SOLUTION

The following problem is the first to be addressed. A plate in a state of plane stress or plane strain is bounded by two concurrent straight edges defining the solid within the angle 2α and by remote boundaries otherwise (Fig. 2.1). In the absence of body forces equilibrium is satisfied identically through the relations

$$\sigma_\theta = \frac{\partial^2 \chi}{\partial r^2} \qquad \sigma_r = \nabla^2 \chi - \sigma_\theta \qquad \tau_{r\theta} = -\frac{\partial}{\partial r}\left(\frac{1}{r}\frac{\partial \chi}{\partial \theta}\right) \tag{2.1}$$

where σ_θ, σ_r, and $\tau_{r\theta}$ are the polar stress components and χ is the Airy stress function (Appendix A). When the plate is linearly elastic, the Airy stress function must satisfy the bipotential equation

$$\nabla^2(\nabla^2 \chi) = 0 \tag{2.2}$$

where the operator ∇^2 is in polar coordinates

$$\nabla^2 \equiv \frac{\partial^2}{\partial r^2} + \frac{1}{r}\frac{\partial}{\partial r} + \frac{1}{r^2}\frac{\partial^2}{\partial \theta^2} \tag{2.3}$$

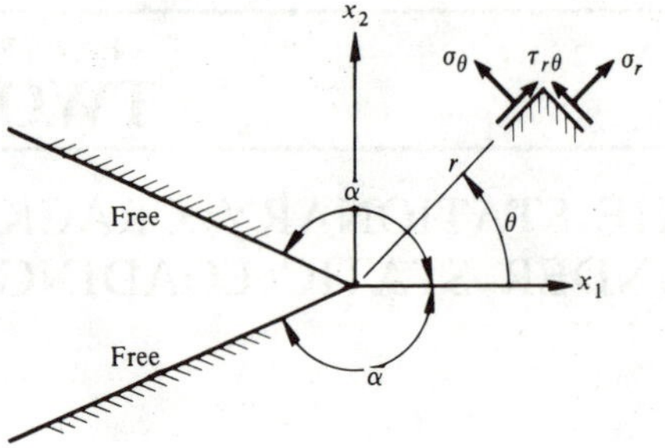

Figure 2.1 Polar stresses at a wedgelike notch.

To solve (2.2) under the present boundary conditions (Fig. 2.1)

$$\sigma_\theta = \tau_{r\theta} = 0 \qquad \text{for } \theta = \pm\alpha \tag{2.4}$$

we follow Williams [9] and assume the product form

$$\chi = r^{\lambda+1} f(\theta) \tag{2.5}$$

This is introduced in (2.2) and (2.4) to give the equations which determine the constant λ and the function f

$$\frac{d^4 f}{d\theta^4} + 2(\lambda^2+1)\frac{d^2 f}{d\theta^2} + (\lambda^2-1)^2 f = 0 \tag{2.6}$$

$$f = \frac{df}{d\theta} = 0 \qquad \text{for } \theta = \pm\alpha \tag{2.7}$$

The general solution of (2.6) is

$$f = C_1 \cos(\lambda-1)\theta + C_2 \sin(\lambda-1)\theta + C_3 \cos(\lambda+1)\theta + C_4 \sin(\lambda+1)\theta \tag{2.8}$$

to be substituted in (2.7). This leads to a homogeneous set of equations for the constants C_1 to C_4. Obviously, (2.8) consists of a symmetric part (first and third terms) and an antisymmetric part (second and fourth); the corresponding form of Eqs. (2.7) after suitable addition and subtraction is

$$\begin{gathered}\begin{bmatrix} \cos(\lambda-1)\alpha & \cos(\lambda+1)\alpha \\ (\lambda-1)\sin(\lambda-1)\alpha & (\lambda+1)\sin(\lambda+1)\alpha \end{bmatrix}\begin{bmatrix} C_1 \\ C_3 \end{bmatrix} = \begin{bmatrix} 0 \\ 0 \end{bmatrix} \\ \begin{bmatrix} \sin(\lambda-1)\alpha & \sin(\lambda+1)\alpha \\ (\lambda-1)\cos(\lambda-1)\alpha & (\lambda+1)\cos(\lambda+1)\alpha \end{bmatrix}\begin{bmatrix} C_2 \\ C_4 \end{bmatrix} = \begin{bmatrix} 0 \\ 0 \end{bmatrix}\end{gathered} \tag{2.9}$$

A nontrivial solution is possible only when the determinants of the equations vanish. This leads to the respective characteristic equations for the eigenvalue λ

$$\begin{gathered}\lambda \sin 2\alpha + \sin 2\lambda\alpha = 0 \\ -\lambda \sin 2\alpha + \sin 2\lambda\alpha = 0\end{gathered} \tag{2.10}$$

To simulate a crack we let α approach π. Then Eqs. (2.10) both reduce to

$$\sin 2\pi\lambda = 0 \tag{2.11}$$

which has only real roots

$$\lambda = \tfrac{1}{2}n \qquad n = \text{integer} \tag{2.12}$$

It will be shown later that the solution will contain unacceptable singularities if n is nonpositive. To each of the remaining eigenvalues $\lambda = \frac{1}{2}n$, $n = 1, 2, 3, \ldots$, there will correspond a relationship between C_{1n} and C_{3n} or C_{2n} and C_{4n} in (2.9). The result is:

$$\begin{aligned} &\text{For } n = 1, 3, 5, \ldots: \qquad C_{3n} = -\frac{n-2}{n+2}C_{1n} \qquad C_{4n} = -C_{2n} \\ &\text{For } n = 2, 4, 6, \ldots: \qquad C_{3n} = -C_{1n} \qquad C_{4n} = -\frac{n-2}{n+2}C_{2n} \end{aligned} \tag{2.13}$$

and thus

$$\begin{aligned} \chi = \sum_{n=1,3,\ldots} r^{1+n/2}\bigg[& C_{1n}\left(\cos\frac{n-2}{2}\theta - \frac{n-2}{n+2}\cos\frac{n+2}{2}\theta\right) \\ & + C_{2n}\left(\sin\frac{n-2}{2}\theta - \sin\frac{n+2}{2}\theta\right)\bigg] \\ + \sum_{n=2,4,\ldots} r^{1+n/2}\bigg[& C_{1n}\left(\cos\frac{n-2}{2}\theta - \cos\frac{n+2}{2}\theta\right) \\ & + C_{2n}\left(\sin\frac{n-2}{2}\theta - \frac{n-2}{n+2}\sin\frac{n+2}{2}\theta\right)\bigg] \end{aligned} \tag{2.14}$$

where the terms multiplied by C_{1n} are symmetric (mode I) and the terms with C_{2n} are antisymmetric (mode II) with respect to $\theta = 0$.

The stresses follow by introducing (2.14) in (2.1). Those written below are the terms corresponding to $n = 1$, which are the only singular ones (as $r^{-1/2}$) in $r = 0$ and are therefore dominant near the crack tip:

$$\begin{aligned} \text{Mode I:} \qquad & \sigma_r = \frac{C_{11}}{4}r^{-1/2}\left(5\cos\frac{\theta}{2} - \cos\frac{3\theta}{2}\right) + \cdots \\ & \sigma_\theta = \frac{C_{11}}{4}r^{-1/2}\left(3\cos\frac{\theta}{2} + \cos\frac{3\theta}{2}\right) + \cdots \\ & \tau_{r\theta} = \frac{C_{11}}{4}r^{-1/2}\left(\sin\frac{\theta}{2} + \sin\frac{3\theta}{2}\right) + \cdots \end{aligned} \tag{2.15}$$

Mode II:
$$\begin{aligned}\sigma_r &= \frac{C_{21}}{4} r^{-1/2}\left(-5\sin\frac{\theta}{2} + 3\sin\frac{3\theta}{2}\right) + \cdots \\ \sigma_\theta &= \frac{C_{21}}{4} r^{-1/2}\left(-3\sin\frac{\theta}{2} - 3\sin\frac{3\theta}{2}\right) + \cdots \\ \tau_{r\theta} &= \frac{C_{21}}{4} r^{-1/2}\left(\cos\frac{\theta}{2} + 3\cos\frac{3\theta}{2}\right) + \cdots\end{aligned} \tag{2.16}$$

The θ-dependence according to (2.15) is illustrated in Fig. 2.26*a*.

Further, the displacements u_r (radially; Fig. 2.2) and u_θ (circumferentially) are found by integrating the kinematical relationships (Appendix A)

$$\varepsilon_r = \frac{\partial u_r}{\partial r} \qquad \varepsilon_\theta = \frac{u_r}{r} + \frac{1}{r}\frac{\partial u_\theta}{\partial \theta} \qquad \gamma_{r\theta} = \frac{1}{r}\frac{\partial u_r}{\partial r} + \frac{\partial u_\theta}{\partial r} - \frac{u_\theta}{r} \tag{2.17}$$

The strains here can be expressed in terms of the stresses by Hooke's law

$$\begin{aligned}&E\varepsilon_r = \sigma_r - \nu\sigma_\theta \qquad E\varepsilon_\theta = \sigma_\theta - \nu\sigma_r \qquad G\gamma_{r\theta} = \frac{E}{2(1+\nu)}\gamma_{r\theta} = \tau_{r\theta} \\ &2G\varepsilon_r = (1-\nu)\sigma_r - \nu\sigma_\theta \qquad 2G\varepsilon_\theta = (1-\nu)\sigma_\theta - \nu\sigma_r \qquad G\gamma_{r\theta} = \tau_{r\theta}\end{aligned} \tag{2.18}$$

in plane stress and plane stress, respectively. All this leads to

Mode I:
$$\begin{aligned}u_r &= \frac{C_{11}}{4G} r^{1/2}\left[(2\kappa - 1)\cos\frac{\theta}{2} - \cos\frac{3\theta}{2}\right] + \cdots \\ u_\theta &= \frac{C_{11}}{4G} r^{1/2}\left[-(2\kappa + 1)\sin\frac{\theta}{2} + \sin\frac{3\theta}{2}\right] + \cdots\end{aligned} \tag{2.19}$$

Mode II:
$$\begin{aligned}u_r &= \frac{C_{21}}{4G} r^{1/2}\left[-(2\kappa - 1)\sin\frac{\theta}{2} + 3\sin\frac{3\theta}{2}\right] + \cdots \\ u_\theta &= \frac{C_{21}}{4G} r^{1/2}\left[-(2\kappa + 1)\cos\frac{\theta}{2} + 3\cos\frac{3\theta}{2}\right] + \cdots\end{aligned} \tag{2.20}$$

where

$$\kappa = \begin{cases} \dfrac{3-\nu}{1+\nu} & \text{plane stress} \\ 3 - 4\nu & \text{plane strain} \end{cases}$$

Additional arbitrary rigid-body displacements may be superposed.

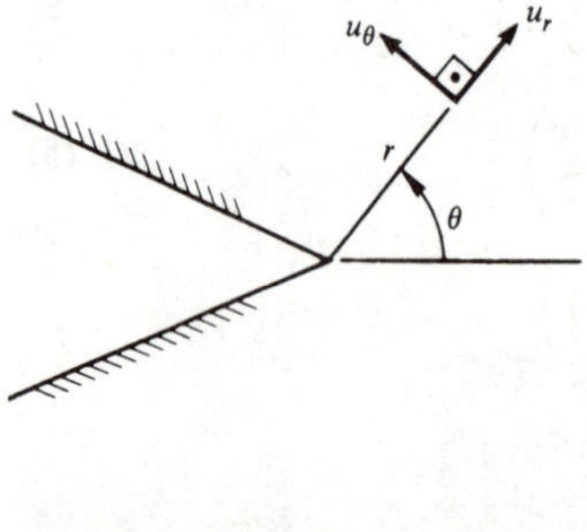

Figure 2.2 Polar displacements at a wedgelike notch.

The functional forms of u_r and u_θ suggest that all roots $\lambda < 0$ would produce infinite displacements in $r = 0$, since the multiplier generalizing $r^{1/2}$ in (2.19) and (2.20) is r^λ. We therefore reject these and concentrate on $\lambda = 0$. This root gives stresses and strains of the types

$$\sigma_{ij} = r^{-1}(\text{functions of } \theta) \qquad \varepsilon_{ij} = r^{-1}(\text{functions of } \theta)$$

and hence for the strain energy density ϕ (Appendix A)

$$\phi = r^{-2}(\text{function of } \theta)$$

But then the total strain energy within any circular area $r < R$ enclosing the crack front would become unbounded

$$\Phi = \int_0^{2\pi}\int_{r_0}^{R} \phi r \, dr \, d\theta \rightarrow \infty \qquad \text{as } r_0 \rightarrow 0$$

which is considered physically unfeasible. For these reasons all roots $\lambda \leq 0$ must be excluded from the solution.

Next we investigate the following situation. A body is bounded by two intersecting planes defining the solid within the angle 2α and by remote surfaces otherwise. The states of stress and strain are independent of the coordinate $x_3 \equiv z$ measured along the line of intersection; this results from the state's being *antiplane* in the sense that $u_1 = u_2 = \sigma_{11} = \sigma_{22} = \sigma_{33} = \sigma_{12} = 0$. The remaining nonzero stress components σ_{13} $(=\sigma_{31})$ and σ_{23} $(=\sigma_{32})$ can be transformed into τ_{rz} and $\tau_{\theta z}$ on the cylindrical coordinate planes in Fig. 2.3. The only nonzero displacement is $u_3 \equiv u_z$. To illustrate, such states may be closely realized in cases of torsion, near shallow notches around the periphery, or along a generator like that indicated in Fig. 2.4.†

Hooke's law and the kinematic constraint are expressed by

$$\tau_{rz} = G\gamma_{rz} = G\frac{\partial u_z}{\partial r} \qquad \tau_{\theta z} = G\gamma_{\theta z} = \frac{G}{r}\frac{\partial u_z}{\partial \theta} \tag{2.21}$$

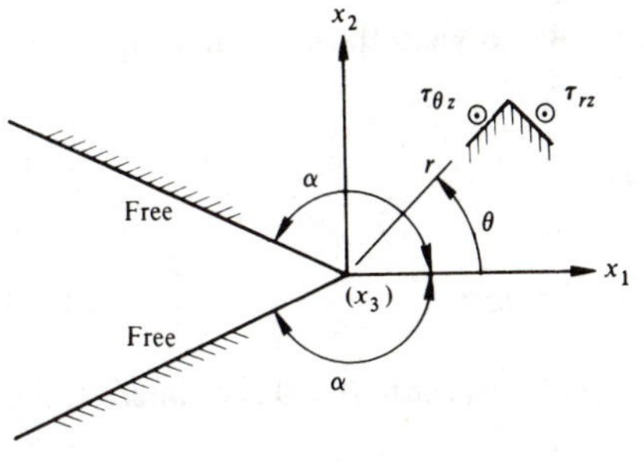

Figure 2.3 Antiplane stress components.

†In Fig. 2.4*b* displacements are also present in the plane of the section, as derived by St. Venant's hypothesis, for example. They can readily be shown to have a negligible influence on the solution in question.

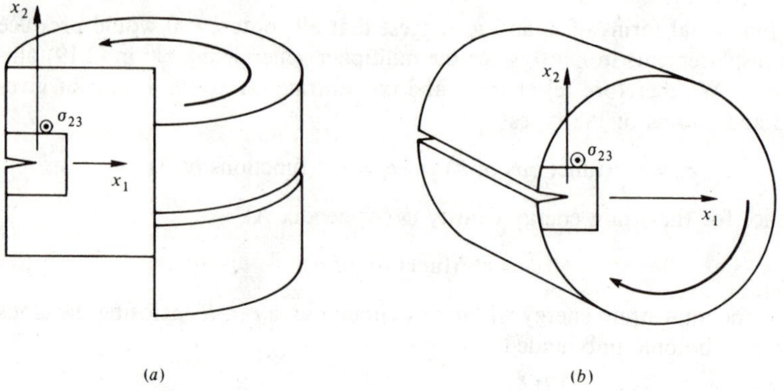

Figure 2.4 Cracks under mode III action by torsion.

Introducing these into the equation of equilibrium (Appendix A)

$$\frac{\partial}{\partial r}(r\tau_{rz}) + \frac{\partial \tau_{\theta z}}{\partial \theta} = 0 \tag{2.22}$$

gives the potential equation for u_z [compare (2.3)]

$$\nabla^2 u_z = 0 \tag{2.23}$$

This equation should be solved under the boundary conditions

$$\tau_{\theta z} = 0 = \frac{\partial u_z}{\partial \theta} \qquad \text{for } \theta = \pm\alpha \tag{2.24}$$

according to Fig. 2.3. The product form

$$u_z = r^{\omega} g(\theta) \tag{2.25}$$

is assumed and introduced in (2.23) and (2.24), to yield the governing equations for ω and g

$$\frac{d^2 g}{d\theta^2} + \omega^2 g = 0 \tag{2.26}$$

$$\frac{dg}{d\theta} = 0 \qquad \text{for } \theta = \pm\alpha \tag{2.27}$$

Since only the antisymmetric displacement with respect to $\theta = 0$ is of interest, the solution of (2.26) is clearly of the type

$$g = D \sin \omega\theta \tag{2.28}$$

where D is a constant. This is inserted in (2.27) to give the characteristic equation for ω

$$\cos \alpha\omega = 0 \tag{2.29}$$

Again moving α toward π to simulate a crack, we obtain the eigenvalues

$$\omega = \tfrac{1}{2}n \qquad n = 1, 3, 5, \ldots \tag{2.30}$$

pertinent to the present physical problem in mode III. The respective constants D_n in (2.28) correspond, so that with (2.25) and (2.21) we finally have the displacement and stresses

$$u_z = D_1 r^{1/2} \sin\frac{\theta}{2} + \cdots \qquad \tau_{rz} = \frac{G}{2} D_1 r^{-1/2} \sin\frac{\theta}{2} + \cdots \qquad \tau_{\theta z} = \frac{G}{2} D_1 r^{-1/2} \cos\frac{\theta}{2} + \cdots \tag{2.31}$$

where only the terms dominant at the crack tip have been written.

Stresses along the extended crack line and displacements along the crack sides will be of particular importance. We first rename the coefficients C_{11}, C_{21}, and D_1 as follows:

$$C_{11} \equiv \frac{K_{\mathrm{I}}}{\sqrt{2\pi}} \qquad C_{21} \equiv \frac{K_{\mathrm{II}}}{\sqrt{2\pi}} \qquad D_1 \equiv \frac{K_{\mathrm{III}}}{G}\sqrt{\frac{2}{\pi}} \tag{2.32}$$

Then we can write the dominant terms of later interest.

Mode I, $x_2 = 0$:

For $x_1 \geq 0$: $$\sigma_{22} = \sigma_\theta(\theta = 0) = \frac{K_{\mathrm{I}}}{\sqrt{2\pi x_1}} = \sigma_r(\theta = 0) = \sigma_{11}$$

For $x_1 \leq 0$: $$u_2^{\pm} = -u_\theta(\theta = \pm\pi) = \pm\frac{\kappa + 1}{2G} K_{\mathrm{I}} \sqrt{\frac{-x_1}{2\pi}}$$

Mode II, $x_2 = 0$:

For $x_1 \geq 0$: $$\sigma_{21} = \tau_{r\theta}(\theta = 0) = \frac{K_{\mathrm{II}}}{\sqrt{2\pi x_1}}$$

For $x_1 \leq 0$: $$u_1^{\pm} = -u_r(\theta = \pm\pi) = \pm\frac{\kappa + 1}{2G} K_{\mathrm{II}} \sqrt{\frac{-x_1}{2\pi}} \tag{2.33}$$

Mode III, $x_2 = 0$:

For $x_1 \geq 0$: $$\sigma_{23} = \tau_{\theta z}(\theta = 0) = \frac{K_{\mathrm{III}}}{\sqrt{2\pi x_1}}$$

For $x_1 \leq 0$: $$u_3^{\pm} = u_z(\theta = \pm\pi) = \pm\frac{2K_{\mathrm{III}}}{G} \sqrt{\frac{-x_1}{2\pi}}$$

$$\kappa = \begin{cases} \dfrac{3-\nu}{1+\nu} & \text{in plane stress} \\ 3-4\nu & \text{in plane strain} \end{cases}$$

referred to Fig. 2.5.†

The three scalars K_I to K_{III} denote the strengths of the singularities and are termed the *stress intensity factors*. They will characterize the dominant elastic fields near the crack tip completely, since the distribution is otherwise fully defined. It is rather easy to demonstrate that the same type of near-tip solution will appear even when body forces are present and when the crack surface is subjected to tractions having finite intensity.

2.2 TWO- AND THREE-DIMENSIONAL ELASTIC FIELDS: SOLUTIONS ON AN ANALYTICAL OR NUMERICAL BASIS

A typical feature of the stress field at any location along an arbitrarily curved crack front is the existence of three possible singularities, associated with relative displacements (1) normal to the crack surface, (2) tangential to the crack front, and (3) normal to those directions.‡ Locally this corresponds to modes I, III, and

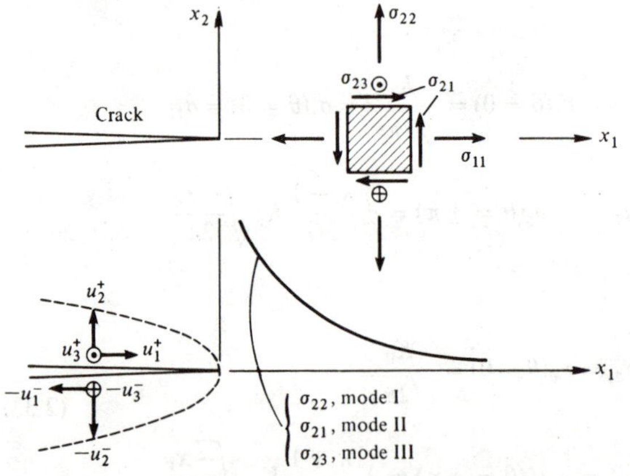

Figure 2.5 Cartesian stress and displacement components at a crack tip.

†These singularities are derived within the frame of the applied first-order small-deformation theory. Instead of being physically defined, they are thus a mathematical consequence, since the results indicate that the displacement gradients near the crack tip should be regarded as large, after all. By including this last alternative in the analysis one might arrive at a more detailed description near the crack tip, with values of the stresses which were finite but still large. In reality however, the onset of yield would invalidate even this elaborate (but elastically based) analysis.

‡This can be shown to hold if the crack surface coincides locally with the osculating plane of the crack front and if isotropy prevails.

II of displacement, as discussed above. Essentially the types of the singularities are also the same. This means that the curvature of the crack front influences the local fields only through the stress intensity factors, as do the geometry of the body and the boundary conditions otherwise.

Therefore the analysis in Sec. 2.1 has already disclosed the general character of the fields which dominate the solution near the crack front, in the plane as well as in space. It is reasonable, as will be discussed later, that this is just that part of the solution which serves to decide whether the state is critical at the crack tip in an essentially elastic body. It remains then to determine the strengths K_{I} to K_{III} when special geometries and types of action are to be considered.

One method of evaluation is based directly on the power-series solution in Sec. 2.1. Then the differential equation and the boundary conditions at the crack sides are satisfied, while the remote boundary conditions still must be met. With a reasonably truncated form of the series and a chosen scheme of approximation (pointwise satisfaction, error minimization, etc.) one then arrives at the set of algebraic equations which determines C_{1n}, C_{2n}, and D_n $(n = 1, 2, 3, \ldots)$. This includes K_{I} to K_{III} through (2.32). Results from such an analysis (in direct or modified versions) are presented in cases 5, 6, 9, and 10 of Appendix C.

In a closed analytical form there exist a few exact solutions applying to cracks in infinitely large bodies (in practice, small cracks). Solutions have been derived on the basis of complex-variable techniques due to Muskhelishvili [17] and Westergaard [18] (compare Appendix B). Further development through the work of Isida [22–24] and others has made it possible to consider finite geometries through the use of conformal mapping. The stress intensity factors are not principal unknowns in such calculations but can easily be deduced. Some typical results are reproduced in cases 11 and 14 to 19 of Appendix C.

To evaluate the strengths K_{I} to K_{III} along the whole length of an arbitrarily curved crack front is quite prohibitive in practice, but results for typical points and simpler geometries are available. In particular, for elliptical internal cracks and edge (thumbnail) cracks we point to classical results as cases 1 and 2 in Appendix C, presented by Irwin [10] on the basis of earlier work by Green and Sneddon [11], [12].

The results in Appendix C must apply equally well if a uniform tension or compression is applied parallel to the plane of the crack. Nor can any other stress field influence the intensity factors if it is superposed with finite intensities. In particular, such a field might be taken as a uniform compression normal to the plane of the crack, having the intensity σ_∞. This explains why expressions such as (C.1) to (C.5) and (C.7) are also valid for the respective geometries if the remote load σ_∞ is replaced by a compressive load σ_∞ acting on the crack faces.

Among numerical methods for determining the stress intensity factors, besides those applying the power series, finite-element methods should be particularly noted. The body is then considered to be discretized as an assemblage of finite elements, and the solution is interpolated between typical primary variables associated with nodal points common to adjacent elements. Two varieties have attracted attention: (1) methods which make use of *singular elements*, which by design include the typical singularities, and (2) energy-based methods which use a relation to be developed between the loss of potential energy (compare $\mathcal{G}$ in the introductory example) and the stress intensity factors. Methods having the latter basis can be grouped together as *compliance methods*. By using methods of type 2 even simple analytical solutions and experimentally based solutions can be established, as discussed later. A related family of methods utilizes Betti's reciprocity theorem to derive stress intensity factors if they are known for other loadings of the considered geometry. As a final alternative of rising importance boundary-integral methods should be mentioned.

Obviously the number of approaches and varieties is large. Those who want to go into the details might start by consulting the discussion in the Bibliography for Chap. 2. Such information is needed if the problem under consideration is not covered by published data.

Finally, it should be mentioned that formulas (C.3), (C.4), (C.7), and (C.8) have been derived by interpolation between results for shallow and very deep notches. Between these extremes one can anticipate errors up to ± 2 percent.

2.3 A QUALITATIVE DISCUSSION OF YIELDING NEAR A CRACK FRONT

Since no real material could carry all the large stresses derived by the elasticity assumption, a plastification extending from the crack tip must take place. In the remainder of this chapter we shall consider various aspects of this yielding, both its internal structure and its extent. First, however, some qualitative feeling for the mode I yielding of plates should be established. Fig. 2.6 is suggestive of the plastic zone if the plate is typically thick or typically thin.

If B is large, we can distinguish between an internal region with approximately

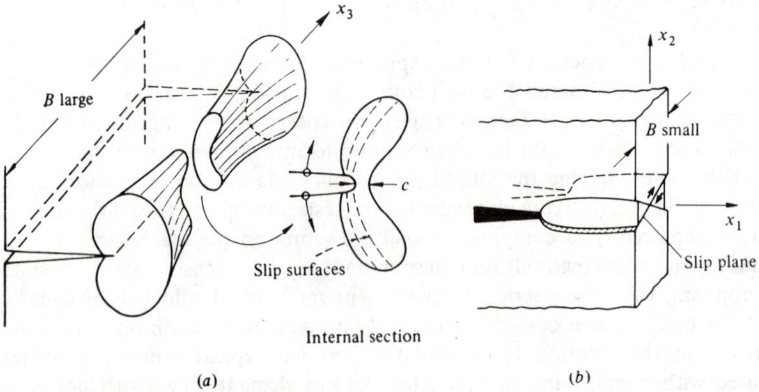

Figure 2.6 Yield zones and mechanisms at the front of a crack through a plate: (*a*) thick; (*b*) thin.

plane strain ($\varepsilon_{33} \approx 0$), the yielding being partly suppressed due to a kinematic constraint from the surrounding elastic material, and an external region where the surface condition ($\sigma_{33} \approx 0$) provides a larger degree of freedom for extended yield. With increasing B it is likely that the relative importance of the external region will be decreasing, as regards fracture strength, for example. Internally it is reasonable for the plastic deformation to be associated with relative motion along curved surfaces parallel to the crack front. This is indicated in Fig. 2.6*a* (the hinge mechanism), together with an approximate shape of the yield zone (discussed later). Altogether the motion corresponds to the indicated opening of the crack.

If B is very small, on the other hand, $\sigma_{33} \approx 0$ will be approximately satisfied through the whole thickness. As $\sigma_{22} \geq \sigma_{11} > 0$ should be expected, this implies (exactly for a material responding to the Tresca criterion) that yielding will be produced essentially by cumulative slip in planes parallel to x_1, making 45° with the plane of the plate. This will limit the height of the yield zone to the order of magnitude of the plate thickness. The total motion has a necking effect in front of the crack as it opens.

2.4 IRWIN'S APPROXIMATION OF THE SIZE OF THE PLASTIC ZONE

It is natural to start this review by referring to an approximate and simple derivation of the yield length along the crack direction introduced by Irwin [93]. An initial observation will be that the yielding must cut off the elastic singular peak of the stress distribution near the crack tip. Considering a plate of elastoplastic material in mode I, this means that σ_{22} in Fig. 2.7 should be represented more realistically by curve 2 and 3 than by the elastic distribution (curve 1). Here σ_Y is the yield stress in uniaxial tension and k is a multiplying constant discussed below.

Equilibrium in the x_2 direction requires that curve 2 be displaced to the right of curve 1. Assume that this is a *pure translation* through the length r_2 and that

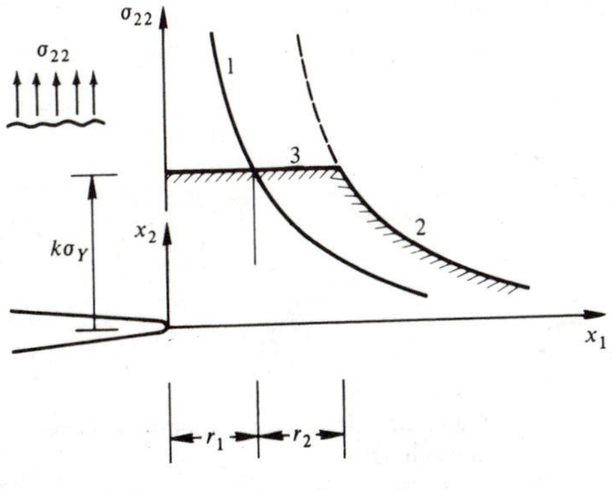

Figure 2.7 Elastic (curve 1) and assumed elastoplastic (curve 2 and 3) stress distribution in front of a crack.

curve 1 can be approximated by the *singular first term* of the elastic power-series solution. Continued equilibrium is then expressed by

$$\int_0^{r_1}\left(\frac{K_I}{\sqrt{2\pi x_1}} - k\sigma_Y\right)dx_1 = k\sigma_Y r_2 \tag{2.34}$$

where r_1 denotes the distance from the crack tip to the point of intersection between curves 1 and 2 and is given by

$$\frac{K_I}{\sqrt{2\pi r_1}} = k\sigma_Y \qquad \text{or} \qquad r_1 = \frac{1}{2\pi}\left(\frac{K_I}{k\sigma_Y}\right)^2 \tag{2.35}$$

Inserting this in (2.34) gives finally

$$r_2 = r_1 \tag{2.36}$$

With $\sigma_{22} \geq \sigma_{11} > 0$ the yield condition (by Tresca) in plane stress $\sigma_{33} = 0$ is

$$\sigma_{22} = \sigma_Y \qquad \text{such that } k = 1 \tag{2.37}$$

This represents the properties of a very thin plate. In the same context $\sigma_{33} > 0$ must appear under conditions of plane strain, when the yield condition would accordingly require $\sigma_{22} > \sigma_Y$. Irwin suggested

$$k = \sqrt{3} \tag{2.38}$$

and found it confirmed by the yield length, observed to be about so much smaller in thick plates. The resulting extent $c = r_1 + r_2$ of the plastic zone has then been proposed as

$$c = \begin{cases} \dfrac{1}{\pi}\left(\dfrac{K_I}{\sigma_Y}\right)^2 & \text{thin plate} \qquad (2.39) \\[2ex] \dfrac{1}{3\pi}\left(\dfrac{K_I}{\sigma_Y}\right)^2 & \text{thick plate} \qquad (2.40) \end{cases}$$

Exactly the same reasoning can be applied to mode III, σ_{22} being replaced by the shear stress σ_{23} and $k\sigma_Y$ being replaced by the shear yield stress τ_Y (Fig. 2.8).

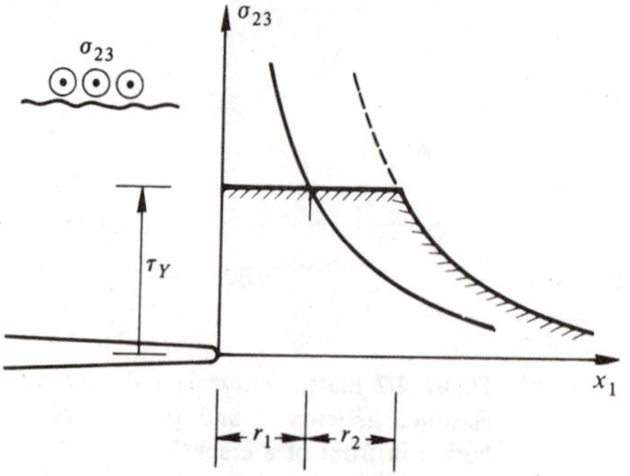

Figure 2.8 Elastic and elastoplastic stresses in mode III.

This leads to the plastic-zone size along the crack direction

$$c = 2r_1 = 2r_2 = \frac{1}{\pi}\left(\frac{K_{\text{III}}}{\tau_Y}\right)^2 \tag{2.41}$$

These results will be compared with exact results as we proceed.† It should be emphasized here that the aim of Irwin analysis can only be to consider a plastification which is highly localized at the crack tip. This is because the analysis has been confined to that region where the singular elastic term would dominate all others, as noted above. Such a type of yielding will also be referred to as *small-scale yielding*.

2.5 THE DUGDALE MODEL

Using the so-called *Dugdale model* [94], we consider the yielding in mode I of a thin plate made from an elastic–ideally plastic material. The analysis thus duplicates part of the Irwin study discussed above and offers a check on that result since it will be exact for a material responding to the Tresca criterion. To begin with, we shall assume a loading case symmetric with respect to the midnormal of the crack and to the crack line itself but otherwise arbitrary.

As discussed in the context of Fig. 2.6*b*, the height of the plastic zone is limited to the plate thickness if the plate is thin and of the Tresca type. To cultivate the assumption of exact plane stress, $\sigma_{33} = 0$, one should further imagine the transition toward an infinitely thin plate and, by consequence, look upon the yield zone as a zero-thickness strip emanating from the crack tip. Thus, the half plane $x_2 > 0$ is to be considered as purely elastic, loaded by the traction $\sigma_{22} = \sigma_Y$ along the yield strip.

Next the length of the yield strip can be determined by the following reasoning (see Fig. 2.9*a*). If $\sigma_{22} = 0$ along the whole length $|x_1| < a + c$, $x_2 = 0$, the loading of the plate will produce positive singularities of strength $K_{\text{I},1}$ in $|x_1| = a + c$ (crack-opening action being assumed). When next we superimpose the traction $\sigma_{22} = \sigma_Y$ along $a < |x_1| < a + c$, this will complete the picture in Fig. 2.9*a* while simultaneously adding a negative singularity strength $K_{\text{I},2}$ to the above value $K_{\text{I},1}$. By manipulating the crack-line traction according to Fig. 2.9*b* we obtain from case 17 of Appendix C

$$K_{\text{I},2} = -\sigma_Y\sqrt{\pi(a+c)} + 2\sigma_Y\sqrt{\frac{a+c}{\pi}}\arcsin\frac{a}{a+c} \tag{2.42}$$

But it is not possible for a singular nonspherical stress state to exist at the border of a plastic zone. A condition to be satisfied if c represents the length of the yield strip is therefore

$$K_{\text{I},1} + K_{\text{I},2} = 0 \tag{2.43}$$

†Exactness is then related to asymptotic states and certain phenomenological assumptions. We mention here that in such an interpretation the result (2.41) turns out to be exact.

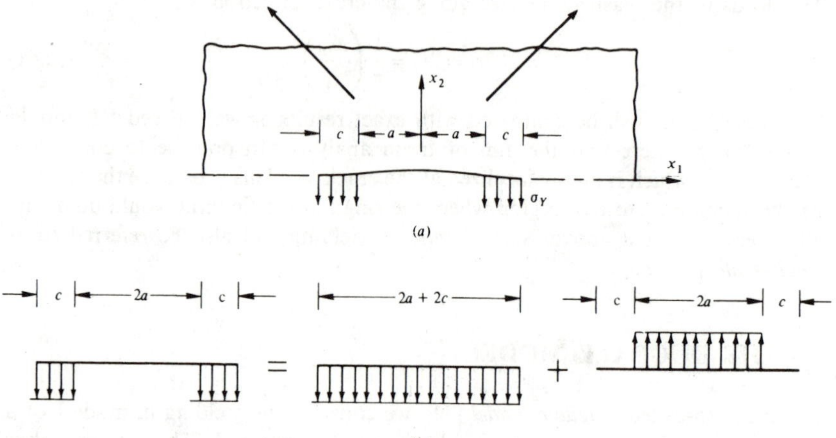

Figure 2.9 The Dugdale model; large arrows symbolize some symmetric external loading.

This is one equation with c as the single unknown, solving our problem. Specifically, if, say $K_{I,1} = \sigma_\infty\sqrt{\pi(a+c)}$ is introduced as for case 3 in Appendix C, $a/W \to 0$, the solution comes out explicitely as

$$\frac{c}{a} = \left(\cos\frac{\pi\sigma_\infty}{2\sigma_Y}\right)^{-1} - 1 \tag{2.44}$$

With decreasing value of c/a the strength $K_{I,1}$ will approach the coefficient K_I applying to the (half) crack length a under the given loading. Simultaneously, (2.42) and (2.43) in general and (2.44) in particular will asymptotically produce the value

$$c = \frac{\pi}{8}\left(\frac{K_I}{\sigma_Y}\right)^2 \tag{2.45}$$

The fact that the crack length no longer appears explicitely must mean that the yielding is so localized as to be governed by K_I alone, i.e., by the singularity of elastic stresses, as in the Irwin study. We then have an asymptotically exact result which is comparable to (2.39). There was an underestimate of about 20 percent in Irwin's prediction.

Reviewing the applied model of analysis, i.e., the elastic half plane of Fig. 2.9*a*, we find that a displacement $u_2 > 0$ may appear at the points $x_1 = \pm a$, $x_2 = 0$. Physically this can be interpreted as a result of the necking ahead of the crack (compare Fig. 2.6*b*). The double value δ in Fig. 2.10 is the so-called *crack-tip opening displacement* (CTOD or COD in fracture language).

From further studies (Appendix B) comes a result which applies in the same

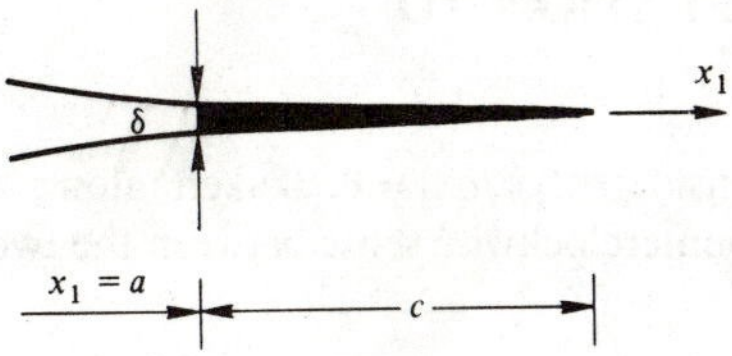

Figure 2.10 Tresca yield zone and COD δ in a thin plate.

context as (2.44) or Fig. 2.11*b*

$$\delta = \frac{8\sigma_Y a}{\pi E} \ln\left[\left(\cos\frac{\pi\sigma_\infty}{2\sigma_Y}\right)^{-1}\right] \tag{2.46}$$

With decreasing σ_∞/σ_Y (and hence c/a) this will asymptotically approach

$$\delta = \frac{K_I^2}{E\sigma_Y} \tag{2.47}$$

which again is a general result for yielding on such a small scale that it is governed by the one external parameter K_I. Results (2.44) to (2.47) are illustrated in Fig. 2.11.

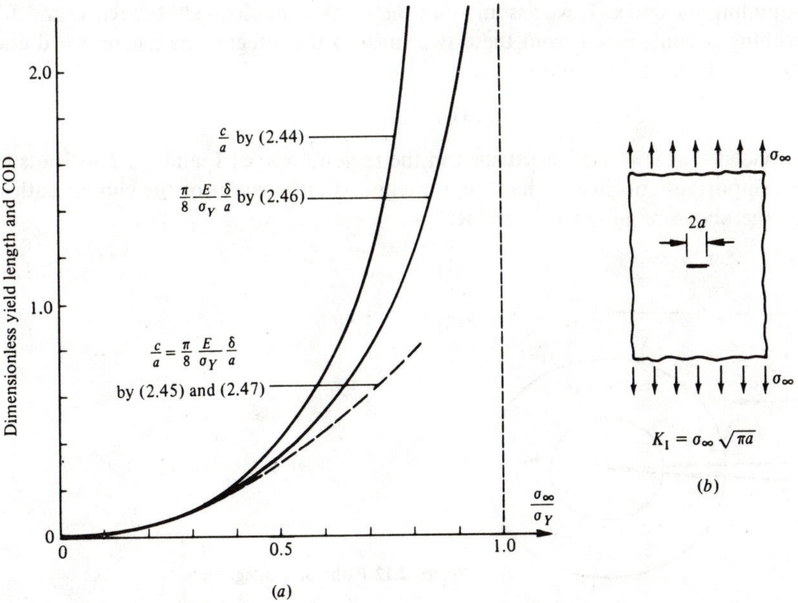

Figure 2.11 (*a*) Yield length c and COD δ in (*b*) a thin plate with a small internal crack.

2.6 THE J INTEGRAL AND ITS RELATION TO CRACK-OPENING DISPLACEMENT

Again, two-dimensional states will be considered: $\partial/\partial x_3 = 0$. Taken along a material curve Γ which is traversed in the counterclockwise sense between the two crack sides (Fig. 2.12), the J integral equals

$$J = \int_{\Gamma} \left(w\, dx_2 - p_i \frac{\partial u_i}{\partial x_1} ds \right) \tag{2.48}$$

by definition [99–102]. Here

$$w = \int_0^{\varepsilon} \sigma_{ij}\, d\varepsilon_{ij} \qquad i, j = 1, 2, 3$$

is the work of deformation per unit of volume, s is the arc length, and p_i is the traction exerted on the body bounded by Γ and the crack surface.† In Appendix D it is shown that the integral of the parenthesis in Eq. (2.48) around the boundary of any simply connected region will vanish if w depends uniquely on ε_{ij} and no singularities occur within that region. The presence of body forces is excluded from this consideration, as are dynamic effects associated with accelerations and kinetic-energy production.

In particular, we look at the closed contour in Fig. 2.12, consisting of the curves Γ and Γ', surrounding the crack tip, and the two straight lines $x_2 = \text{const}$ bounding the crack. If we assume that the crack is nonloaded between Γ and Γ', nothing is contributed from these two lines to the integral since here $p_i = 0$ and $dx_2 = 0$. Then there remains

$$J(\Gamma) - J(\Gamma') = 0$$

provided that $w = w(\varepsilon_{ij})$ is satisfied in the region between Γ and Γ'. This leads to the important conclusion that the J integral is *independent* of the chosen path if all the above requirements are met.

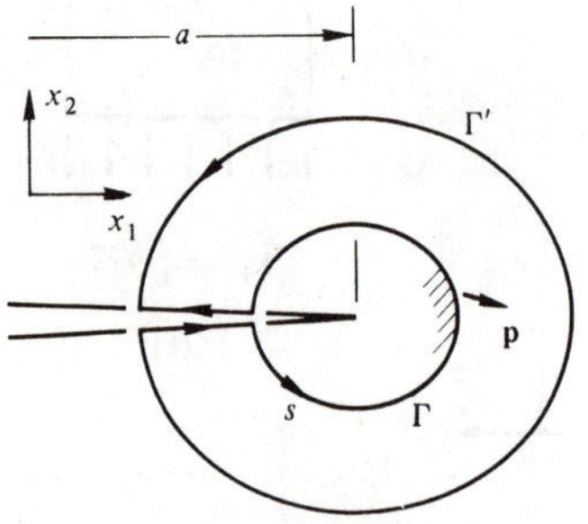

Figure 2.12 Paths of J integration.

†Summation over the range of repeated indices is understood.

Trivially, the last condition is satisfied for purely elastic (hyperelastic) states when w is identical to the strain energy density ϕ (Appendix A). It might be satisfied even under elastoplastic conditions if the straining had been *proportional* in all material points within the enclosed region. This means that the strain (and stress) components have been increased monotonically and in a fixed ratio at each material point, which is never realized exactly but sometimes approximately in a cracked body *before* the crack has started to move. The (approximate) path independence is the most important property of the J integral; special physical interpretations may also be observed. With the invariance it follows (1) that J must express a typical scalar measure of the state of straining at the crack tip since its value is conserved if Γ is shrunk arbitrarily close to the crack tip and (2) that J can be evaluated along a path chosen to give the greatest computational advantage.

An example illustrating properties 1 and 2 is provided by the Dugdale model. The COD δ indicated in Fig. 2.10 will now be expressed by the J integral. Since a path of integration can be arbitrarily chosen in the elastic regime, we take the curve Γ around the yield strip boundary, from the lower side at $x_1 = a$, past $x_1 = a + c$, to the upper side at $x_1 = a$. Since here $dx_2 = 0$ and $\sigma_{22} = \sigma_Y$, $\sigma_{21} = 0$, there remains

$$J = -\int_a^{a+c} \sigma_Y \frac{\partial}{\partial x_1}(u_2^+ - u_2^-)\,dx_1 = \int_0^{\delta} \sigma_Y\,d(u_2^+ - u_2^-) \tag{2.49}$$

where u_2^+ (u_2^-) is the displacement of the upper (lower) side. This implies that

$$J = \sigma_Y \delta \tag{2.50}$$

which shows that a simple material constant relates the J integral to the COD δ in a thin plate, ideal plasticity and Tresca properties being considered.

For small-scale yielding (2.47) can be inserted in (2.50) to provide

$$J = \frac{K_{\mathrm{I}}^2}{E} \tag{2.51}$$

considering the case of plane stress, mode I. This is an important specialization of a more general result to be derived later, relating J to the stress intensity factors. A direct way of obtaining (2.51) would be to insert the elasticity solution (2.15), (2.19), and (2.32) in the defining expression (2.48), then shrinking Γ toward the crack tip to exclude all nonsingular terms from consideration.

In Sec. 2.8 the relation

$$\delta \approx 0.6 \frac{K_{\mathrm{I}}^2}{E\sigma_Y}$$

is pointed out for plane strain (or a thick plate) when ideal plasticity and small-scale yielding are assumed. Invoking $J \approx K_{\mathrm{I}}^2/E$ [compare the plane-stress result (2.51)], we must then have

$$J \approx 1.6\sigma_Y \delta \tag{2.52}$$

under the circumstances. As will be discussed later, there are indications that (2.52) may have a relevance even in plane-strain cases more generally met.

2.7 FURTHER ELASTOPLASTIC ANALYSIS IN MODE III

The analysis of antiplane states, corresponding to cracking in mode III, may have an importance both on its own and as an indicator of results to be expected from the more elaborate considerations in mode I.

For the case in Fig. 2.8 we found, using Irwin's simple device, that the extent of yielding in front of the crack tip is equal to

$$c = \frac{1}{\pi}\left(\frac{K_{III}}{\tau_F}\right)^2$$

[compare (2.41)]. This corresponds to the yield lengths (2.39) and (2.40) in mode I. What makes solution (2.41) particularly interesting, however, is the fact that it is *exact* when the yielding in mode III is so localized that it is governed by the singularity of the elastic solution.

The preceding equations were not presented in historical order. Equation (2.41) derives from an analysis by Hult and McClintock [103], the first exact analysis for any elastoplastic stress field in a cracked body. Irwin subsequently saw that this result implied a simple parallel displacement of the elastic stresses and carried the observation over to mode I to provide a fair (though not exact) prediction.

We summarize here a specialized version of the exact analysis, aiming directly at small-scale yielding in mode III. To save indices, we adopt the following notation:

For stresses:	$\sigma_{31} \equiv \tau_x$	$\sigma_{32} \equiv \tau_y$
For shear strains:	$2\varepsilon_{31} \equiv \gamma_x$	$2\varepsilon_{32} \equiv \gamma_y$
For displacement:	$u_3 \equiv u$	

the crack front being parallel to x_3. The remaining cartesian components are zero according to the assumption (Sec. 2.1). Basic requirements are respectively the equilibrium condition, the yield condition, and the flow rule

$$\frac{\partial \tau_x}{\partial x} + \frac{\partial \tau_y}{\partial y} = 0 \tag{2.53}$$

$$\tau_x^2 + \tau_y^2 = \tau_Y^2 \tag{2.54}$$

$$\dot{\gamma}_x = \lambda \tau_x \qquad \dot{\gamma}_y = \lambda \tau_y \qquad \lambda \ (\propto \dot{\varepsilon}_e) \geq 0 \tag{2.55}$$

together with Hooke's law for elastic strains

$$\gamma_x = \frac{\tau_x}{G} \qquad \gamma_y = \frac{\tau_y}{G} \tag{2.56}$$

The strains or strain rates are related respectively to the displacements or particle velocities (Appendix A)

$$\gamma_x = \frac{\partial u}{\partial x} \qquad \gamma_y = \frac{\partial u}{\partial y} \tag{2.57}$$

$$\dot\gamma_x = \frac{\partial \dot u}{\partial x} \qquad \dot\gamma_y = \frac{\partial \dot u}{\partial y} \tag{2.58}$$

Equations (2.54) and (2.55) assume isotropic and isochoric yielding (as by the Tresca or von Mises assumption), and $\tau_Y = \text{const}$ is assumed with a view to ideal plasticity.

For convenience, we reformulate the problem in terms of arc length s_α along lines (characteristics) α which are normal to the resulting shear stress in the plane of $x_1 \equiv x$ and $x_2 \equiv y$. ϕ is now understood to be the angle between the α line and the x axis, as in Fig. 2.13. With

$$\tau_x = -\tau_Y \sin\phi \qquad \text{and} \qquad \tau_y = \tau_Y \cos\phi \tag{2.59}$$

implying (2.54), (2.53) becomes

$$0 = \cos\phi \frac{\partial \phi}{\partial x} + \sin\phi \frac{\partial \phi}{\partial y} = \frac{\partial x}{\partial s_\alpha}\frac{\partial \phi}{\partial x} + \frac{\partial y}{\partial s_\alpha}\frac{\partial \phi}{\partial y} \equiv \frac{\partial \phi}{\partial s_\alpha}$$

as expressed in a single independent variable s_α. Similarily, we have

$$\frac{\partial \dot u}{\partial s_\alpha} = \cos\phi \frac{\partial \dot u}{\partial x} + \sin\phi \frac{\partial \dot u}{\partial y} = \lambda(\tau_x \cos\phi + \tau_y \sin\phi) = 0$$

where (2.58), (2.55), and (2.59) are invoked in that order. These results signify that the α lines are straight and that the yielding takes place with a constant velocity $\dot u$ on each α line. Similarly, it follows by (2.57), (2.56), and (2.59) that the elastically related displacement u is a constant on each α line. Consequently, the total integrated displacement u must be a constant of the α line.

The intercept of an α line and a nonloaded boundary must be at right angles, since the resulting shear stress has to be tangent to the bounding curve. Therefore,

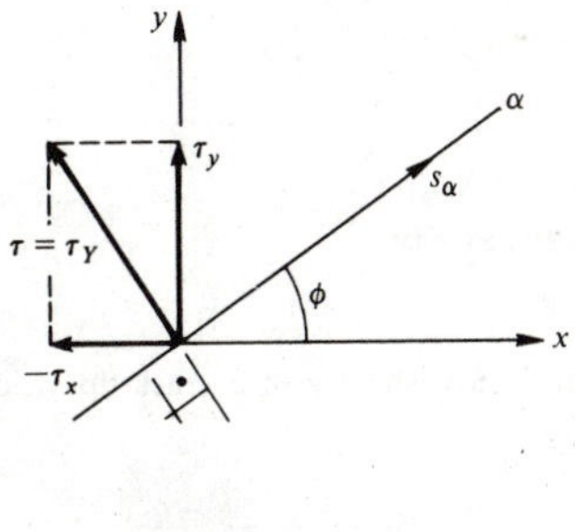

Figure 2.13 A characteristic direction α in mode III yielding.

the structure of the field at the tip of a nonloaded crack will be like that of Fig. 2.14, i.e., a fan centered at the crack tip connecting α lines perpendicular to the crack sides. Since the displacements in the fan are constant along each line, they may be imagined as representing the different strata if the blades of a physical fan were hinged at the crack tip. A further evaluation of their magnitudes and the extent of the yield zone is necessary. We then aim at connecting the elastoplastic near field to the surrounding elastic far field along the bounding curve $r = R(\theta)$, as indicated in Fig. 2.15.

With polar coordinates the following must hold within the fan:

$$\tau_r \, (\equiv \tau_{rz}) = 0 \qquad \tau_\theta \, (\equiv \tau_{z\theta}) = \tau_Y \qquad u \, (\equiv u_z) = u(\theta)$$

and hence [compare (2.21)]

$$\gamma_\theta = \frac{1}{r}\frac{du}{d\theta} \qquad \gamma_\theta[R(\theta)] = \frac{1}{R(\theta)}\frac{du}{d\theta} = \frac{\tau_Y}{G} \tag{2.60}$$

applying to the shear strain generally and on the boundary where Hooke's law is valid. $du/d\theta$ can be eliminated from both parts of (2.60) to give

$$\gamma_\theta = \frac{R(\theta)}{r}\frac{\tau_Y}{G} \tag{2.61}$$

The last equality in (2.60) further leads to the displacement

$$u = \frac{\tau_Y}{G}\int_0^{\theta_1} R(\theta)\, d\theta \tag{2.62}$$

at $\theta = \theta_1$ when $u(0) = 0$ is assumed.

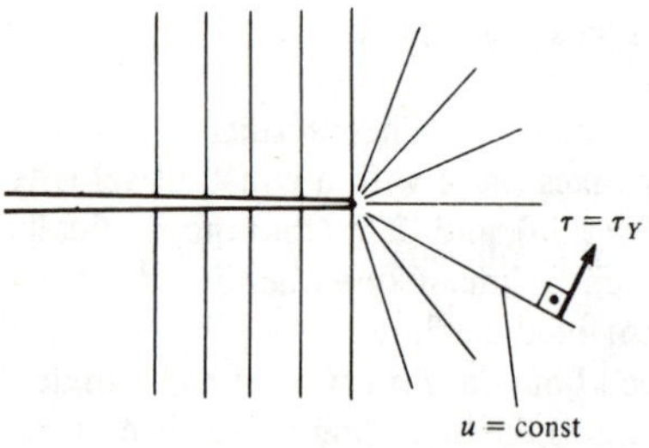

Figure 2.14 Near-tip structure in mode III yielding.

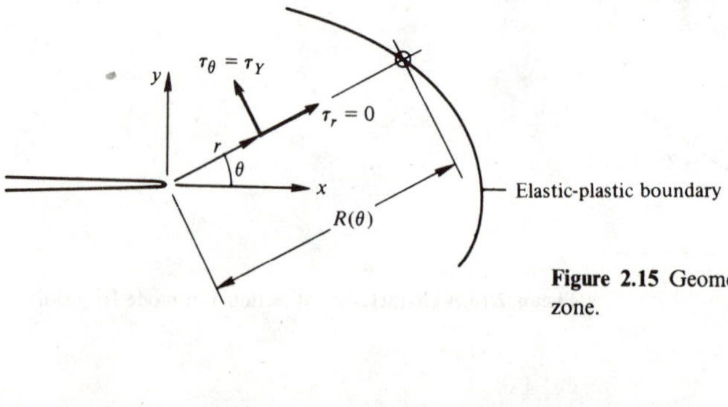

Figure 2.15 Geometry of a crack-tip yield zone.

Equation (2.61) indicates a singular shear strain at $r = 0$ when yielding is taken into account, the singularity being even harder than in the elastic case (when it was characterized by $r^{-1/2}$). A similar experience was related to the Dugdale model through a COD δ which represented a discontinuity in the total displacement u_2 and hence a singularity in ε_{22}. By analogy, we might also expect a displacement discontinuity in the present case; it does appear as a crack opening displacement normal to the xy plane (the distance between the upper and the lower stratum) and follows mathematically from (2.62) as†

$$\delta = u\left(\frac{\pi}{2}\right) - u\left(-\frac{\pi}{2}\right) = \frac{\tau_Y}{G}\int_{-\pi/2}^{\pi/2} R(\theta)\, d\theta \qquad (2.63)$$

It clearly remains only to determine the boundary $R(\theta)$, which will be approached from the elastic side by some simple use of complex-variable techniques. Since u outside the yield zone satisfies the potential equation [compare (2.23)], it must be a harmonic function of $z \equiv x + iy$‡ and we are therefore at liberty to express it as the imaginary part of an analytic function $\Omega(z)/G$

$$u = \frac{1}{G}\,\mathrm{Im}\,[\Omega(z)] \qquad (2.64)$$

It follows, by (2.56) and (2.57), that

$$\tau_y + i\tau_x = \left(\frac{\partial}{\partial y} + i\frac{\partial}{\partial x}\right)\mathrm{Im}\,[\Omega(z)] = \Omega'(z) \qquad (2.65)$$

which means that the combination $\tau_y + i\tau_x$ is also an analytic function of z. The converse is that z must be an analytic function of ξ, defined by

$$\xi = \frac{\tau_y + i\tau_x}{\tau_Y} \qquad (2.66)$$

hence

$$z = F(\xi) \qquad (2.67)$$

This is an effective formulation with a view to including the boundary condition at the interface $r = R(\theta)$. Here (2.54) should be satisfied, or equivalently

$$|\xi| = 1 \qquad (2.68)$$

which means that (2.54) will be mapped on the unit circle in the ξ plane.

The planes z and ξ are shown in Fig. 2.16. To a point

$$z = x + iy = R(\theta)e^{i\theta}$$

will correspond the stresses $\tau_x = -\tau_Y \sin\theta$ and $\tau_y = \tau_Y \cos\theta$, and so, by (2.66),

$$\xi = e^{-i\theta}$$

†More appropriate might be the term crack warping displacement.
‡Obviously z is not used here in the meaning of x_3.

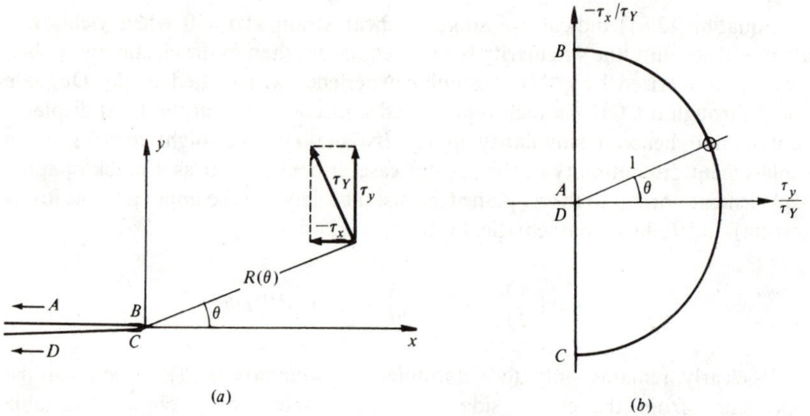

Figure 2.16 (*a*) The physical z plane and (*b*) the ξ plane with stresses at yield on the unit circle.

The radius vector of the boundary follows next from (2.67), as

$$R(\theta) = e^{-i\theta}F(e^{-i\theta}) \tag{2.69}$$

where F still has to be determined.

We note that the crack sides $x < 0$, $y = 0$ must be mapped on the imaginary axis in the ξ plane, since here $\tau_y = 0$. Stated inversely, $z = F(\xi)$ has to be real and negative when $\tau_y/\tau_Y = \mathrm{Re}\,\xi = 0$ (condition 1). Further, $R(\theta)$ is obviously real, implying by (2.69) that $\mathrm{Im}\,[e^{-i\theta}F(e^{-i\theta})] = 0$ (condition 2). With the assumption of yielding governed by the elastic singularity, it also follows that the singular elastic-stress distribution should be asymptotically attained away from the crack tip. By (2.31) and (2.32) this means that

$$\tau_x = \tau_r \cos\theta - \tau_\theta \sin\theta \to -\frac{K_{\mathrm{III}}}{\sqrt{2\pi r}} \sin\frac{\theta}{2}$$

and

$$\tau_y = \tau_r \sin\theta + \tau_\theta \cos\theta \to +\frac{K_{\mathrm{III}}}{\sqrt{2\pi r}} \cos\frac{\theta}{2}$$

as $r \to \infty$, or, equivalently,

$$\tau_y + i\tau_x = \xi\tau_Y \to \frac{K_{\mathrm{III}}}{\sqrt{2\pi z}} \qquad \text{as } |z| \to \infty \tag{2.70}$$

The inverse statement is

$$z = F(\xi) \to \frac{K_{\mathrm{III}}^2}{2\pi\tau_Y^2\xi^2} \qquad \text{as } |\xi| \to 0$$

which is the last of the three boundary conditions to be imposed on the analytic function $F(\xi)$.

It is easily verified that the three conditions are all satisfied by the solution

$$z = F(\xi) = \frac{K_{\text{III}}^2}{2\pi\tau_Y^2}\left(1 + \frac{1}{\xi^2}\right) \tag{2.71}$$

which combines with (2.69) into

$$R(\theta) = \frac{K_{\text{III}}^2}{\pi\tau_Y^2}\cos\theta \tag{2.72}$$

and leads to the stresses outside this boundary, by (2.66),

$$\tau_y + i\tau_x = \frac{K_{\text{III}}}{\sqrt{2\pi}}\left(z - \frac{K_{\text{III}}^2}{2\pi\tau_Y^2}\right)^{-1/2} \tag{2.73}$$

The interpretation of the result (2.72) is that the yield zone is a circle tangent to the crack tip, having the diameter

$$c = R(0) = \frac{K_{\text{III}}^2}{\pi\tau_Y^2} \tag{2.74}$$

as shown in Fig. 2.17. This confirms result (2.41). Under purely elastic conditions $\tau_y + i\tau_x = K_{\text{III}}/\sqrt{2\pi z}$ would apply throughout the field [compare (2.70)], and (2.73) can therefore be taken to indicate that the stress field outside the yield zone is the same as the elastic one, except for translation by the length $K_{\text{III}}^2/2\pi\tau_Y^2 = c/2$ in the x direction. This is what Irwin recognized and carried over to mode I. We complete the present deductions by introducing (2.72) in Eqs. (2.61) to (2.63) to obtain the strain and displacement fields

$$\gamma_\theta = \frac{K_{\text{III}}^2}{\pi G\tau_Y}\frac{\cos\theta}{r} \qquad u = \frac{K_{\text{III}}^2}{\pi G\tau_Y}\sin\theta \tag{2.75}$$

within the yield zone and the COD

$$\delta = \frac{2}{\pi}\frac{K_{\text{III}}^2}{G\tau_Y} \tag{2.76}$$

As it turns out, there is a striking resemblance between expressions (2.47) and (2.76) for the two types of crack opening displacement.

The above outline of the specialized Hult–McClintock solution largely follows a presentation by Rice [21], who has also made many original contributions to

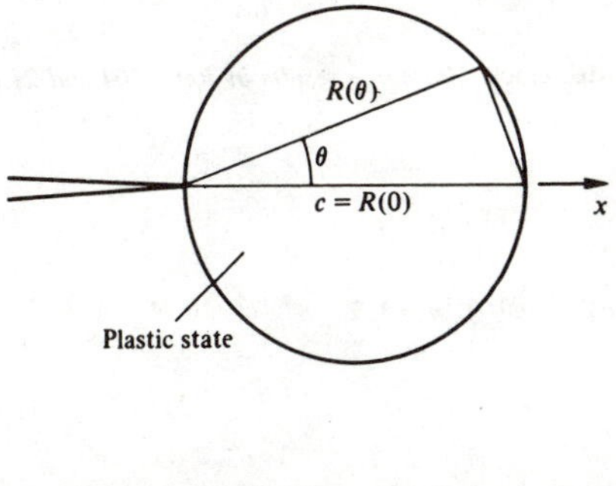

Figure 2.17 Mode III small-scale yield zone for ideal plasticity.

fracture mechanics. In the present context we first relate his results for an edge crack in an elastic–ideally plastic body. When the plastic zone has grown enough to be comparable to the crack length a (while remaining small compared with the other dimensions), the analysis provides the yield length c and the COD δ through

$$\frac{c}{a} = \frac{2}{\pi}\frac{1+s^2}{1-s^2} E_2\left(\frac{2s}{1+s^2}\right) - 1 \tag{2.77}$$

$$\frac{\delta}{a} = 2\frac{\tau_Y}{G}\left[\frac{2}{\pi}(1+s^2)E_1(s^2) - 1\right] \tag{2.78}$$

where $s = \tau_\infty/\tau_Y$ and E_1 and E_2 are the complete elliptic integrals of the first and second kind, respectively.† The results are shown in Fig. 2.18, which has obvious features in common with Fig. 2.11. In particular, we note that (2.74) and (2.76) are asymptotically approached by (2.77) and (2.78) when yielding is localized, $\tau_\infty/\tau_Y \to 0$.

Next, a result is discussed which concerns a material obeying a nonlinear

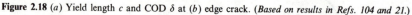

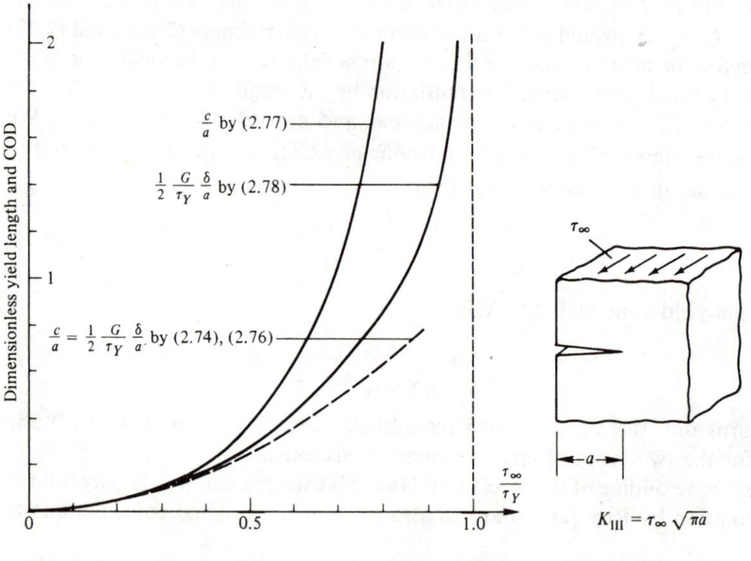

Figure 2.18 (*a*) Yield length c and COD δ at (*b*) edge crack. (*Based on results in Refs. 104 and 21.*)

†

$$E_n(q) = \int_0^{\pi/2} (1 - q^2 \sin^2\theta)^{n-3/2}\, d\theta \qquad n = 1, 2$$

stress–strain relation of the type

$$\gamma = \begin{cases} \gamma_Y\left(\dfrac{\tau}{\tau_Y}\right) & \text{if } \tau \le \tau_Y \quad (2.79a) \\ \gamma_Y\left(\dfrac{\tau}{\tau_Y}\right)^m & \text{if } \tau \ge \tau_Y \quad (2.79b) \end{cases}$$

This may correspond to elastic behavior up to $\gamma = \gamma_Y \equiv \tau_Y/G$ and to strain-hardening elastoplastic response when $\gamma > \gamma_Y$; $m = 1$ and $m = \infty$ are limiting cases related to the elastic and elastic–ideally plastic response, respectively. For the shear strain along the extended crack line the following solution was obtained for small-scale yielding

$$\gamma_y(x > 0, y = 0) = \gamma_Y\left[\frac{mK_{\text{III}}^2}{(m+1)\pi\tau_Y^2 x}\right]^{m/(m+1)} \tag{2.80}$$

Equation (2.80) contains the transition from the elastic solution (2.33), having a singularity of the type $r^{-1/2}$ (or $x^{-1/2}$), to the elastoplastic solution (2.75) with the singularity r^{-1}. The type of the singularity will be reflected in the displacements of the crack faces near the tip. A distinct COD appears in the ideally plastic case, whereas a continuous change from one side to the other is exhibited if the material is purely elastic or strain-hardening (Fig. 2.19). This final point is a matter of some interest in later discussions.

2.8 FURTHER ELASTOPLASTIC ANALYSIS IN MODE I

2.8.1 Ideal Plasticity and Plane Strain

The Dugdale model dealt with ideal plasticity (Tresca type) and plane stress. We are now going to look at the same kind of problem in plane strain, approximately realized in thick plates. Unfortunately it is not possible to present any simple analytical total solution here, but we can still establish certain typical properties of the fields near the crack tip and then complement the evaluation by numerical or roughly approximate analyses.

The plane state of deformation is assumed to be dominated by isotropic and isochoric yielding (such as by Mises or Tresca) near the crack tip; it can then be analyzed by slip-line theory. The field on one side of the crack is indicated in Fig. 2.20. The plane of symmetry is a principal one and should therefore be

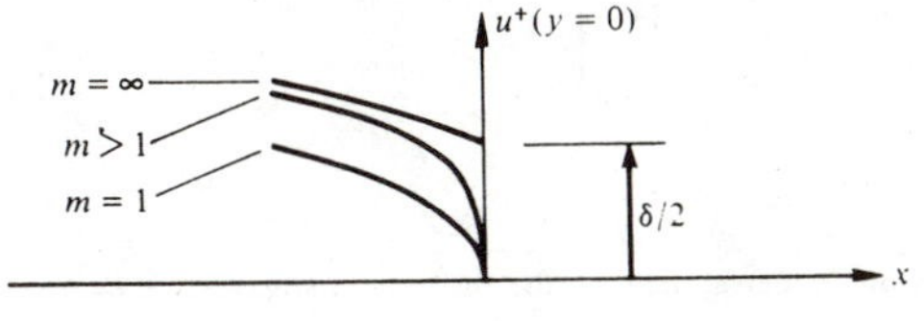

Figure 2.19 Crack-tip profiles for varying material properties.

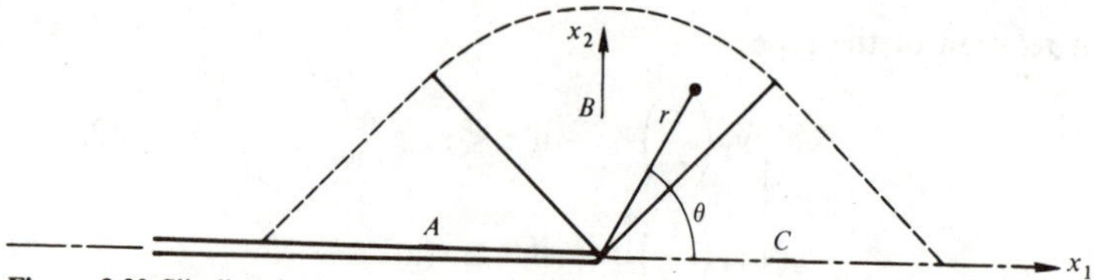

Figure 2.20 Slip-line field at a crack tip in mode I.

intersected at 45° by the slip lines. Further, the maximum stress is likely to be parallel to x_1 along the crack sides and normal to x_1 ahead of the crack. This has been achieved by the fields A and C, respectively, the α lines being solid and the β lines dashed. Connecting these fields must be a circular fan centered on the crack tip, field B. The figure shows the structure of the total field near the crack tip but tells nothing about the extent of the plastic zone.

Acting on the α and β lines are the mean stress $\tilde{\sigma} = (\sigma_{11} + \sigma_{22})/2$ and the yield shear stress τ_Y (Fig. 2.21). In region A, therefore,

$$\sigma_{22} = \sigma_{\min} = \tilde{\sigma} - \tau_Y = 0 \qquad \sigma_{11} = \sigma_{\max} = \tilde{\sigma} + \tau_Y = 2\tau_Y \tag{2.81}$$

where $\sigma_{22} = 0$ is the boundary condition on the crack side. Since along a β line one has to satisfy

$$\tilde{\sigma} + 2\tau_Y\theta = \text{const} \tag{2.82}$$

[compare (A.66)] and since θ is larger by $\pi/2$ in region A than in region C, in C we must have

$$\tilde{\sigma} = \tau_Y + 2\tau_Y\frac{\pi}{2} = (1 + \pi)\tau_Y \tag{2.83}$$

so that here

$$\sigma_{11} = \sigma_{\min} = \tilde{\sigma} - \tau_Y = \pi\tau_Y \qquad \sigma_{22} = \sigma_{\max} = \tilde{\sigma} + \tau_Y = (2 + \pi)\tau_Y \tag{2.84}$$

If Mises properties are dominant in zone C, then $\tau_Y = \sigma_Y/\sqrt{3}$ and $\sigma_{33} = \tilde{\sigma}$, whereby

$$\sigma_{11} = 1.81\sigma_Y \qquad \sigma_{22} = 2.97\sigma_Y\dagger \qquad \sigma_{33} = 2.39\sigma_Y \tag{2.85}$$

while for Tresca properties $\tau_Y = \sigma_Y/2$ and

$$\sigma_{11} = 1.57\sigma_Y \qquad \sigma_{22} = 2.57\sigma_Y\dagger \qquad \sigma_{11} \le \sigma_{33} \le \sigma_{22} \tag{2.86}$$

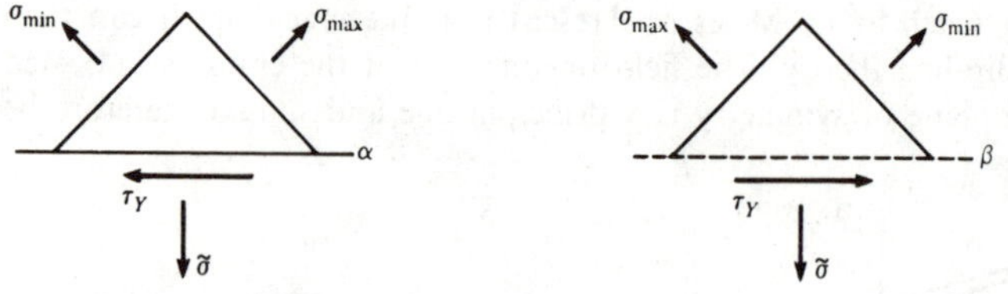

Figure 2.21 Orientation of α and β slip lines.

†The Irwin factor $k = \sqrt{3}$ is an obvious underestimate of the theoretical stress value σ_{22}.

In both cases the results indicate a very high spherical or mean-stress level ahead of the crack tip in a thick plate. This derives from the kinematic constraint $\varepsilon_{33} \approx 0$, which is exerted by the far less strained surrounding material. By the constraint the stress relief through yielding is suppressed compared with the plane-stress situation; this favors a more brittle type of fracture in the global sense.

The velocity field in fan B must be of the type

$$\dot{u}_r = f'(\theta) \qquad \dot{u}_\theta = -f(\theta) + g(r)$$

in order to conform with the vanishing of the strain rates on the α lines (radially) and β lines (circumferentially)

$$\dot{\varepsilon}_r = \frac{\partial \dot{u}_r}{\partial r} = 0 \qquad \dot{\varepsilon}_\theta = \frac{\dot{u}_r}{r} + \frac{1}{r}\frac{\partial \dot{u}_\theta}{\partial \theta} = 0$$

by (A.22). Further

$$\dot{\gamma}_{r\theta} = \frac{1}{r}\frac{\partial \dot{u}_r}{\partial \theta} + \frac{\partial \dot{u}_\theta}{\partial r} - \frac{\dot{u}_\theta}{r} = \frac{f''(\theta)}{r} + g'(r) + \frac{f(\theta)}{r} - \frac{g(r)}{r}$$

By choice we can put $\dot{u}_r = \dot{u}_\theta = 0$ in $r = 0$ on the boundary of region C. This implies that $g(0) = f(\pi/4)$; by invoking continuity in the normal velocity and a finite strain rate in C we can conclude that $g'(0)$ has a finite value. After integration in time there remains a singular shear strain of the type [21]

$$\gamma_{r\theta} = \frac{F(\theta)}{r} \tag{2.87}$$

This intense shear deformation must correspond to such motion at the crack tip as indicated in Fig. 2.22, roughly confirming the hinge mechanism in Fig. 2.6. We can therefore expect that (2.87) will lead to a COD δ in mode I (analogous to the singular shear strain producing a crack opening in mode III).

It is no simple matter to evaluate this crack opening exactly, however. Numerical analyses have been undertaken by Tracey [108], Sorensen [109], and

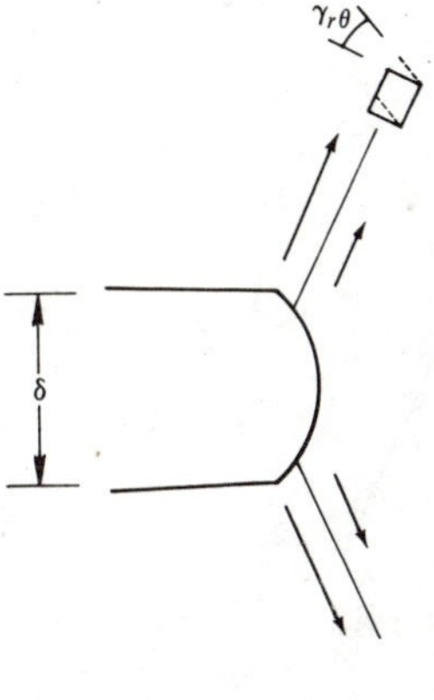

Figure 2.22 COD δ by intense shearing (hinge mechanism).

several others based on finite elements incorporating the $\gamma \propto 1/r$ singularity at the crack tip. This will determine the function $F(\theta)$ in principle, and the COD follows by

$$\frac{\delta}{2} = \int_{\pi/4}^{3\pi/4} F(\theta) \sin\theta \, d\theta \tag{2.88}$$

Before examining the results, however, we shall look into an estimation suggested by Rice [106], in which all yielding was assumed to be concentrated along two straight lines emanating from the crack tip.

The yield lines are shown to different scales in Fig. 2.23. We consider each line as the extension of an imaginary equivalent crack operating in mode II and being characterized in the elastic state by the stress intensity factor

$$K'_{\mathrm{II}} = K_{\mathrm{I}} \frac{\max[\tau_{r\theta}(0,\theta)]}{\sigma_\theta(0,0)} = \frac{K_{\mathrm{I}}}{4}\left(\sin\frac{\theta}{2} + \sin\frac{3\theta}{2}\right)_{\max} = \frac{2}{3\sqrt{3}} K_{\mathrm{I}} \tag{2.89}$$

where K_{I} is the factor for the actual crack and $\tau_{r\theta}$ and σ_θ are given by (2.15). It is natural to assume that yielding will take place along that direction which maximizes $\tau_{r\theta}$; by simple differentiation this turns out to be

$$\theta = \theta_0 = \arccos\tfrac{1}{3} = 70.6° \tag{2.90}$$

From now on the procedure is analogous to that of the Dugdale model. A highly localized yielding with $\tau_{r\theta} = \tau_Y$ assumed along the strip will cancel the singularity K'_{II} when the strip length is

$$c' = \frac{\pi}{8}\left(\frac{K'_{\mathrm{II}}}{\tau_Y}\right)^2 \tag{2.91}$$

[compare (2.45)] and lead to a tangential crack opening

$$\delta' = (1 - \nu^2)\frac{(K'_{\mathrm{II}})^2}{E\tau_Y} \tag{2.92}$$

[compare (2.47)]. The multiplier $1 - \nu^2$ reflects the present state of plane strain.

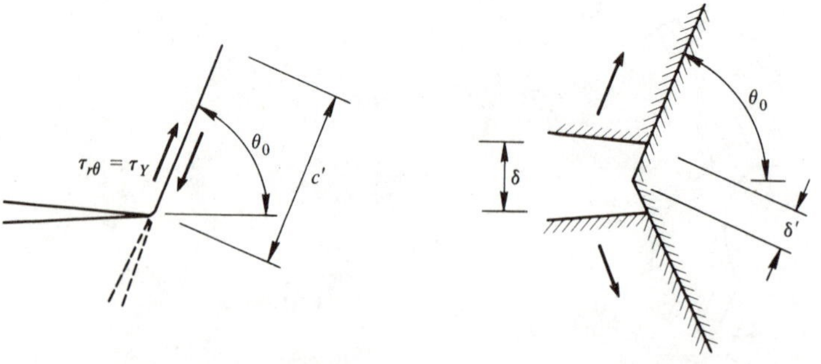

Figure 2.23 Symmetric sliding mechanism approximating plane-strain yield.

Taking $\nu = 0.3$ and $\tau_Y = \sigma_Y/\sqrt{3}$ and inserting from (2.89) and (2.90), we obtain

$$c' = 0.17\left(\frac{K_\mathrm{I}}{\sigma_Y}\right)^2 \tag{2.93}$$

and

$$\delta = 2\delta' \sin\theta_0 = 0.44\frac{K_\mathrm{I}^2}{E\sigma_Y} \tag{2.94}$$

where the last result is the desired COD (compare Fig. 2.23).

The finite-element solution has shown that small-scale yielding will occur within a zone approximately as indicated in Fig. 2.24, the largest distance from the crack tip being about $0.15(K_\mathrm{I}/\sigma_Y)^2$ in a direction of about 71°. The COD has been derived, by various authors, within 0.4 to 0.7 times $K_\mathrm{I}^2/E\sigma_Y$.† Apart from the reduction to yielding in two distinct directions, the estimation above has given reasonable values. The yield length c *along* the actual crack is even more difficult to determine; it is strongly dependent on ν, decreasing when ν increases, and possibly vanishing for $\nu = 0.5$. For $\nu = 0.3$, $c \approx 0.03(K_\mathrm{I}/\sigma_Y)^2$ has been recorded, which is somewhat less than the Irwin approximation $c \approx 0.1(K_\mathrm{I}/\sigma_Y)^2$.

2.8.2 Strain-Hardening Plasticity

We first investigate the stress-strain fields at a crack tip when the material obeys a nonlinear constitutive relation of the type

$$\varepsilon_{ij} = \varepsilon_Y\left(\frac{\sigma_e}{\sigma_Y}\right)^m \frac{3}{2}\frac{s_{ij}}{\sigma_e} \tag{2.95a}$$

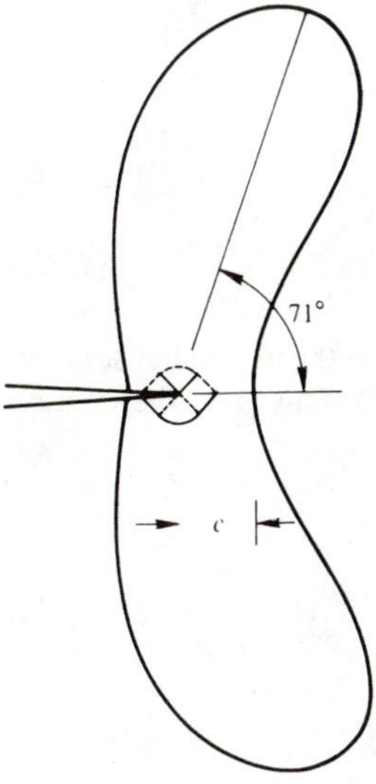

Figure 2.24 Shape of the ideal-plastic small-scale yield zone in plane strain, mode I.

†With the latest investigations a multiplier somewhat above 0.6 seems to have emerged.

where m, ε_Y, and σ_Y are constants,

$$s_{ij} = \sigma_{ij} - \frac{\delta_{ij}\sigma_{kk}}{3} \tag{2.95b}$$

are the deviatoric stresses, and

$$\sigma_e = (\tfrac{3}{2}s_{ij}s_{ij})^{1/2} \tag{2.95c}$$

is the effective stress associated with von Mises properties (compare Appendix A). Subsequently, the results will be discussed in relation to real elastoplastic behavior.† Equation (2.95) defines an incompressible isotropic material and generalizes the uniaxial stress - strain relation

$$\varepsilon = \varepsilon_Y\left(\frac{\sigma}{\sigma_Y}\right)^m \tag{2.96}$$

or the simple shear relation (2.79b). The case $m = 1$ can therefore be identified as linearly elastic, assuming $\nu = \frac{1}{2}$, while $m = \infty$ will have a bearing on ideal plasticity.

The analysis referred below was undertaken by Hutchinson [111] in complete analogy with Williams' power-series solution (Sec. 2.1). Accordingly, the Airy stress function χ was first introduced by (2.1) to enforce equilibrium and substituted for the stresses in (2.95). Next, these strains were substituted into the compatibility equation (A.23) to provide the equation governing χ. This is the analog of (2.2) for plane-strain or plane-stress conditions. Just to indicate its complexity we show the resulting form for plane stress

$$\begin{aligned}
&\frac{\partial^2}{\partial r^2}\left[\sigma_e^{m-1}\left(2r\frac{\partial\chi^2}{\partial r^2} - \frac{\partial\chi}{\partial r} - \frac{1}{r}\frac{\partial^2\chi}{\partial\theta^2}\right)\right] + \frac{6}{r}\frac{\partial^2}{\partial\theta\,\partial r}\left[\sigma_e^{m-1}r\frac{\partial}{\partial r}\left(\frac{1}{r}\frac{\partial\chi}{\partial\theta}\right)\right]\\
&\quad + \frac{\partial}{\partial r}\left[\sigma_e^{m-1}\left(-\frac{2}{r}\frac{\partial\chi}{\partial r} - \frac{2}{r^2}\frac{\partial^2\chi}{\partial\theta^2} + \frac{\partial^2\chi}{\partial r^2}\right)\right]\\
&\quad + \frac{1}{r}\frac{\partial^2}{\partial\theta^2}\left[\sigma_e^{m-1}\left(-\frac{\partial^2\chi}{\partial r^2} + \frac{2}{r}\frac{\partial\chi}{\partial r} + \frac{2}{r^2}\frac{\partial^2\chi}{\partial\theta^2}\right)\right] = 0
\end{aligned} \tag{2.97}$$

Something just as bad is encountered in plane strain. Again the method of solution is to assume for χ the power and product expression (2.5). This gives the stresses, by (2.1),

$$\begin{aligned}
\sigma_r &= r^{\lambda-1}\left[(\lambda+1)f + \frac{d^2f}{d\theta^2}\right] \equiv r^{\lambda-1}\bar{\sigma}_r\\
\sigma_\theta &= r^{\lambda-1}(\lambda+1)\lambda f \equiv r^{\lambda-1}\bar{\sigma}_\theta\\
\tau_{r\theta} &= -r^{\lambda-1}\lambda f' \equiv r^{\lambda-1}\bar{\tau}_{r\theta}
\end{aligned} \tag{2.98}$$

†Formally, (2.95) is a nonlinear elastic relation, or a total-strain description as related in (A.47) to (A.51). Only when the straining is proportional in all points can (2.95) describe real plastic behavior.

and the effective stress, by (2.95c)

$$\sigma_e = r^{\lambda-1}(\bar{\sigma}_r^2 + \bar{\sigma}_\theta^2 - \bar{\sigma}_r\bar{\sigma}_\theta + 3\bar{\tau}_{r\theta}^2)^{1/2} \equiv r^{\lambda-1}\bar{\sigma}_e \tag{2.99}$$

or

$$\sigma_e = r^{\lambda-1}\left[\frac{3(\bar{\sigma}_r - \bar{\sigma}_\theta)^2}{4} + 3\bar{\tau}_{r\theta}^2\right]^{1/2} \equiv r^{\lambda-1}\bar{\sigma}_e \tag{2.100}$$

for plane stress or plane strain, respectively. With (2.5) we have the same r dependence in all terms of the compatibility equation. This reduces (2.97) to the ordinary, nonlinear differential equation

$$\begin{aligned}\left[m(\lambda-1) - \frac{d^2}{d\theta^2}\right]&\left\{\bar{\sigma}_e^{m-1}\left[(\lambda+1)(\lambda-2)f - 2\frac{d^2f}{d\theta^2}\right]\right\} \\ &+ [m(\lambda-1)+1]m(\lambda-1)\bar{\sigma}_e^{m-1}\left[(\lambda+1)(2\lambda-1)f - \frac{d^2f}{d\theta^2}\right] \\ &+ 6[m(\lambda-1)+1]\lambda\frac{d}{d\theta}\left(\bar{\sigma}_e^{m-1}\frac{df}{d\theta}\right) = 0\end{aligned} \tag{2.101}$$

with $\bar{\sigma}_e$ as defined by (2.98) and (2.99).

With the boundary conditions (2.7) the interesting eigenvalue λ which solves (2.101) (or the corresponding equation in plane strain) was accurately determined to be

$$\lambda = \lambda_1 = \frac{m}{m+1} \tag{2.102}$$

For $m = 1$ this result confirms $\lambda = \frac{1}{2}$ in (2.12). The smaller values of λ would give an infinite total work of deformation within any finite area including the crack tip and were discarded for that reason; the larger values resulted in nonsingular stresses and strains. The root λ_1 leads to the stresses, by (2.98),

$$\{\sigma_r, \sigma_\theta, \tau_{r\theta}\} = \sigma_Y\tilde{K}r^{-1/(m+1)}\{\tilde{\sigma}_r, \tilde{\sigma}_\theta, \tilde{\tau}_{r\theta}\} \tag{2.103}$$

where the constant $\sigma_Y\tilde{K}$ has been included, $\tilde{K}$ having here the dimension of $(\text{length})^{1/(m+1)}$, to make $\tilde{\sigma}_{ij} = \bar{\sigma}_{ij}/\sigma_Y\tilde{K}$ dimensionless. Equation (2.103) is the analog of (2.15), with f and $\bar{\sigma}_{ij}$ as determined for any m value by (2.101) and (2.102) using (2.7). Introduction of (2.103) into (2.95) gives the strains

$$\{\varepsilon_r, \varepsilon_\theta, \gamma_{r\theta}\} = \varepsilon_Y\tilde{K}^m r^{-m/(m+1)}\{\tilde{\varepsilon}_r, \tilde{\varepsilon}_\theta, \tilde{\gamma}_{r\theta}\} \tag{2.104}$$

where $\tilde{\varepsilon}_{ij}$ are dimensionless functions of θ, related to the value of m being considered. Typical distribution functions are reproduced in Fig. 2.25 for the parameter value $m = 3$. The analysis indicated a continuous transition from the elastic-stress distribution when $m = 1$ to the ideally plastic stress distribution when $m \to \infty$, (see Fig. 2.26).† The strain distribution should be indeterminate in the

†For plane strain we have just considered the slip-line solution, Eqs. (2.81) to (2.85). In plane stress the solution when $m \to \infty$ will converge toward Mises results as given in Eq. (A.67) of Appendix A. The Dugdale model is for Tresca material, as already observed.

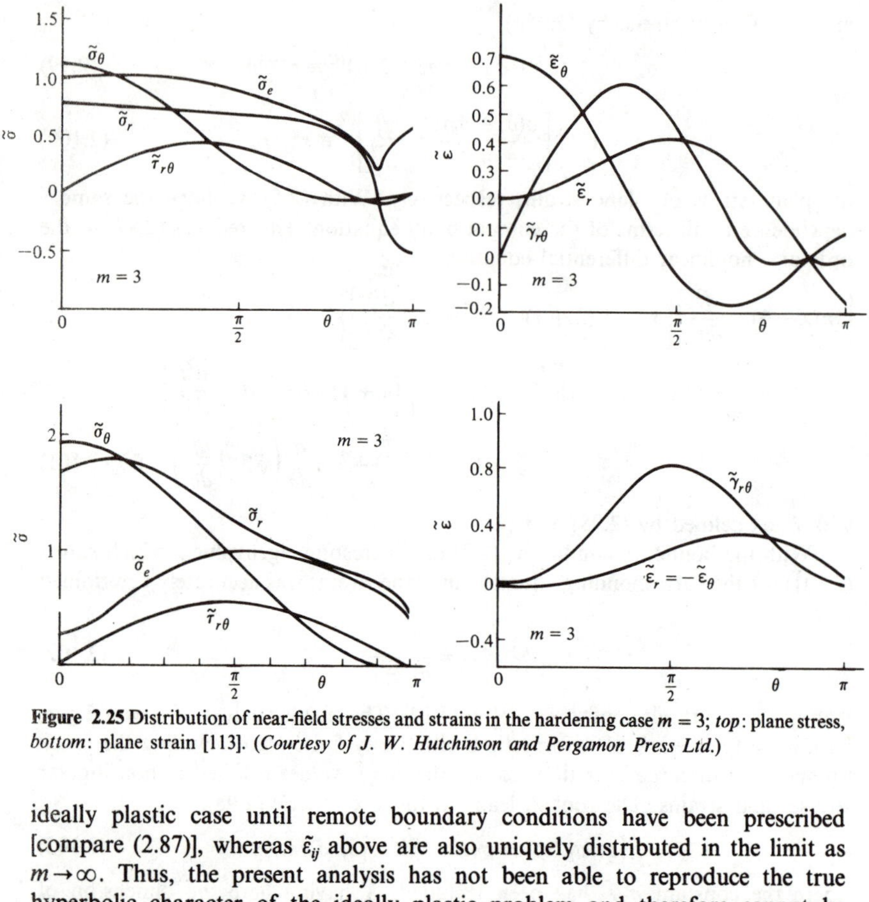

Figure 2.25 Distribution of near-field stresses and strains in the hardening case $m = 3$; *top*: plane stress, *bottom*: plane strain [113]. (*Courtesy of J. W. Hutchinson and Pergamon Press Ltd.*)

ideally plastic case until remote boundary conditions have been prescribed [compare (2.87)], whereas $\tilde{\varepsilon}_{ij}$ above are also uniquely distributed in the limit as $m \to \infty$. Thus, the present analysis has not been able to reproduce the true hyperbolic character of the ideally plastic problem and therefore cannot be unconditionally applied to this case by taking m to infinity.

Disregarding ideal plasticity, it remains to determine only the scalar $\tilde{K}$, the strength of the singularity in analogy to the linearly elastic case. Before going into this problem, however, we may well question the validity of the above results in relation to actual elastoplastic behavior of the material. Near the crack tip, where the present singular solution dominates, the stress or strain components are indeed proportional in all points considered. These predictions are therefore consistent with a truly (flow theory, isotropic strain-hardening) plastic generalization of (2.96). A theorem of Ilyushin may extend this region of validity even further by asserting that the state would be proportional throughout the body if the boundary tractions were increased in the same ratio and (2.95) were applicable everywhere. However, the last assumption is not acceptable, since instead the response to smaller straining should be linearly elastic and compressible ($\nu < \frac{1}{2}$);

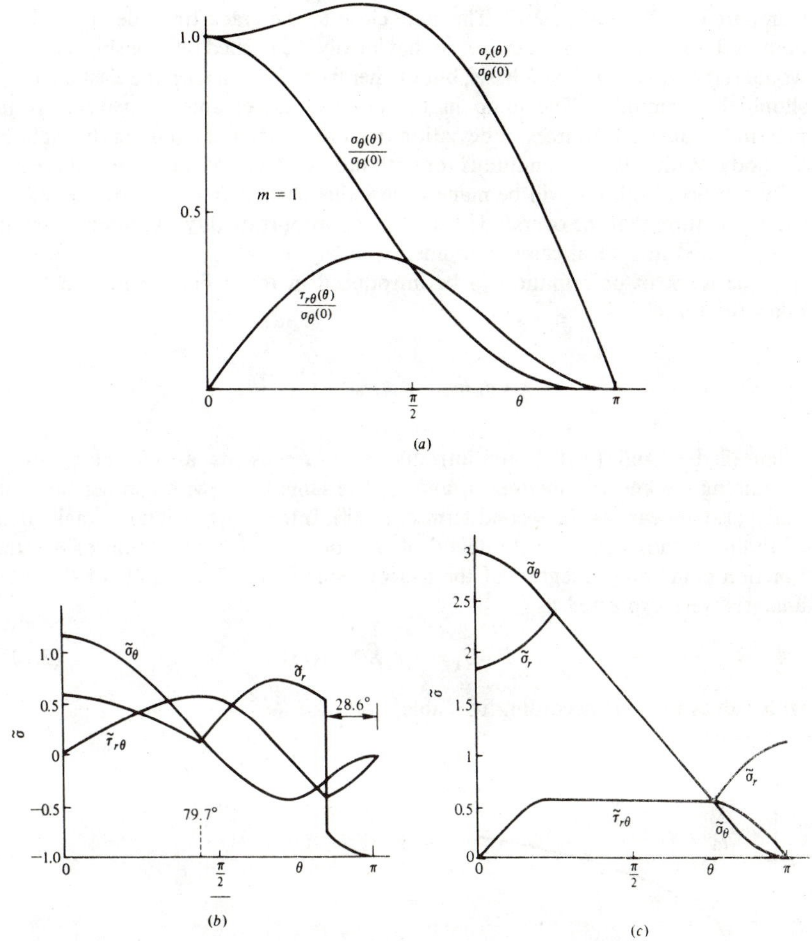

Figure 2.26 Distribution of near-field stresses in the limiting cases of (*a*) linear elasticity ($m = 1$) and (*b*) and (*c*) ideal plasticity ($m \to \infty$); (*b*) plane stress; (*c*) plane strain [113]. (*a*) derives from Eqs. (2.15) with $r = \text{const.}$ [(*b*) *and* (*c*) *courtesy of J. W. Hutchinson and Pergamon Press Ltd.*]

uniaxially this would correspond to something like

$$\varepsilon = \begin{cases} \varepsilon_Y\left(\dfrac{\sigma}{\sigma_Y}\right) & \text{if } \sigma \le \sigma_Y \\ \varepsilon_Y\left(\dfrac{\sigma}{\sigma_Y}\right)^m & \text{if } \sigma \ge \sigma_Y \end{cases} \tag{2.105}$$

Compare (2.79) and Fig. 2.27. The state close to the crack tip is described by a point well to the right on the curve, so that locally the dashed relationship assumed would serve equally well as a basis; but further from the crack tip the elastic regime should be dominant. The jump in the uniaxial slope, and the varying compressibility are both sources of deviation from proportional straining throughout the body. With this, the conditions for path independence of the J integral are not fully satisfied. Still, use will be made of just this particular tool below, based on the expectation that the overall deviation from proportionality will often be small, as confirmed in several cases by numerical analyses.

The work of deformation to be introduced in the defining equation (2.48) takes the form

$$w = \int_0^{\varepsilon} \sigma_{ij}\, d\varepsilon_{ij} = \sigma_Y \varepsilon_Y \tilde{K}^{m+1} r^{-1} g_1(\theta) \tag{2.106}$$

when (2.103) and (2.104) are introduced; g_1 represents a sum of products containing the known functions $\tilde{\sigma}_{ij}$ and $\tilde{\varepsilon}_{ij}$. The same form [with another function $g_2(\theta)$] also appears as the second term of (2.48). Integrating around a small circle of radius r leads next to a J integral of the form $\sigma_Y \varepsilon_Y \tilde{K}^{m+1}$(function of m), the function containing integrals of the θ-dependent terms. The results of the total analysis were expressed as

$$J = J_1 = \sigma_Y \varepsilon_Y \tilde{K}^{m+1} I(m) \tag{2.107}$$

with values of $I(m)$ according to Table 2.1.†

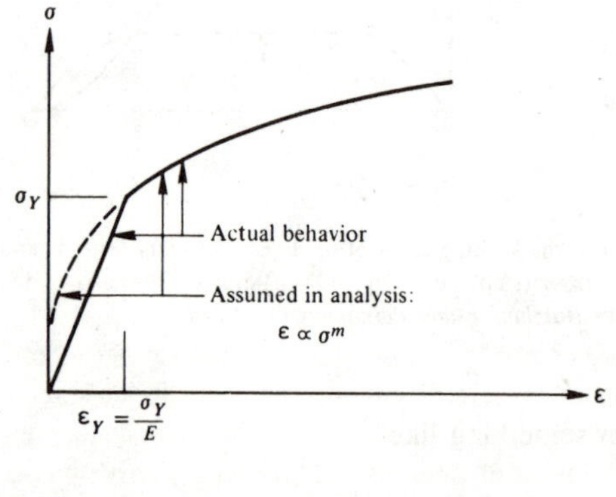

Figure 2.27 Strain versus uniaxial stress (idealizations).

†As expected, J is path- or r-independent in this near-tip region. This property might have been anticipated as a starting point for deriving the root λ_1: if the stress and the strain expressions had been based directly on (2.1), (2.5), and (2.95), an expression of J would ensue which would lose its r dependence only if $\lambda = \lambda_1$. This way of reasoning was followed by Rice and Rosengren [110] who, simultaneously with, and independently of Hutchinson undertook a study restricted to plane-strain situations.

Table 2.1 Values of $I(m)$†

m	1	3	5	9	13	∞
Plane stress	2π	3.86	3.41	3.03	2.87	~2.8
Plane strain	$2\pi(1-\nu^2)$	5.51	5.01	4.60	4.40	~4.3

†The reported values of $I(m)$ correspond to such scaling of the stresses that $\tilde{\sigma}_e$ attains the maximum value of unity.

When J is invariant throughout the body, it should follow that

$$J_1 = J_2 = J \tag{2.108}$$

where J_2 is the integral as derived well away from the crack tip, possibly in the elastic region. With (2.107) and (2.108) the strength of the plastic singularity ("plastic stress intensity factor") then follows as

$$\tilde{K} = \left[\frac{E}{\sigma_Y^2}\frac{J}{I(m)}\right]^{1/(m+1)} \qquad E \equiv \sigma_Y/\varepsilon_Y \tag{2.109}$$

and the singular stress-strain field has been completely determined for a given geometry and external action. The final result, together with (2.103) and (2.104), adds detail to the earlier assertion that J must represent a scalar measure of the state of straining at the crack tip. If yielding is highly localized (or small-scale) at the crack tip, expression (3.35) in particular applies (with $\beta = 1$ in plane stress or $\beta = 1 - \nu^2$ in plane strain) so that from (2.109)

$$\tilde{K} = \left(\frac{K_I}{\sigma_Y}\right)^{2/(m+1)}\left[\frac{\beta}{I(m)}\right]^{1/(m+1)} \tag{2.110}$$

In principle, the problem has now been solved within the frame of certain assumptions and expectations, for cracked bodies of a material displaying a power-law hardening. Again we have seen the transition in the strain singularities [compare (2.104) with (2.81)] from the type $r^{-1/2}$ for linear elasticity to r^{-1} for ideal plasticity. In the latter limit the stresses turned out to be finite, according to (2.103), and their values coincided with the results derived by methods of characteristics (slip-line theory in the case of plane strain).

Finally, we discuss briefly the derived results [111] for *linear* hardening, when the curve $\sigma > \sigma_Y$ in Fig. 2.27 is replaced by a straight line having the slope E_t. The near-tip solution turns out to be identical to the linearly elastic one as far as radial ($r^{-1/2}$) and circumferential distribution of stresses and strains are concerned. Only the strengths have to be reduced [by a factor of $(E_t/E)^{1/2}$ in plane stress and $1.15(1-\nu^2)^{1/2}(E_t/E)^{1/2}$ in plane strain] when yielding is small-scale and geometry and external actions are given. The final result presumes that E_t is small compared with E. The fact that the singularity of w (including the strain energy density ϕ) has turned out to be r^{-1} in all cases considered is merely a reflection of the fact that J is locally path-independent.

2.9 PLASTIC INSTABILITY, BIFURCATIONS, AND LIMIT LOAD

Even for bodies where (hypothetically) cracks do not appear or do not grow, there is a limit to the action which can be exerted. On the assumption that loads are controlled, such critical action may be linked to "plastic instability," i.e., load maxima associated with geometry changes in essentially tensioned members, if the material hardens and to the limit load if the material can be viewed as ideally plastic. Both types are relevant as limiting cases in the analysis of combined cracking and extensive yielding. They will be discussed briefly in the order mentioned, the first only in qualitative terms. Related to such collapse may be bifurcation patterns through the localization of plastic deformation into necking regions where final fracture takes place. In addition, there is, of course, the important class of collapse mechanisms related to buckling, which is not considered in this book.

Take first a homogeneous cylindrical tension bar of length L (initially L_0) and cross-sectional area S (initially S_0), which carries a load P given by

$$P = \sigma S$$

where σ is the "true" (or Cauchy-defined) stress referred to the deformed section. Load maximum or plastic instability following monotonic loading will then be characterized by

$$dP = S\,d\sigma + \sigma\,dS = 0 \qquad (2.111)$$

as indicated in Fig. 2.28. If the ductility is high, so that elastic strains are relatively negligible, the deformation can be viewed as isochoric, i.e.,

$$d(LS) = S\,dL + L\,dS = 0 \qquad (2.112)$$

whereupon a combination of (2.111) and (2.112) leads to

$$\frac{d\sigma}{\sigma} = \frac{dL}{L} \equiv d\varepsilon \qquad \text{or} \qquad \sigma = \frac{d\sigma}{d\varepsilon} \qquad (2.113)$$

Thus, the carrying capacity has been exhausted when the true stress equals the tangent modulus $d\sigma/d\varepsilon$ in the relation between σ and the true or logarithmic strain, $\varepsilon = \int d\varepsilon = \ln(L/L_0)$. Modifications of this result also apply to other tensioned structures, e.g., pressure vessels of certain regular shapes. The observation aims

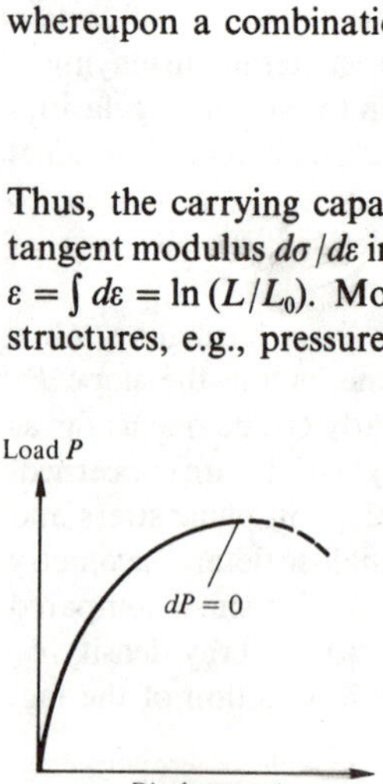

Figure 2.28 Plastic instability of a tension bar.

principally at a homogeneous body and is approximately valid if the net section is not significantly reduced by cracking. This kind of restriction may be hypothetical, as indicated initially, but serves to provide an upper bound of the collapse load resulting from any combined assessment.

Experience indicates that there is some relation between the unstable state as determined for a homogeneous or smooth deformation and the bifurcation into alternative critical patterns. Thus, in our particular example it is generally observed that near the load maximum (theoretically somewhat later) a localized contraction or necking will occur. At the same location the final separation takes place (Fig. 2.29*a* to *c*). The absolutely ductile rupture in metals, normally realizable only in such extremes as pure lead or gold, will proceed by contracting into a zero section as in Fig. 2.29*b*, whereas the normal event is that of Fig. 2.29*c*. The neck has had a doubly adverse effect, partly by locally increasing the axial stress σ and partly as the site of radial normal stresses which contribute to a high level of triaxial tension near the axis. This may further promote a cracking from the interior of the neck, related to the growth and coalescence of voids, until final separation occurs at the surface in the mode of slanted shearing. In the smaller dimension even the breaking of the ligaments between the voids may have the character of plastic instability.

As this suggests, crack growth is normally involved in the final failure, even if it happens at a load level which can be determined without having to take cracking specifically into account.

Further examples of the determination of critical loads are provided by the limit-load analysis. With the assumption of ideal plasticity and small changes in geometry, the limit load is defined as that stationary load value which can take the body into a mechanism of unbounded displacements (P_0 in Fig. 2.30). We shall review a few results with particular bearing on fracture-mechanical events, adding some comments to the individual cases. Being concerned in each case with a plate of thickness B, we make use of the notation

$$\sigma_Y = \text{yield stress in uniaxial tension}$$

$$\sigma'_Y = \begin{cases} \dfrac{2\sigma_Y}{\sqrt{3}} & \text{for Mises material} \\ \sigma_Y & \text{for Tresca material} \end{cases}$$

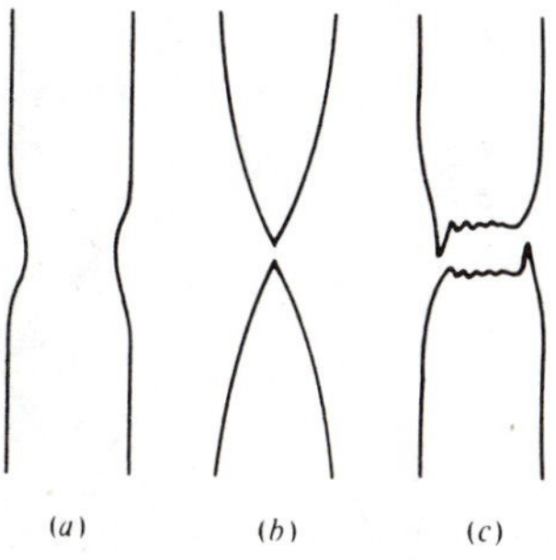

Figure 2.29 Necking and types of ultimate separation in a tension bar of circular section.

Table 2.2 Limit loads and yield mechanisms of some typical cracked geometries

Case 1: Plate with internal through-crack

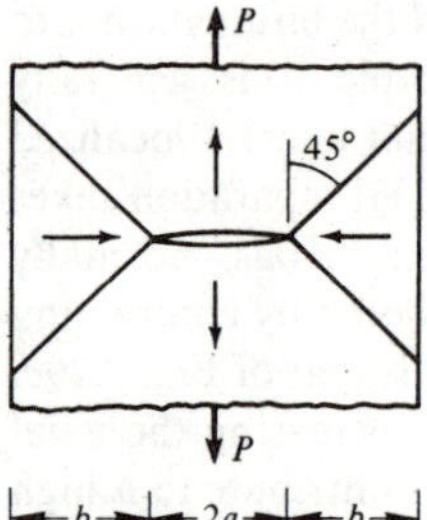

1*a*. Plane strain

$$P_0 = 2bB\sigma'_Y$$

yield lines (shear lines) at 45°

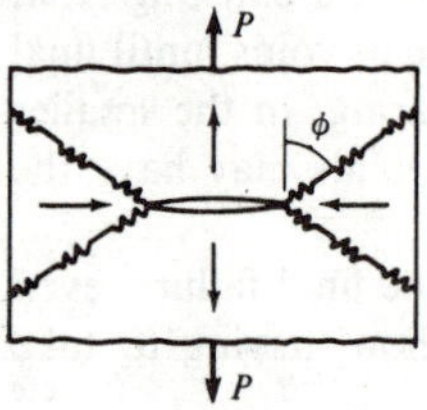

1*b*. Plane stress

$$P_0 = 2bB\sigma_Y$$

yield lines (necking) at $\phi = 54.7°$ (Mises) or $45° \leq \phi \leq 90°$ (Tresca)

Case 2: Plate with edge cracks

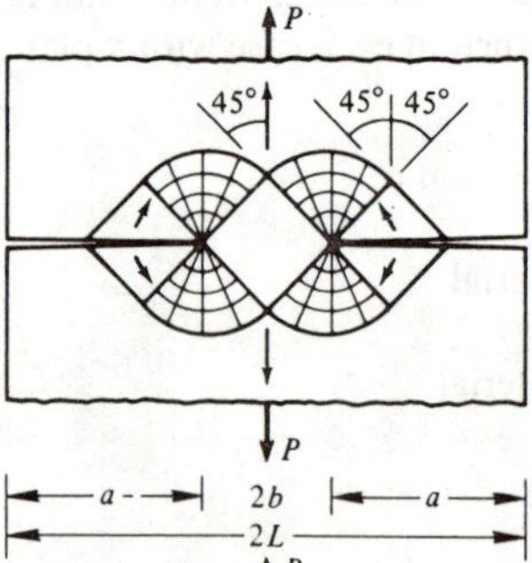

2*a*. Plane strain

$$P_0 = \begin{cases} \left(1 + \dfrac{\pi}{2}\right) 2bB\sigma'_Y & \dfrac{L}{b} \geq 8.6 \\ \left(1 + \ln \dfrac{1 + L/b}{2}\right) 2bB\sigma'_Y & \dfrac{L}{b} \leq 8.6 \end{cases}$$

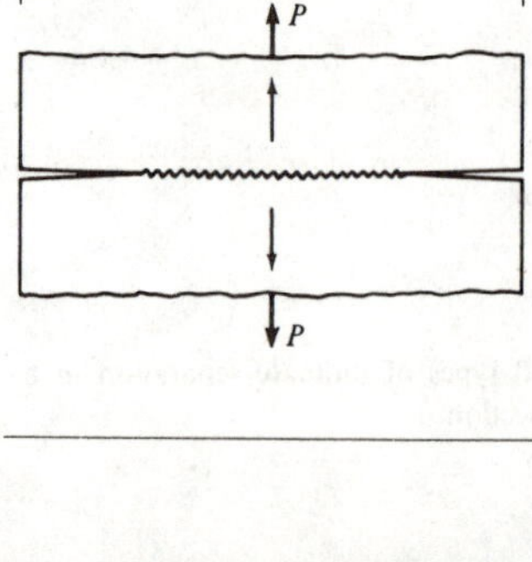

2*b*. Plane stress

$$P_0 = 2bB\sigma'_Y \qquad \frac{L}{b} > \text{about } \frac{4}{3}$$

Case 3: Double cantilever beam

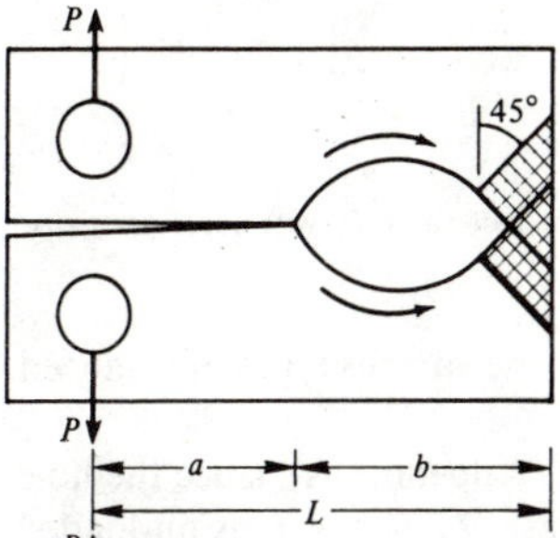

3*a*. Plane strain

$$P_0 = \beta b B \sigma'_Y$$

β given by

$$1.26\sqrt{\beta^2 + \frac{2L}{b}\beta} - \beta = 1$$

($L/b \to \infty$; $PL \to M$ gives limit moments M_0 for case 4)

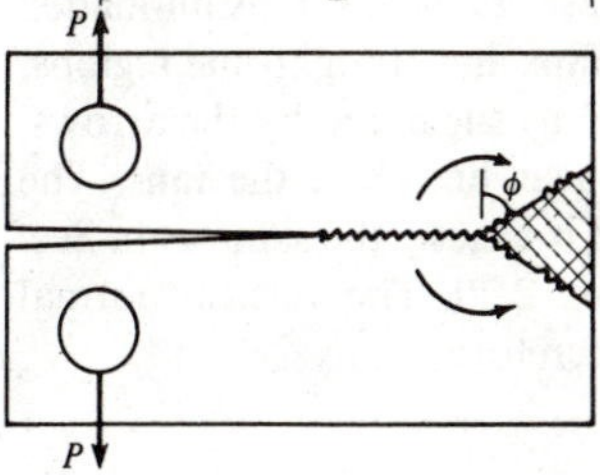

3*b*. Plane stress

$$P_0 = \beta b B \sigma'_Y$$

β given by

$$\beta^2 + 2\beta\left[\left(1 + \frac{\sigma_Y}{\sigma'_Y}\right)\frac{L}{b} - 1\right] = \frac{\sigma_Y}{\sigma'_Y}$$

ϕ as in case 1*b*

Case 4: Four-point bending and edge crack

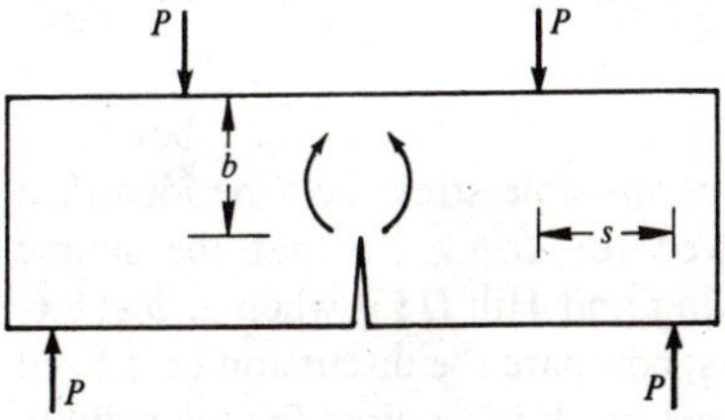

4*a*. Plane strain

$$M_0 = P_0 s = 0.315\sigma'_Y b^2 B$$

4*b*. Plane stress

$$M_0 = P_0 S = \frac{\frac{1}{2}\sigma'_Y b^2 B}{1 + \sigma'_Y/\sigma_Y}$$

For kinematic fields at the crack tip see case 3

Case 5: Three-point bending and edge crack

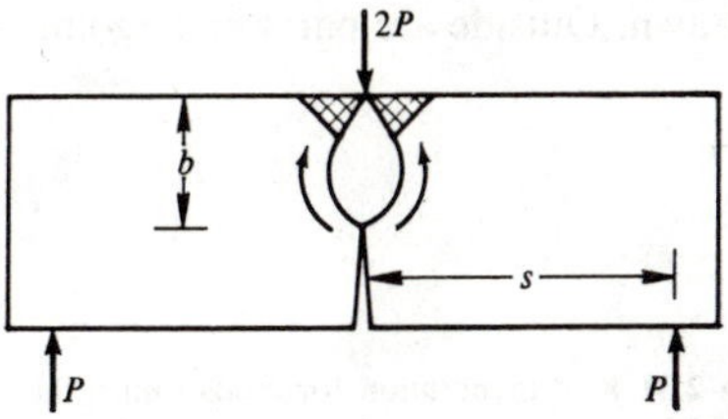

5*a*. Plane strain (see the figure)

$$M_0 = P_0 s \approx 0.31\sigma'_Y b^2 B$$

An equality similar to 4*b* may apply in plane stress

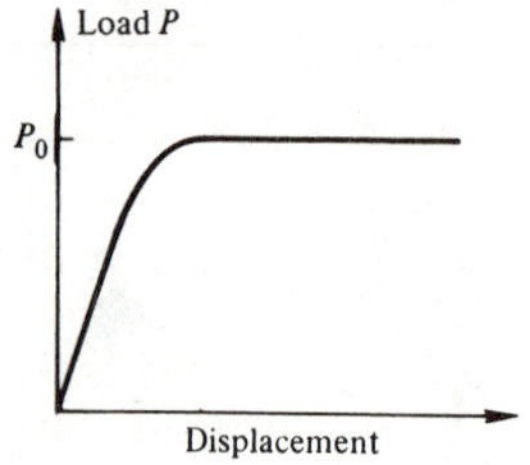

Figure 2.30 The limit-load plateau value P_0 with ideal plasticity.

and the symbolism in Fig. 2.31 to indicate (Table 2.2) the various types of localized or distributed yielding.

Case 1 is the same as dealt with in Example A.3 (Appendix A), since the hole considered there may have an arbitrary shape. Case 2*a* is the tension-loaded *Prandtl field*, a slip-line system consisting of central fans and straight-line regions. In the latter part the material moves as rigid (elastic) as suggested by the arrows, while the plastic deformation is restricted to the shear lines and the fans. The detailed derivation of the stresses in the slip-line field is exactly the same as in Sec. 2.8.1, the upper left quadrant corresponding to Fig. 2.20. The vertical normal stress in the central section (compare zone C) is therefore

$$\sigma_{22} = (2+\pi)\tau_Y \equiv \left(1+\frac{\pi}{2}\right)\sigma'_Y$$

giving the load

$$P = 2bB\sigma_{22} = \left(1+\frac{\pi}{2}\right)2bB\sigma'_Y$$

For this to be interpreted as the limit load (and not merely an upper bound to it) it is also necessary that a corresponding permissible stress field be identified throughout the body. This has been achieved for $L/b \geq 8.6$, but the simple solution must be replaced by one due to Ewing and Hill [135] when $L/b < 8.6$. Case 2*b* is a pure neck between the edge cracks; compare the discussion on a fixed width in Example A.1. Again, a sufficiently deep crack is required for the validity of the solution, approximately $L/b \geq \frac{4}{3}$ [123].

Bending cases 3 to 5 were treated by Ewing and Richards [136], Green [137], and Green and Hundy [138], respectively. As in case 2*a*, there are zones where distributed yielding may take place, within which the zero-stretching trajectories or the characteristics of the field have been drawn. Outside are parts undergoing

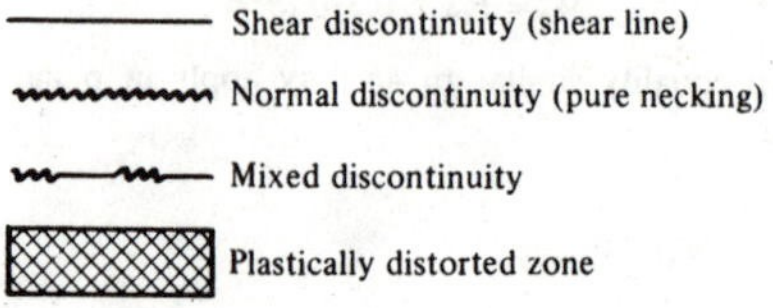

Figure 2.31 Key to notation for fields under the limit load.

rigid-body rotations, as the arrows indicate. In case 3*a*, particularly, the rotation occurs by gliding along circular shear lines. The difference between the limit moments M_0 for three- and four-point bending is very small in plane strain and probably also in plane stress if buckling is prevented.

We bring this discussion of stationary cracks to a close with a brief review. Sections 2.1 and 2.2 were concerned with purely elastic analyses, concluding with singular stresses at the crack tip and indicating that in reality yielding has to take place. On this basis the analysis in Secs. 2.3 to 2.8 was largely restricted by the assumption that yielding is highly localized at the crack tip. Cases of more extensive yield are mentioned (partly without derivations) in Secs. 2.5 to 2.7 and formally included in Sec. 2.8.2. The larger the extent of yielding realized the higher must have been the resistance or toughness toward separation at the crack front. The most extreme (infinite) toughness is necessary to permit yielding to develop to such levels as are considered in this section, when the load extrema are given by the plastic deformation of the total body.

PROBLEMS

2.1 Reversing the procedure in Sec. 2.1, verify that the solution

$$\chi = C_{11} r^{3/2}\left(\cos\frac{\theta}{2} + \frac{1}{3}\cos\frac{3\theta}{2}\right)$$

is statically and kinematically permissible in an elastic plate. Derive the polar stresses σ_θ and $\tau_{r\theta}$, and find that they correspond to the (crack) surfaces $\theta = \pm\pi$ being nonloaded. Compare (2.15).

2.2 A large plate has a small internal crack through the thickness. The crack is oriented normal to the direction of a remote tension $\sigma_\infty = 1000$ MPa (see $\alpha = 0$ in Prob. 2.4). For the crack length $2a = 10$ mm, consult Appendix C to find the stress intensity factor

$$K_I = \sigma_\infty\sqrt{\pi a} = 125 \text{ MPa m}^{1/2}$$

2.3 A cracklike slit in a large plate is loaded by a central wedge force P per unit of plate thickness. Consult Appendix C to find the stress intensity factor

$$K_I = \frac{P}{\sqrt{\pi a}}$$

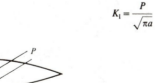

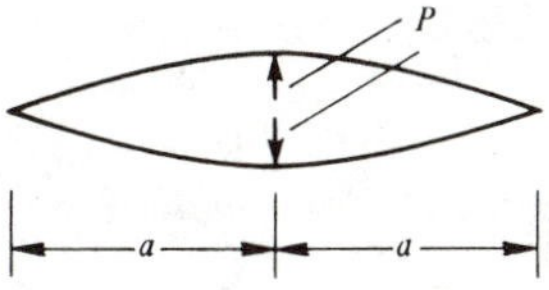

Problem 2.3

2.4 Problem 2.2 is generalized into the crack being rotated through the angle α with respect to the normal orientation. Derive the stress intensity factors

$$K_I = \sigma_\infty\sqrt{\pi a}\cos^2\alpha$$

$$K_{II} = \sigma_\infty\sqrt{\pi a}\sin\alpha\cos\alpha$$

by transforming the remote stress onto the crack direction and its normal.

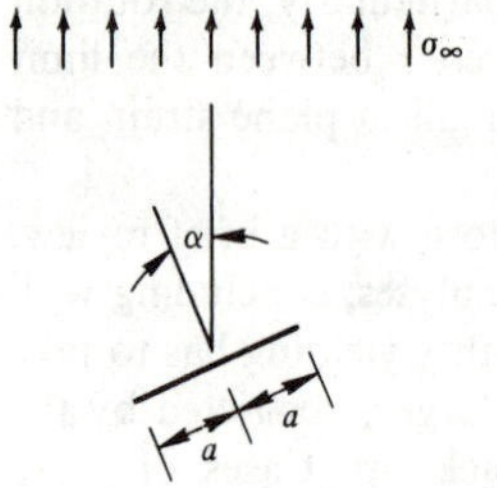

Problem 2.4

2.5 A large plate contains a wedge-loaded cracklike slit, as described in Prob. 2.3. The material is elastic and ideally plastic with yield stress σ_Y. The plate is so thin that a plane-stress condition will be assumed. Local buckling may be disregarded.

(*a*) Write formulas for the length c of the yield zone extending from one end of the slit, assuming that the yielding is sufficiently localized to be governed by the elastic singularity K_I. Apply (1) the Irwin approximation and (2) the Dugdale solution.

(*b*) Using the Dugdale model, derive an equation which determines c valid even outside the small-scale yielding range considered in part (*a*).

(*c*) With the data $P = 20$ MN/m, $a = 10$ mm, and $\sigma_Y = 1000$ MPa verify the solutions $c = 4.0$ mm and $c = 5.0$ mm for (1) and (2) of part (*a*) and $c = 3.5$ mm for part (*b*). Obviously, with this high ratio of c/a only the last solution should be considered valid.

2.6 A historical prelude to the Dugdale-model evaluation of the COD is the following prescription. Consider the Irwin "equivalent crack," extending the length $c/2$ by (2.39) beyond the actual crack tip in an imaginary elastic material. For this imaginary crack and material evaluate the separation of the crack sides at the location of the actual crack tip. This separation $\hat{\delta}$ is taken to be an approximation of the true COD δ.

For small-scale yielding behavior, use (2.33) to derive in this way

$$\hat{\delta} = \frac{4}{\pi}\frac{K_I^2}{E\sigma_Y} = \frac{4}{\pi}\delta$$

under mode I plane-stress conditions.

2.7 For the compact-tension test specimen shown, verify the stress intensity factor

$$K_I = 6.82\frac{P}{B\sqrt{a}}$$

and the limit load in a plane-strain condition

$$P_0 = 0.257Ba\sigma_Y$$

by consulting Appendix C and Sec. 2.9.

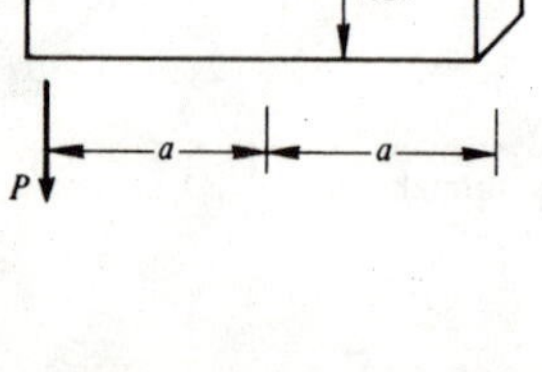

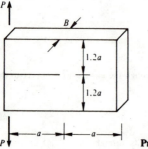

Problem 2.7

CHAPTER

THREE

ENERGY BALANCE AND CRACK GROWTH

3.1. LOCAL AND GLOBAL FORMS OF SEPARATION WORK

We focus our attention on an existing crack or cracklike notch, which may extend to the outer surface, as in Fig. 3.1, or be totally embedded in the body. The body is loaded by surface tractions $\mathbf{p} = \{p_i\}$ and body forces per unit of volume $\mathbf{b} = \{b_i\}$ $(i = 1, 2, 3)$. The displacements $\mathbf{u} = \{u_i\}$ and their gradients are assumed to be small before and during the growth of the crack.

Crack growth is shown in Fig. 3.2. The dashed area represents the fracture process zone, which by assumption (Chap. 1) has a negligible thickness. The boundary between this zone and the surrounding continuum can then be regarded as the surface of the crack. The crack area is assumed to increase from A in stage a of development to $A + \Delta A$ in stage b. During this event a new crack surface S_c^+ is created. Together with the existing crack surface S_c and the exterior surface S_e of the body, this forms the boundary of the continuum during the transition from stage a to stage b. In addition to the forces on the existing surface already mentioned (including $p_i = 0$ on S_c if the crack is not loaded) tractions p_i will also act on S_c^+. They vary during the transition from p_i^a in stage a through a complex intermediate history to p_i^b $(= 0$ if the crack is not loaded) in stage b.

Since now, by definition, the traction exerted through S_c^+ from the continuum on the process zone is $-\mathbf{p}$, the work done on this zone during the transition will be

$$-\int_{S_c^+} \left(\int_{(a)}^{(b)} p_i \, du_i \right) dS$$

(sum over repeated indices) where dS denotes the differential area of surface and

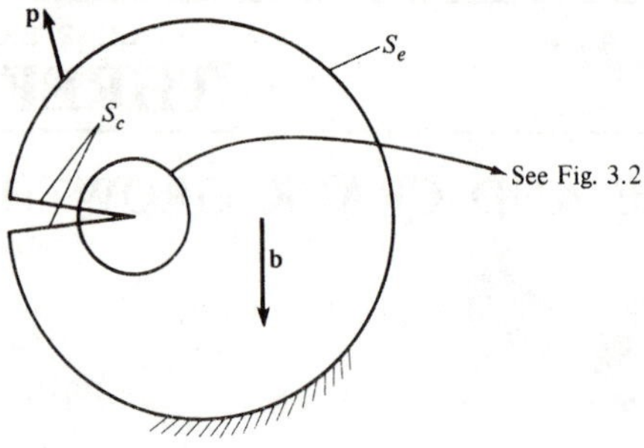

Figure 3.1 A cracked specimen under body forces, surface tractions, and kinematic constraints.

du_i are the displacement increments at that location.† Accordingly, work is done per unit of crack area generated subsequent to stage a equal to

$$C = -\lim_{\Delta A \to 0} \frac{1}{\Delta A} \int_{S_c^+} \left(\int_{(a)}^{(b)} p_i \, du_i \right) dS \qquad (3.1)$$

which will be called the *separation work.*

Alternatively, a global expression for the separation work can be obtained by applying the principle of virtual work to the body. With the displacement field $du_i(x_j)$ corresponding to a differential crack motion within $t^a < t < t^b$, this implies the identity

$$\int_{S+S_c^+} p_i \, du_i \, dS + \int_V (b_i - \rho \ddot{u}_i) \, du_i \, dV = \int_V \sigma_{ij} \, d\varepsilon_{ij} \, dV \qquad (3.2)$$

where S has been introduced as the union of S_e and S_c. The right-hand integral expresses the work of deformation, the strain increments $d\varepsilon_{ij}$ deriving from the

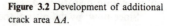

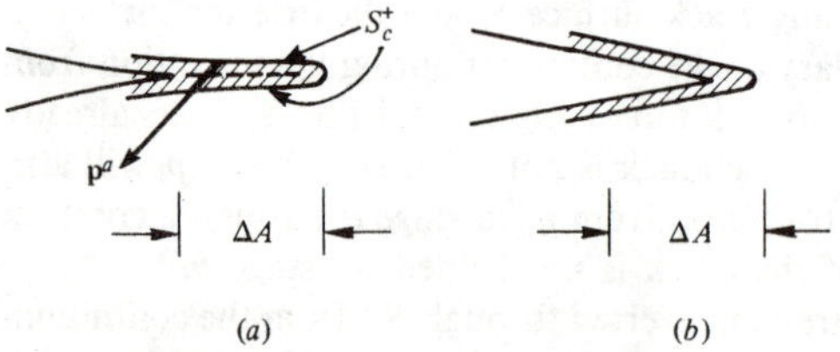

Figure 3.2 Development of additional crack area ΔA.

†The inner integral should be strictly interpreted as

$$\int_{t^a}^{t^b} p_i \dot{u}_i \, dt \qquad \dot{u}_i = \frac{du_i}{dt}$$

where $\dot{u}_i$ are the velocities and t^a and t^b are the times bounding the transition period. Note further that under the small-displacement assumption the order of integrations in space and time is arbitrary.

field du_i. Integrating (3.2) through the transition yields

$$\int_S \left(\int_{(a)}^{(b)} p_i \, du_i \right) dS + \int_V \left[\int_{(a)}^{(b)} (b_i - \rho \ddot{u}_i) \, du_i \right] dV - \int_V \left(\int_{(a)}^{(b)} \sigma_{ij} \, d\varepsilon_{ij} \right) dV = - \int_{S_c^+} \left(\int_{(a)}^{(b)} p_i \, du_i \right) dS \tag{3.3}$$

In the first integral, which does not include the new crack surface, the variation of p_i may be neglected in the limit as $\Delta A \to 0$, and similarly the b_i then appear as constants. The volume integrals further contain the complete differential

$$\ddot{u}_i \, du_i = \tfrac{1}{2} d(\dot{u}_i \dot{u}_i) \tag{3.4}$$

and also

$$\sigma_{ij} \, d\varepsilon_{ij}^e = d\phi \tag{3.5}$$

when the deformation is split into an elastic and a plastic part,

$$d\varepsilon_{ij} = d\varepsilon_{ij}^e + d\varepsilon_{ij}^p \tag{3.6}$$

ϕ being the strain energy density. With the notation

$$f^b - f^a = \Delta f$$

(f representing u_i, T, or Φ below) the left-hand side of (3.3) can then be expressed as

$$\int_S p_i \, \Delta u_i \, dS + \int_V b_i \, \Delta u_i \, dV - \Delta T - \Delta \Phi - \int_V \left(\int_{(a)}^{(b)} \sigma_{ij} \, d\varepsilon_{ij}^p \right) dV$$

in view of the limit as $\Delta A \to 0$. Here

$$T = \frac{1}{2} \int_V \rho \dot{u}_i \dot{u}_i \, dV \tag{3.7}$$

is the total kinetic energy, and

$$\Phi = \int_V \phi \, dV \tag{3.8}$$

is the total elastic-strain energy.

We assume finally that the progress of the fracture surface is well enough defined to be described by the area A as a single parameter. In the limit, then, (3.3) turns into

$$W - \frac{dT}{dA} - \frac{d\Phi}{dA} - D = C \tag{3.9}$$

where

$$W = \int_S p_i \frac{du_i}{dA} \, dS + \int_V b_i \frac{du_i}{dA} \, dV \tag{3.10a}$$

is the external work done by p_i and b_i and

$$D = \lim_{\Delta A \to 0} \frac{1}{\Delta A} \int_V \left(\int_{(a)}^{(b)} \sigma_{ij} \, d\varepsilon_{ij}^p \right) dV \tag{3.11}$$

is the continuum dissipation, both referred to the unit of new crack area; $df/dA \equiv \lim (\Delta f/\Delta A)$ as $\Delta A \to 0$ is understood above.

Note that if the work is done only by a number of concentrated forces P_k $(k = 1, 2, 3, \ldots, n)$ through associated displacements u_k, (3.10a) will simplify into

$$W = P_k \frac{du_k}{dA} \tag{3.10b}$$

The separation work has been expressed by quantities localized at the crack front in (3.1) and through a global energy balance in (3.9). Both forms will find later application. The second is suggestive in that the separation work C appears as the surplus of expended work over the dissipation D and the increase in kinetic and elastic energy.

During the growth of a crack, forces and displacements may change. Some of these are controlled actions while others follow as dependent variables (reactions). That kind of control is considered possible (by some assumption on the material's behavior) which permits a quasi-static crack growth, i.e., a development through successive stages of equilibrium such that $dT/dA = 0$. By a *virtual* crack growth we shall understand any infinitesimal motion which takes place quasi-statically when actions (a traction *or* displacement component in each direction at all boundary points) are kept constant.

Alternatively, (3.9) can be written

$$\mathscr{G} = C + D \tag{3.12}$$

as the balance between the net expenditure of mechanical energy

$$\mathscr{G} \equiv W - \frac{dT}{dA} - \frac{d\Phi}{dA} \tag{3.13}$$

and the total dissipation $C + D$ within the cracked body. Both sides of (3.12) will be discussed; we begin with the left-hand side in the case of elastic response.

3.2. QUASI-STATIC CRACK GROWTH IN AN ELASTIC BODY

In the context of essentially elastic response $\mathscr{G}$ by (3.13) is traditionally termed the *crack driving force*; this is contrasted by the right-hand *resistance* in (3.12) if the crack is actually growing. Under the present quasi-static conditions $\mathscr{G}$ is given by

$$\mathscr{G} \, dA = W \, dA - d\Phi \tag{3.14}$$

with

$$W \, dA = \int_S (p_i \, du_i) \, dS + \int_V (b_i \, du_i) \, dV \tag{3.15a}$$

or

$$W\,dA = P_k\,du_k \tag{3.15b}$$

the strain energy Φ depending uniquely on the crack area A and the current forces p_i, b_i, and P_k.† Alternatively, we can consider Φ as depending on A and the displacements u_i, u_k at the loading locations. These relationships imply such great simplifications that fracture mechanics has been extensively aimed at essentially elastic behavior.

3.2.1 The Crack Driving Force in Terms of Energy Rates

Useful interpretations of (3.14) are sought. It is profitable to start here by looking into the effects of a single of coupled load (Fig. 3.3*b*). Equation (3.14) with (3.15*b*) then takes the form

$$\mathscr{G}\,dA = P\,du - d\Phi \tag{3.16}$$

In particular, for *linear* elasticity we have

$$\Phi = \tfrac{1}{2}Pu \tag{3.17}$$

as can be deduced from (3.16) on condition that A be kept constant. The terms of (3.16) have been illustrated in Fig. 3.3*a*. Lines OF and OF' reflect loading and unloading for the given body with crack areas A and $A + dA$, respectively, while the trajectory $P(u, A)$ indicates a possible relationship between the load P and the displacement u as they interact while the crack is moving. The term $P\,du$ then appears as the area $EE'F'F$, while $d\Phi$ is the difference between the areas below OF' and OF, thus

$$d\Phi = \Phi(A + dA) - \Phi(A) = OE'F' - OEF$$

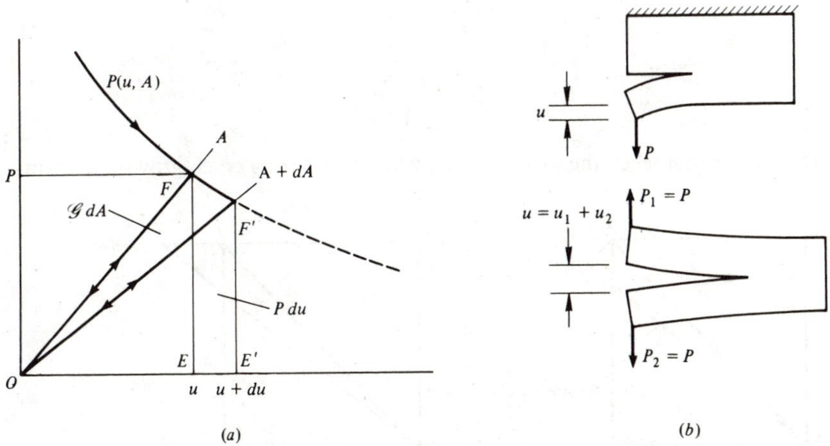

Figure 3.3 A possible load trajectory for a growing crack. Identification of the crack driving force $\mathscr{G}$.

†Recall that such uniqueness depends on conditions being either isothermal or adiabatic.

By (3.16) we then have the sum

$$\mathscr{G}\,dA = EE'F'F + OEF - OE'F' = OF'F$$

equal to the area between the loading curves for A and $A + dA$ and the tractory $P(u, A)$. Having measured a curve $P(u, A)$ and knowing the loading curves for neighboring crack areas, we can thus follow the development of the crack driving force as a crack is growing. This graphical interpretation can be carried over to a mathematical form in two different ways, related to Fig. 3.4*a* and 3.4*b*, respectively. First, $\mathscr{G}\,dA$ can be viewed as the reduction of Φ as the crack area increases if u is kept constant. Thus,

$$\mathscr{G} = -\left(\frac{\partial \Phi}{\partial A}\right)_u \tag{3.18}$$

because the small triangle to the right of $u = \text{const}$ is negligible for infinitesimal increments of A. Second, $\mathscr{G}\,dA$ can be viewed as the increase in complementary energy

$$\Omega = Pu - \Phi \tag{3.19}$$

if P is kept constant while the crack grows. Thus,

$$\mathscr{G} = \left(\frac{\partial \Omega}{\partial A}\right)_P \tag{3.20}$$

or

$$\mathscr{G} = P\left(\frac{\partial u}{\partial A}\right)_P - \left(\frac{\partial \Phi}{\partial A}\right)_P$$

If for *linear* elasticity the equality $\Omega = \Phi$ is introduced [compare (3.17) and (3.19)], it further follows from (3.20) that

$$\mathscr{G} = \left(\frac{\partial \Phi}{\partial A}\right)_P \tag{3.21}$$

or

$$\mathscr{G} = \tfrac{1}{2}P\left(\frac{\partial u}{\partial A}\right)_P$$

This shows that when the loading is kept constant during crack growth, the work

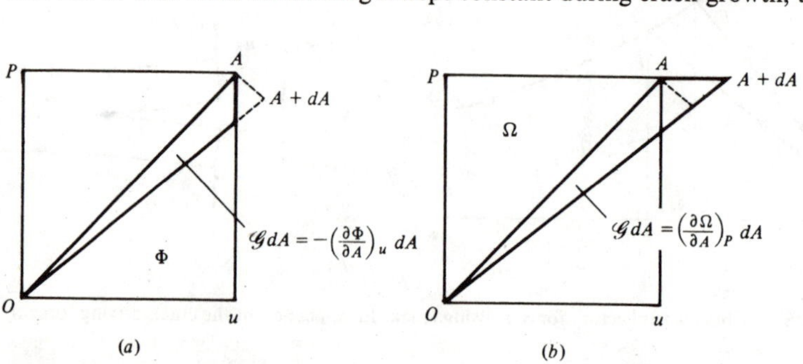

Figure 3.4 Alternative forms of the crack driving force.

done on such a body is divided equally between crack driving force and strain energy rate.

A generalization of (3.17) and (3.19) such that (3.18) and (3.20) hold for any distributed loading or some system of concentrated loads, can be expressed by

$$\Omega = \int_S p_i u_i \, dS + \int_V b_i u_i \, dV - \Phi \tag{3.22a}$$

$$\Omega = P_k u_k - \Phi \tag{3.22b}$$

combined with $\Omega = \Phi$ in the case of linear elasticity.

The relations (3.18), (3.20), and (3.22) are easily demonstrated by invoking the uniqueness of the strain energy function (Appendix E). One must satisfy $b_i = 0$ for (3.18) to apply. This form holds if displacements are viewed as controlled, while (3.20) assumes a fictitious load control. One might even imagine some mixed type of control, to include both displacements and forces. The general result can be expressed in terms of potential energy Π

$$\mathcal{G} = -\frac{\partial \Pi}{\partial A} \qquad \text{keeping actions constant} \tag{3.23}$$

$$\Pi = \Phi - \int_{S_p} p_i u_i \, dS - \int_V b_i u_i \, dV$$

where S_p denotes the surface where p_i is controlled. The form (3.23) includes those above, as $\Pi = \Phi$ for assumed displacement control (when $b_i = 0$ is required) while $\Pi = -\Omega$ for load control. We can then make the general statement that the crack driving force in an elastic body equals the rate of potential-energy decrease during a virtual crack growth.

In the context of (3.17) we can carry the discussion a bit further by introducing the linear relation

$$u = ZP$$

where $Z(A)$ is the compliance function, here assumed to be uniquely related to A. This gives

$$\Phi = \tfrac{1}{2}P^2 Z = \frac{\tfrac{1}{2}u^2}{Z} \tag{3.24}$$

and then from (3.21) or (3.18)

$$\mathcal{G} = \tfrac{1}{2}P^2 \frac{dZ}{dA} = -\tfrac{1}{2}u^2 \frac{d(Z^{-1})}{dA} \tag{3.25}$$

The compliance methods mentioned in Sec. 2.2 have their background in the above considerations. As is seen, the derivation of the crack driving force may be tantamount to finding the compliance function $Z(A)$. This can be done analytically (in simple cases; see Examples 3.1 and 3.2 below), numerically, or experimentally.

Example 3.1 The double-cantilever-beam (DCB) specimen finds application in experimental fracture mechanics. Assuming the beams to be slender, $a \gg h$ in the figure, we recall the elementary theory of cantilevered beams to write

$$u = 2\frac{Pa^3}{3EI} \qquad I = \frac{Bh^3}{12}$$

for plane-stress conditions. [$E/(1-\nu^2)$ replaces E in the case of plane strain, i.e., larger width of the beam.] This implies

$$Z = \frac{2a^3}{3EI}$$

and thus, by the use of (3.25),

$$\mathscr{G} = \frac{P^2}{2B}\frac{dZ}{da} = \frac{P^2}{2B}\frac{2a^2}{EI} = \frac{u^2}{2B}\frac{9EI}{2a^4}$$

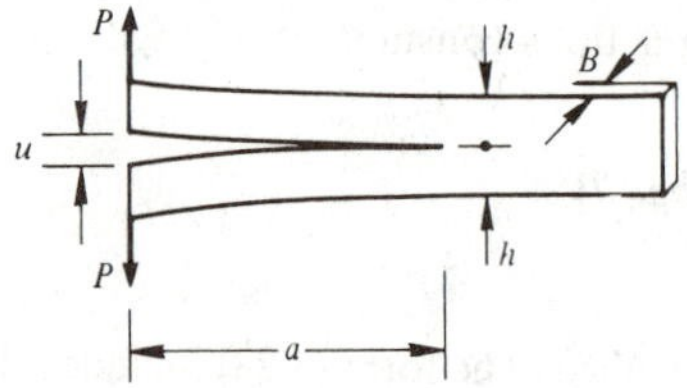

Example 3.1

Example 3.2 Example 1.1 involved

$$Z = \frac{H}{\eta Eb} = \frac{H}{\eta E(A_t - A)}$$

where the total area $A_t = A + b \cdot 1$ has been introduced. This is seen from (1.1), where $\sigma = P/b$. Accordingly,

$$\mathscr{G} = -\frac{u^2}{2}\frac{d(Z^{-1})}{dA} = \frac{\eta E}{2H}u^2$$

in agreement with (1.4).

3.2.2 The Crack Driving Force in Terms of Stress Intensity Factors

For the purely elastic body $\mathscr{G} = C$ applies, since then $D = 0$ in (3.12); we can therefore resort to (3.1) when a relation is sought between $\mathscr{G}$ and the near-tip field parameters. The expression

$$-\mathscr{G}\,\Delta A = \int_{S_c^+}\left(\int_{(a)}^{(b)} p_i\,du_i\right)dS \qquad \Delta A \to 0 \tag{3.26}$$

is now the work done on the elastic continuum to bring it quasi-statically from a state a, when tractions p_i^a are carried across the future path of the crack, to a state b, when $p_i^b = 0$ are carried after the passage. We have here a nonloaded crack in mind and shall return later to this assumption. The stress history is a successive unloading behind the crack front as it progresses, while simultaneously the state of stress in material points ahead of the crack is changed. However, (3.18) also shows that this work equals the change $\Delta\Phi$ of a state variable, which can be derived as for any other chosen path between states a and b. Therefore, instead of considering the actual, complex history, we could profitably approach b by a proportional relaxation from p_i^a to $p_i^b = 0$ in the whole surface S_c^+. This implies finding the local work when S_c^+ is subjected to additional forces varied from 0 to $-p_i^a$, such that the crack is gradually opened along ΔA to its final shape as included in the displacement field u_i^b.

We principally consider two-dimensional problems, the fields being independent of one coordinate (x_3) in a cartesian system x_1, x_2, x_3. Then the crack area can be replaced by the crack length a, considering a unit length along x_3, while x_1 is taken along the assumed codirectional path of the crack tip (Fig. 3.5). Introducing in (3.26) $p_i = -\sigma_{2i}$ $(+\sigma_{2i})$ on the upper (lower) crack side, and having $u_i = u_i^+$ above and $u_i = u_i^-$ below, we can then write the equation as

$$\mathcal{G} = \lim_{\Delta a \to 0} \frac{1}{\Delta a} \int_0^{\Delta a} \left[\int_{(a)}^{(b)} \sigma_{2i} d(u_i^+ - u_i^-) \right] dx_1 \tag{3.27}$$

For the chosen path of simultaneous relaxation and for *linear* elasticity this turns into

$$\mathcal{G} = \lim_{\Delta a \to 0} \frac{1}{2\Delta a} \int_0^{\Delta a} \sigma_{2i}^a (u_i^+ - u_i^-)^b \, dx_1 \tag{3.28}$$

since the additional loads and displacements are proportional in all points of the fracture surface.

It is further assumed that the origin is at the crack tip, as in Fig. 3.5*a*. For the case of symmetry (mode I, Fig. 3.5*a*) or antisymmetry (modes II and III) with respect to $x_2 = 0$, the stresses and displacements of concern are those considered

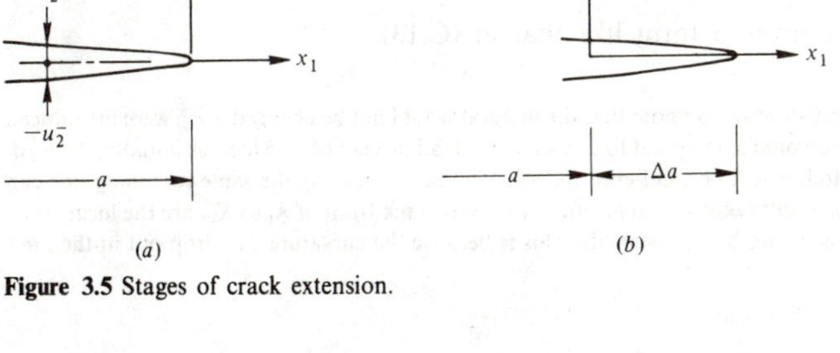

Figure 3.5 Stages of crack extension.

in (2.33), K_{I} to K_{III} being the stress intensity factors. These functions will dominate the fields at the crack tip and will govern the integral (3.28) uniquely in the limit as $\Delta a \to 0$.†

Also, let Fig. 3.5*a* represent an existing crack in the state *a* (compare Fig. 3.2); then σ_{2i} according to (2.33) are directly to be interpreted as σ_{2i}^a. On the other hand, u_i^b denote the crack geometry after continued progress into state *b* (Fig. 3.5*b*). In the limit as $\Delta a = 0$ this will be a translation of the fields at the crack front such that u_i^b will be the displacements according to (2.33) when $-x_1$ is replaced by $\Delta a - x_1$. This leads to

$$\int_0^{\Delta a} \sigma_{2i}^a (u_i^+ - u_i^-)^b \, dx_1 = \frac{2}{\pi G} \left[\frac{\kappa + 1}{4} (K_{\mathrm{I}}^2 + K_{\mathrm{II}}^2) + K_{\mathrm{III}}^2 \right] \int_0^{\Delta a} \sqrt{\frac{\Delta a - x_1}{x_1}} \, dx_1$$

where the last integral turns out to be $\pi \, \Delta a/2$. Finally, with the identity $\beta \equiv (\kappa + 1)E/8G$ we obtain from (3.28)

$$\mathcal{G} = \frac{\beta}{E} K_{\mathrm{I}}^2 + \frac{\beta}{E} K_{\mathrm{II}}^2 + \frac{1+\nu}{E} K_{\mathrm{III}}^2 \qquad \beta = \begin{cases} 1 & \text{for plane stress, } \sigma_{33} = 0 \\ 1 - \nu^2 & \text{for plane strain, } \varepsilon_{33} = 0 \end{cases} \tag{3.29}$$

provided the crack moves in the plane of the existing crack. In a pure opening mode ($K_{\mathrm{II}} = K_{\mathrm{III}} = 0$)

$$\mathcal{G} = \frac{\beta}{E} K_{\mathrm{I}}^2 \tag{3.30}$$

whereas the second or third term in (3.29) would appear singly if mode II or mode III were isolated.

Example 3.3 The stress intensity factor of the DCB specimen in Example 3.1 is sought. With (3.30) we find

$$K_{\mathrm{I}} = \sqrt{\frac{E\mathcal{G}}{\beta}} = \frac{Pa}{\sqrt{BI}} = \sqrt{\frac{12}{h^3}} \frac{Pa}{B}$$

as reproduced in (C.12) of Appendix C.

For the plate in Example 3.2 and Fig. 1.2 we similarly obtain

$$K_{\mathrm{I}} = \sqrt{\frac{\eta}{2H\beta}} Eu$$

which can be given a form like that in (C.13).

†With this in mind we may also note that the integral would not be changed if σ_{2i}^b were introduced with finite values. This would correspond to a given crack load instead of zero load as initially assumed. Thus, the results to follow will be valid even for a loaded new crack. By the same reasoning one can also conclude that the result holds for an arbitrarily curved crack front, if K_{I} to K_{III} are the local stress intensity factors (plane strain being assumed). This is because the curvature will drop out in the limit as $\Delta a \to 0$.

3.2.3 The Crack Driving Force as Related to the *J* Integral

Let Σ represent the area enclosed by the curve Γ in Fig. 2.12 and assume that the curve is shrunk toward the crack tip ($\Sigma \to 0$). Within this area the gradients are so large (toward singularities at the crack tip) that they dominate all local derivatives with respect to the crack length. Thus, the fields within $\Sigma \to 0$ will be "stationary" in the sense that they mainly translate with the crack tip during a differential crack motion. (This property was also anticipated in the transition related to Fig. 3.5.) Given the external action, when the crack tip moves a small step forward, the changes observed at a fixed location in Σ will therefore be the same as when the observer moves the same length back toward the stationary crack. Briefly, this is formulated by the substitution

$$\frac{\partial}{\partial x_1} = -\frac{\partial}{\partial a} \tag{3.31}$$

applying to some function of x_1 and a with x_1 measured from a fixed origin. Then the second right-hand term of (2.48) equals

$$-\int_\Gamma p_i \frac{\partial u_i}{\partial x_1}\, ds = \int_\Gamma p_i \frac{\partial u_i}{\partial a}\, ds \tag{3.32}$$

which can be interpreted as the rate of work exerted per unit thickness by the outside material on the material inside Γ as the crack moves. Likewise, $\int_\Gamma w\, dx_2\, 1$ can be seen as the total strain energy carried by particles into Σ per unit thickness and crack advance when that region moves with the crack tip. The sum J will therefore represent a net expenditure of mechanical energy per unit crack area during virtual growth, which again equals the crack driving force.† We have thus arrived at a simple relation and an important physical interpretation of the J integral

$$J = \mathscr{G} \tag{3.33}$$

provided that elasticity prevails and body forces are negligible.

Finally invoking (3.29) or (3.30), we note that the relation between J and the stress intensity factor for linear elasticity is

$$J = \frac{\beta}{E} K_{\mathrm{I}}^2 + \frac{\beta}{E} K_{\mathrm{II}}^2 + \frac{1+\nu}{E} K_{\mathrm{III}}^2 \tag{3.34}$$

†The conservation

$$\int_\Gamma w\, dx_2 = -\frac{\partial}{\partial a}\int_\Sigma w\, d\Sigma = -\frac{\partial}{\partial a}(\Phi \text{ within } \Sigma)$$

is a consequence of Gauss' theorem and (3.31). Equations (3.20) and (3.22*a*) can therefore be invoked to yield the same conclusion as in (3.33).

in combined modes, while in pure mode I

$$J = \frac{\beta}{E} K_{\mathrm{I}}^2 \tag{3.35}$$

For plane stress ($\beta = 1$) the last result was derived as (2.51) from other considerations. To repeat: a more direct (but certainly more elaborate) way of relating J to the stress intensity factors would be to insert the near-tip elasticity solution in the defining integral (2.48).

Example 3.4 We turn back to Example 3.2 (Fig. 1.2) attempting now to evaluate $\mathcal{G}$ by going through the J integral. The path of integration can be profitably chosen as in the figure, horizontally along the edges and vertically well removed from the crack tip. Along the horizontal branches we have $dx_2 = 0$, $u_1 = 0 = p_3$, $u_2 = \text{const}$, and along the left-hand vertical branch the stresses can be assumed to vanish, so that $\phi = p_i = 0$. Contributions to the integral (2.48) are therefore limited to the right-hand branch; here the displacements can be assumed to be x_1-independent and vertical, corresponding to the strain component $\varepsilon_{22} = u/H$. What remains, then, is the term containing the energy density

$$w = \phi = \frac{\sigma_{22}\varepsilon_{22}}{2} = \frac{\eta E \varepsilon_{22}^2}{2} = \frac{\eta E (u/H)^2}{2}$$

This leads to the same result as before

$$\mathcal{G} = J = H\phi = \frac{\eta E}{2H} u^2$$

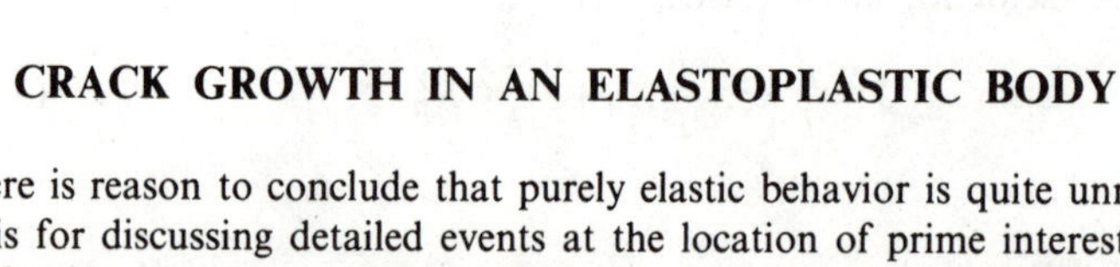

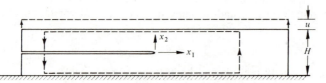

Example 3.4

3.3 CRACK GROWTH IN AN ELASTOPLASTIC BODY

There is reason to conclude that purely elastic behavior is quite unrealistic as a basis for discussing detailed events at the location of prime interest in fracture mechanics, i.e., the crack front. Still, this assumption has produced qualitative notions and quantitative results which turn out to be very useful in evaluating the more complex elastoplastic states. Specifically, in Chap. 2 we have seen quantities of plastic origin, e.g., yield lengths or CODs, being expressed in terms of the elastic stress intensity factors K in cases where yielding is so localized or small-scale as to be *governed* by the surrounding singular elastic solution. Similar effects apply to the energy-related quantities $\mathcal{G}$ or J, considered next. Eventually, even the fracture process as such should reasonably be K-governed when yielding is sufficiently localized. However, further difficulties have to be faced if extensive yielding and a more detailed fracture event call for explicit attention.

3.3.1 Significance of Singular Stresses

We have seen (Sec. 3.2.2) that stress singularities of a certain hardness (the type $r^{-1/2}$, with strengths K_{I} to K_{III}) were necessary to explain a nonzero separation work, $C = \mathcal{G} > 0$, in a purely elastic body. These singularities were formal consequences of the linearization, both in material characterization and in geometric description. If, alternatively, a higher-order description allowing for large displacement gradients had been invoked, the notion of a finite separation work might have to be abandoned in favour of more detailed field considerations. It is fortunate that lumping such alternative field characteristics into the derived scalar quantity C has provided a practical working concept and, intuitively, an appealing macroscopic counterpart to the ideal micro work. As it turns out, a similar kind of first-order assessment cannot be successfully combined with all types of material behavior. A generalization of the above requirements has been pointed out [141], to the effect that singular stresses are always necessary within a continuum description to predict nonzero separation work. Such formal singularities have also been documented for elastoplastic materials with nominal work hardening, both for stationary (Sec. 2.8.2) and steadily growing cracks. They exist as well in time-dependent (viscoelastic or viscoplastic) models, which may be appropriate under given thermal conditions. However, it has even been suggested that singularities as hard as the above elastic ones are necessary to give $C > 0$, [144, 145]. By conventional models this would seem to exclude the separation work as a characterizing property for all time-independent materials but the linearly elastic one. Contradictory views and results are being claimed on this issue, however, and augmentations have even been suggested to ensure the existence of C or C-like measures. Without such precautions it is at any rate quite clear that $C = 0$ will be the only result by assumptions which imply that stresses are bounded at the crack tip. This property will be illustrated below.

3.3.2 Quasi-Static Relationships during Small-Scale Yielding

From its definition as external work over stored elastic energy [compare (3.13) assuming $dT = 0$] it is evident that the crack driving force approaches its elastic value when yielding becomes increasingly localized. Thus

$$\mathcal{G} \to \mathcal{G}_{\mathrm{el}} = \frac{\beta}{E} K_{\mathrm{I}}^2 + \frac{\beta}{E} K_{\mathrm{II}}^2 + \frac{1+\nu}{E} K_{\mathrm{III}}^2 = J_{\mathrm{el}} \tag{3.36}$$

the last three expressions referring to total elasticity. Furthermore, the difference between the J values for small-scale yielding and for total elasticity should also go to zero, since the integral could be evaluated along remote paths where small crack-tip yielding does not interfere.† Therefore

$$J \to J_{\mathrm{el}}$$

and we have by (3.12) and (3.36)

$$\mathscr{G} = J = \frac{\beta}{E} K_{\mathrm{I}}^2 + \frac{\beta}{E} K_{\mathrm{II}}^2 + \frac{1+\nu}{E} K_{\mathrm{III}}^2 = C + D \tag{3.37}$$

in the limit of localized yielding. The three left-hand expressions are equal in such cases [as also anticipated in (3.33)], each the same as the total dissipation $C + D$. In particular, the expressions will represent the separation work C alone if the material is purely elastic and the continuum dissipation D alone if the stresses are bounded by a nonhardening level.

To illustrate the last property, confirming the observation in Sec. 3.3.1, we look back to (2.49), which represents the J integral of the Dugdale model. The substitution (3.31) remains valid if such yielding is sufficiently localized and leads to

$$J = \int_a^{a+c} \sigma_Y \frac{\partial}{\partial a} (u_2^+ - u_2^-)\, dx_1 \tag{3.38}$$

Since this expression can be interpreted as the plastic work done on the yield zone ahead of the crack per unit crack advance and within a unit thickness, it should be identified as the dissipation D. Thus, for the Dugdale model we have seen that

$$J = \sigma_Y \delta = D \qquad \text{and} \qquad C = 0 \tag{3.39}$$

observing (2.50) and (3.37). The fact that some if not all of the supplied energy $\mathscr{G}$ is absorbed as dissipation D in the continuum is sometimes referred to as *screening* of the fracture process zone by its surrounding yield region.

Furthermore, we recall that the formulation with J in (3.37) assumes two-dimensional relations, that the one with K_{I} to K_{III} assumes linearity in the elastic behavior and that both assume a further growth coplanar to the original crack. With regard to the final form we may find it reasonable that the three scalars K_{I} to K_{III}, which define an elastic near-tip field, govern even the dissipating mechanisms when they are sufficiently localized.

Finally, it would be profitable to define quantitatively what has been somewhat vaguely described as small-scale or localized yielding. Two typical lengths which can be compared are the extent of the yield zone and the crack length a. The first is of the order of $0.1(K_{\mathrm{I}}/\sigma_Y)^2$ in mode I, where σ_Y represents the (initial) yield stress; then a measure proportional to their ratio

$$\alpha = \frac{K_{\mathrm{I}}^2}{a\sigma_Y^2} \tag{3.40}$$

†Total path independence would demand evaluation outside the plastic zone anyway.

(or a similar dimensionless yield length in another mode) will express the relative extent of yielding. Empirically, yielding so localized that (3.37) is valid has certainly been realized for α in the order 0.1 or less. Incidentally, the limit $\alpha = 0$ of an elastoplastic state should not be confused with total elasticity (as sometimes happens). The singularity of the last case would obviously be modified by local yielding however small the load at a finite yield stress.

3.3.3 Graphical-Experimental Interpretation of the *J* Integral

Generally, i.e., for extensive yielding, the J integral cannot be associated with any continued crack growth.† This is because the condition $w = w(\varepsilon_{ij})$ for invariance is certainly not satisfied in the unloaded material behind the crack front. Still, J may characterize the state at the tip of a *stationary* crack, as mentioned in Sec. 2.6, and may therefore possibly serve as a critical quantity if the initial crack motion is considered. The value of J for a given elastoplastic material would then be the same, assuming proportional straining, as that of an imaginary nonlinearly elastic material having an identical stress-strain relation during loading. For the last material we know the connection $J = \mathscr{G}$ by (3.33), where $\mathscr{G}$ has the property of being related to crack growth as in Fig. 3.3 (linearity being no condition for validity). By this indirect reasoning we have arrived at the same graphical interpretation of J, that is, $J\,\Delta A$ representing the area between the loading curves for crack areas A and $A + \Delta A$ as $\Delta A \to 0$ (see Fig. 3.6).‡ This interpretation is the basis of experimental techniques to determine the J integral in large-scale yielding situations. The result can also be formulated as

$$J = -\left(\frac{\partial U}{\partial A}\right)_{u_1} = -\int_0^{u_1} \left(\frac{\partial P}{\partial A}\right)_u du = \int_0^{P_1} \left(\frac{\partial u}{\partial A}\right)_P dP \tag{3.41}$$

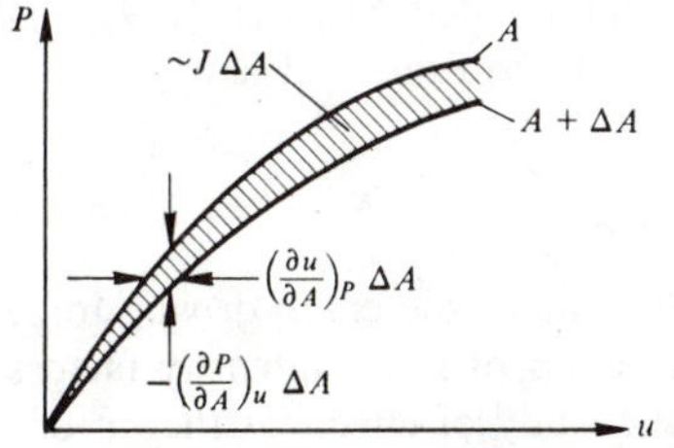

Figure 3.6 Two-specimen interpretation of the J integral.

†Such interpretations in Sec. 3.3.2 were due to the fact that the yielding was so localized. A possible connection may exist for a very limited crack advance (Chapter 4).

‡The difference between the two materials lies only in the fact that the curve for $A + \Delta A$ must be determined by loading from below for a new crack (new test piece) in the elastoplastic case, whereas it can be produced by unloading from above after a ΔA crack growth in the imaginary elastic case.

with the external work

$$U = \int_0^{u_1} P\,du \tag{3.42}$$

replacing Φ of (3.18) and index 1 denoting the level of an effective load P and its displacement u.

As special applications of (3.41), consider platelike specimens of thickness B and uncracked ligament lengths b, as in cases 1 to 5 of Table 2.2. If yielding is extensive in a low-hardening material, P values close to the limit load may be realized, i.e.,

$$P \approx C_1 b$$

for essential stretching of the net section and

$$P \approx C_2 b^2$$

for essential bending, C_1 and C_2 being constants.† Observing that $\partial P/\partial A = -(1/B)(\partial P/\partial b)$, we then derive from the second form of (3.41)

$$J = \begin{cases} \dfrac{C_1}{B}u = \dfrac{Pu}{bB} = \dfrac{U}{bB} & \text{stretching} \qquad (3.43) \\[2ex] \dfrac{2C_2 b}{B}u = \dfrac{2Pu}{bB} = \dfrac{2U}{bB} & \text{bending} \qquad (3.44) \end{cases}$$

$U = Pu$ has been introduced here by rigid, ideal-plastic idealization.

It is interesting to note that (3.43) also holds for the deep-crack specimen of Fig. 1.2 (thickness B) regardless of the constitutive properties of the material. We see this by repeating the steps of Example 3.4, finally replacing ϕ by U/bHB instead of its linear representation. A similar validity of (3.44) for deep cracks and unspecified elastoplastic behavior was first demonstrated by Rice [149]. The importance of these results in testing situations will be examined later.

3.3.4 Crack Growth and the Separation Work

For the totally elastic body the separation work equals the crack driving force, which is available by various means including tables of stress intensity factors, when conditions are quasi-static. In general, the study of effects at the tip of a crack moving in a nonelastic solid is far more cumbersome, with the result that certain aspects of the problem have not yet been fully explored.

Analytically, the elastoplastic modeling of unloading in the wake of the crack tip is difficult, and the solutions available so far are for ideal plasticity and linear

†Note that β in case 3 will be asymptotically proportional to b as L/b increases.

hardening only. They indicate a softening of the stress singularity, from the type $r^{-1/2}$ when the crack is stationary to r^s, $-\frac{1}{2} < s < 0$, when the crack moves steadily in a linearly hardening material [119]. This may have an implication on the separation work, as recorded in Sec. 3.3.1. We mention also a smoothening of the crack-tip profile, predicted for ideal plasticity and steady motion [21]. This reflects a logarithmic strain singularity and replaces the discrete, well-defined tip opening of the stationary crack.

Numerically, it is easy to suggest the basis of an approach which simulates crack growth. Explicit use can then be made of the definition (3.1) to derive the separation work. Applications will be indicated below and remaining uncertainties pointed out.

The procedure is to discretize the structure as finite elements (Sec. 2.2) and to simulate crack growth by relaxing the interconnection along the prospective crack path between adjacent elements. This is illustrated in Fig. 3.7 for a plane geometry and mode I. The crack might first have been modeled over a length a by prescribing free surfaces at $x_2 = 0\pm$ and $x_1 < 0$. Next, relaxing the nodal force Q as shown, we simulate the further growth of the crack over the length Δa. The hatching in the figure is to symbolize the fracture process zone: infinitely thin but still a physical concept which may support the arguments. When the nodal force Q is relaxed such that a relative displacement u occurs, work is done negatively on the continuum and positively on the process zone (Fig. 3.7b). The latter contribution can be interpreted as the separation work C multiplied by Δa and

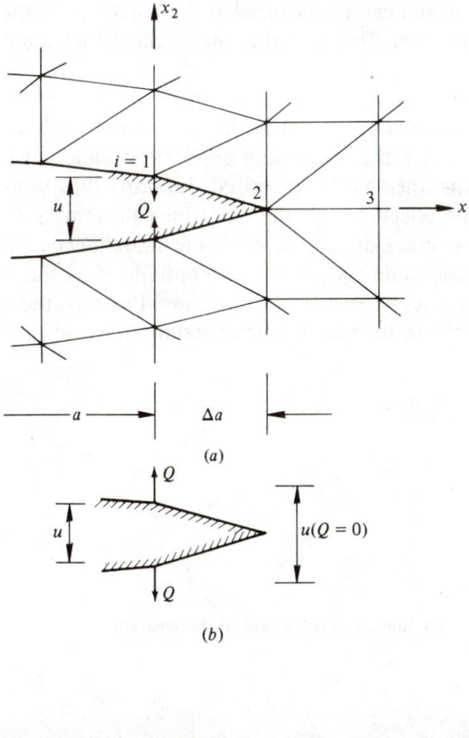

Figure 3.7 Finite-element modeling of the separation work.

the thickness B. In this first approximation expression (3.1) is therefore modeled by

$$C = \frac{1}{B\,\Delta a} \int_0^{u(Q=0)} Q\,du \equiv C^{(1)} \tag{3.45}$$

as in Fig. 3.8, while a further refinement may be the averaging of the integral over a series of consecutive nodes $i = 1, 2, \ldots, n$. That is,

$$C = \frac{1}{Bn\,\Delta a} \sum_{i=1}^{n} C_i^{(1)} \equiv C^{(2)} \tag{3.46}$$

convergence as $n\,\Delta a$ goes to zero being anticipated.

As an alternative to this local evaluation, one can go into the bookkeeping of loads and displacements, as well as of the strain-energy variation and the dissipation

$$\int_{V_e} \left(\int \sigma_{ij}\, d\varepsilon_{ij}^{p} \right) dV_e$$

in each element volume V_e during the relaxation. Summation over the grid then leads to an approximation of the left-hand side of (3.9) if a quasi-static situation $dT = 0$ is considered. This is an evaluation of C by global energy balance. As it appears, C is a kind of dissipation which can be localized at the crack tip, at the boundary of the continuum, while the term D is related to the plastic deformation within the continuum.

The direct method (3.45) has some application in elastostatics. Here it can be shown, via Betti's theorem, to be numerically related to the compliance methods for deriving the crack driving force. A rather promising application under such conditions is in predicting crack paths when modes are mixed. Tentative directions are then assumed and a critical one selected, e.g., the one which maximizes C.

Further uses of the method are in crack dynamics. While the unloading curve, as in Fig. 3.8, is uniquely given under static conditions, assumptions of its shape have been rather arbitrarily made in cases when inertia contributes. Possibly, there may be guidelines from basic principles to the best or correct assumption, but such

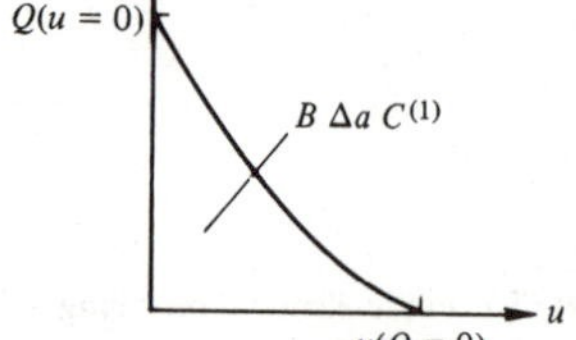

Figure 3.8 Numerical equivalent of the separation work.

indication has not yet appeared. Nevertheless, much promising work using the relaxation technique has been done in elastodynamics.

More severe difficulties have been met in static or dynamic cases where yielding is so developed that it demands explicit attention. The problem is related here to convergence of (3.45) or (3.46) as Δa is reduced in a numerical simulation, and even (in view of Sec. 3.3.1) to the existence of C. A practical approach which may be successful in the study of extended cracking as well as initiation is in *comparing* neighboring values of C, which may have converged to a fixed ratio long before the separate final values could be attained. A modification to ensure the existence of nonzero separation work and to improve the convergence has been suggested by Kanninen et al. [167] in that a process zone occupying a finite area is explicitly modeled to the fracture analysis. This is physically attractive and has conformed well in macroscopic prediction but only after considerable computational effort. We return later to predictive capabilities in the context of stable cracking, a subject which will also be addressed on nonenergetic premises.

3.3.5 Cohesive Zones

By a cohesive zone here we understand a narrow lamina extending from the crack front, across which closing tractions are transferred depending on some parameter(s) recording the fracture process. Specifically, an *extended-crack model* in mode I will be considered, where the cohesive zone is formed as a nonmaterial, hypothetical extension of the opened crack (Fig. 3.9).

In the context of brittle fracture the existence of the cohesive zone was advocated by Barenblatt [95], who (by implication) anticipated some coupling to exist between the closing stress $\sigma_{22}(a < x < a + c)$ and the separation $\bar{u}_2(a < x < a + c) = u_2^+ - u_2^-$ of the hypothetical crack faces. The length c of the zone enters a condition that the resulting singularity at the leading edge $x = a + c$

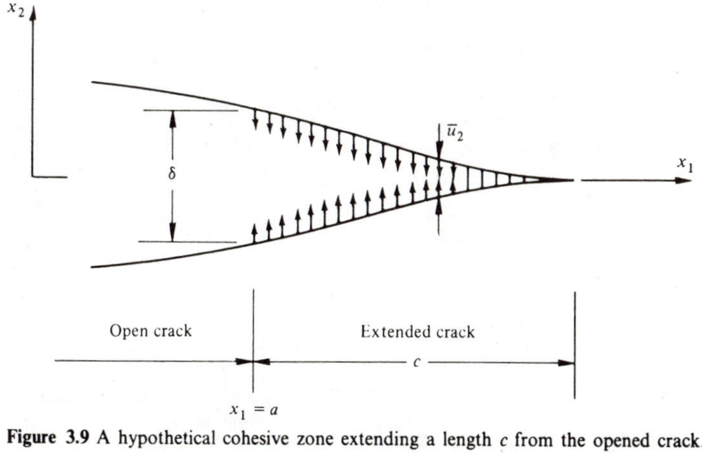

Figure 3.9 A hypothetical cohesive zone extending a length c from the opened crack.

of the zone will be canceled. Thereby the crack faces will join smoothly at $x = a$, as indicated in the figure. Formally, this is an obvious generalization of the Dugdale model where $\sigma_{22} = \sigma_Y$ acted ahead of the crack. However, the physical reference of Barenblatt was some interatomic separation taking place on a very small length scale with large but finite stresses occurring. By contrast, the Dugdale yield length is macroscopic, and yielding not fracture is the physical reference.

The work hypothetically required to separate a material element situated in the crack path will be

$$\int_0^\delta \sigma_{22}\, d\bar{u}_2 \equiv J_z \tag{3.47}$$

as measured per unit or traversed area, where δ is the opening displacement at the trailing edge of the zone. The work is recognized as the J integral for the contour adjacent to the zone as in (2.49) and Fig. 3.10. A similar but not identical energy measure is the work exerted on the cohesive zone through its boundaries per unit width and hypothetical advance of the crack tip

$$\int_a^{a+c} \sigma_{22} \frac{\partial \bar{u}_2}{\partial a}\, dx_1 \equiv C_z \tag{3.48}$$

A prerequisite in both cases is that σ_{22} be nonsingular within $a \le x_1 \le a + c$. If the cohesive zone is small enough to have relatively stationary local properties, the transport substitution (3.31) can be invoked to give the equality

$$C_z = -\int_a^{a+c} \sigma_{22} \frac{\partial \bar{u}_2}{\partial x_1}\, dx_1 = J_z \tag{3.49}$$

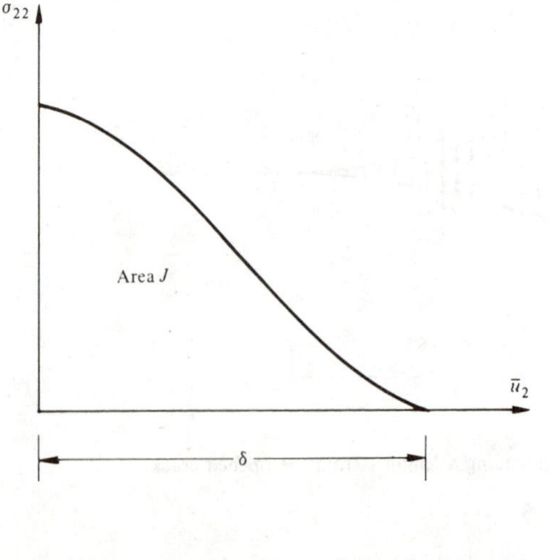

Figure 3.10 A postulated cohesive relation $\sigma_{22}\,(\bar{u}_2)$, identifying as the work of individual separation the J integral for the contour adjacent to the cohesive zone.

A global energy balance like (3.9) is readily formulated, with the above work C_z of crack-tip translation replacing the earlier energy sink C. Thus

$$W - \frac{dT}{dA} - \frac{d\Phi}{dA} - D = C_z \tag{3.50}$$

We reconcile this with all details in Sec. 3.1 by temporarily viewing the cohesive stress as an external load on part of the crack surface S_c and by introducing $C = 0$. The latter observation anticipates that singularities which might contribute to $C > 0$ have been canceled, as assumed above.

Given a body under external action, and considering now the elastic case $D = 0$, it is evident that the left-hand terms in (3.50) will become those of (3.9) as c goes to zero. A further argument to this effect may follow in the static two-dimensional case through (3.49) and path independence of J. $C = C_z$ and $\mathcal{G} = C_z$ is then the asymptotic result, which means that the models of Griffith and Barenblatt are equivalent from the point of view of energy balance.

To summarize, in a cohesive model the work of individual separation has been identified as the area J_z below the cohesive curve, by (3.47), while the local work of crack-tip translation equals C_z by (3.48). Equality of the two measures is achieved as the length c approaches zero. In an otherwise elastic solid this common limiting value equals the elastic crack driving force $\mathcal{G}$ according to (3.23) or (3.30).

The extended notion of separation work J_z or C_z may be the key to energy considerations outside the frame of simple continuum theories. Indeed, promising steps have already been made in applications to concrete and fibrous composites [180–183]. A cohesive zone is introduced to model partial cracking or pullout of fibers ahead of the open crack, with empirical data to help estimate critical cohesive properties. Further applications may go through a defined damage parameter governing the effective closing stress, as proposed by Refs. 184 and 185 in a related context.

While the above experiences were restricted to an elastic unbroken material, it remains a possibility of this energetic approach to assess the combined effect of a finite cohesive zone and a nonelastic adjacent material. It may provide a way to discuss a realistic fracture process, and supplement the modeling mentioned at the end of Sec. 3.3.4. The approach may be viewed as physically reasonable though perhaps numerically discouraging because of the very fine computational mesh often needed in a detailed study. No systematic investigation of such applications has yet come to the author's attention.

The topic of cohesive zones will not be pursued in the present text. Further insight can be obtained in the references.

PROBLEMS

3.1 A slender rectangular elastic bar has a central longitudinal slit through the thickness, as shown. A pair of opening loads P acts at the midspan. Assuming $h \ll a$, use the compliance method to derive the crack driving force

$$\mathcal{G} = \frac{3}{4} \frac{P^2 a^2}{E B^2 h^3}$$

referring to one crack tip in a plane-stress condition, then find the stress intensity factor

$$K_I = \frac{1}{2} \sqrt{\frac{3}{h} \frac{Pa}{Bh}}$$

Hint: To preserve symmetry, admit the simultaneous advance of both crack tips while associating with each the energy of half the bar.

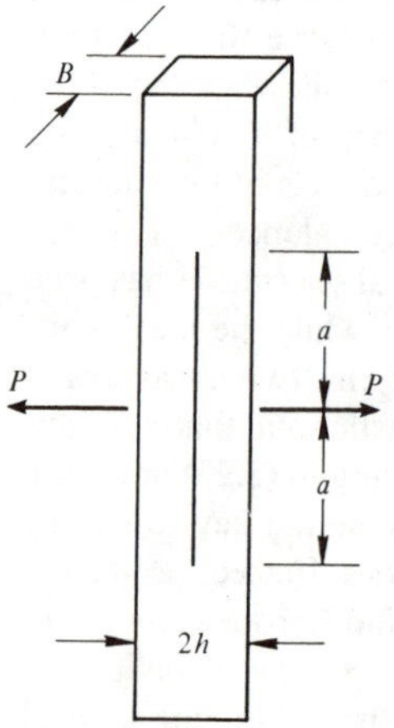

Problem 3.1

3.2 A rectangular elastic strut has the thickness B and a slit of length a starting from the top. Assuming $h \ll a$, use the compliance method to derive the stress intensity factors

$$K_I = \sqrt{\frac{3}{h} \frac{P}{B}} \quad \text{and} \quad K_I = \frac{1}{2} \sqrt{\frac{3}{h} \frac{P}{B}}$$

respectively, for the two alternative loadings.

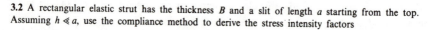

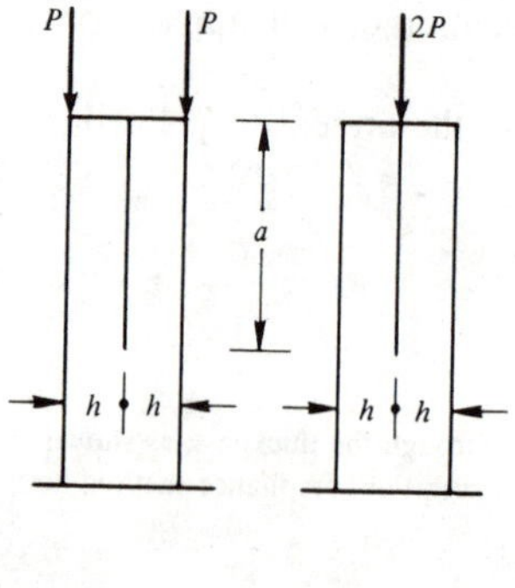

Problem 3.2

3.3 The J integral can be approximated by expressions like

$$J = -\frac{1}{B}\frac{\Delta U}{\Delta a} \qquad \text{or} \qquad J = \frac{2U}{Bb}$$

the latter applying to deep-crack, bending-type specimens. Considering the three-point bending specimen shown and the (imaginary) loading curves for three neighboring crack lengths, derive in either way the J value of 0.1 MN m^{-1} for $a/W = 0.6$, $u = 1$ mm.

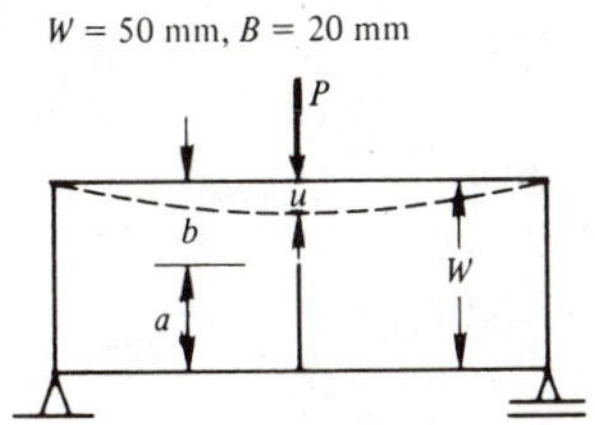

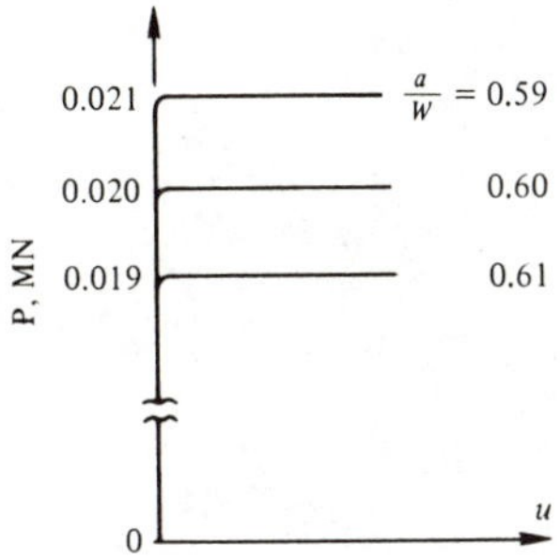

Problem 3.3

CHAPTER
FOUR

CRACK INITIATION AND GROWTH: CRITERIA AND STATIC ANALYSIS

As mentioned in Chap. 1, a number of hypotheses may serve to define the state at the crack tip as critical. In a formal sense the situation is similar to that of discussing the transition from an elastic to an elastoplastic state at a material point. Some measure or measures defining the state should exist; for the yield transition it might be an invariant function of the stresses, such as the "effective" stress, while for the crack growth it may be the separation work, the strength of a singularity, the COD, or others. In both cases, the expectation is that one or more of these candidates can be compared with one or more typical constants pertaining to the material. Equality of the two (or of functions on both sides) then defines the critical state. Thus, in the crack problem the material property will be related to the loss of a continuity, much as the yield stress is related to the transition from an elastic state.

4.1 GRIFFITH ANALYSIS

Both in a historical frame and along the lines of the energy balance discussed in Chap. 3, it is natural to begin our systematic evaluation of critical states by referring once more to Griffith. He was primarily concerned with materials that might be considered purely elastic (specifically glass); then by (3.12) and (3.13)

$$\mathscr{G} \equiv W - \frac{dT}{dA} - \frac{d\Phi}{dA} = C \tag{4.1}$$

assuming $D = 0$. In this context his proposition [2] was that the separation work C will take on a typical constant value 2γ when the crack moves in a given material under given external conditions. The hypothesis implies that

$$\mathscr{G} = 2\gamma = \text{const} \tag{4.2}$$

where γ alone is to be interpreted as the *surface energy* assigned to one side of the fracture surface. As it stands, condition (4.2) applies to the transfer of energy *when* the crack is growing. It must be sharpened to provide a sufficient criterion for the initiation of crack motion. A common interpretation is that cracking will start if (4.2) is satisfied with $\mathcal{G}$ referred to a small quasi-static motion, or a virtual crack growth in the preceding notation. Then, by (3.23),

$$\mathcal{G} = -\frac{\partial \Pi}{\partial A} \tag{4.3}$$

where any of the forms (3.18), (3.20), (3.29), or (3.33) may be equivalent under the pertinent conditions.

We can appreciate the consequences of this assumption by looking at a related example. A block lying on a table is loaded actively by a tangential load P and, in consequence, by an opposing frictional force f. For a displacement ds of the block the work done equals the increment in kinetic energy

$$(P - f)\, ds = dT$$

or

$$P - \frac{dT}{ds} = f \tag{a}$$

where f can be viewed as the frictional work per unit of displacement. This expression is comparable to Eq. (4.1). It is presumed that when the block does slide, the frictional work is at a constant critical value; that is, $f = F = \text{const}$ (compare $C = 2\gamma$). The load P to start motion is sought.

As formulated above, the problem has no solution; but if we assume as an additional hypothesis that $f = F$ is to be satisfied as we try a virtual displacement with $dT = 0$, the result by (a) is

$$P = F$$

which is the trivially expected answer, provided that no initial frictional barrier adds to the resistance. On the other hand, if a gap existed between the static and the kinetic friction, we know that a higher load $P_0 > F$ would be necessary to start the motion. Equation (a) might serve again, to indicate a dynamic event in the next instant

$$\frac{dT}{ds} = P_0 - F > 0 \tag{b}$$

Obviously, some additional information would then be necessary to provide the value of the initial load.

By analogy, we expect the energy balance (4.2) and (4.3) to predict an initiation load which is realistic or, if a barrier to the initiation (Chap. 1) exists, a lower bound to the true value. Dynamic effects would actually be involved in the latter case. Such a barrier might be geometrically conditioned, e.g., by the crack tip's being blunted in some way, or a product of material response and environmental interplay. Once the separation has been initiated, the surface energy γ can be physically interpreted as the work it takes to overcome the attractive forces in the atomic lattice. If this takes place when the crack front is slowly moving, $\mathscr{G}$ retains the forms derived in Sec. 3.2. Therefore, Eqs. (4.2) and (4.3) represent a relationship that is valid for smooth initiation as well as for a continued slow motion.†

Assume further that the slow type of growth is governed by some load or displacement factor $\mu(A)$, such that $d\mu/dA > 0$. This corresponds to increasing action, following initiation at some level $\mu = \mu_0$ with the precrack area A_0 (Fig. 4.1). While that growth proceeds, $\mathscr{G} = \mathscr{G}(\mu, A)$ satisfies (4.2), and thus

$$\frac{d\mathscr{G}}{dA} = \left(\frac{\partial \mathscr{G}}{\partial \mu}\right)_A \frac{d\mu}{dA} + \left(\frac{\partial \mathscr{G}}{\partial A}\right)_\mu = 0 \tag{4.4}$$

The ending of the period with increasing μ, that is,

$$\frac{d\mu}{dA} = 0 \tag{4.5}$$

designates the transition into its quasi-static continuation with $d\mu/dA < 0$, or a dynamic development. We must conclude that the first alternative is an *unstable* branch (dashed line in Fig. 4.1), it being practically impossible to ward off any small jump of the crack by a corresponding reduction of the controlled action such that acceleration is suppressed. We define the crack as being *stable* when $d\mu/dA > 0$, which is tantamount to a controlled quasi-static growth, and unstable when $d\mu/dA \leq 0$. The equality includes the optimal point in Fig. 4.1 and also the shortcut $\mu = \text{const} = \mu_0$ shown. In the last case $\mathscr{G}(\mu, A)$ is independent of A according to (4.4) and (4.5), and the lack of stability is encountered in an arbitrary quasi-static crack jump taking place under a constant action. The notion of metastability is also applied to such a crack; Example 1.1 serves to illustrate the case. For the unstable crack it follows from (4.4) that

$$\left(\frac{\partial \mathscr{G}}{\partial A}\right)_\mu \geq 0 \tag{4.6}$$

when $(\partial \mathscr{G}/\partial \mu)_A > 0$ is reasonably assumed. As Fig. 4.1 suggests, this conclusion might be satisfied even from the beginning; the inequality of (4.6) then corresponds to a crack motion which is totally dynamic.

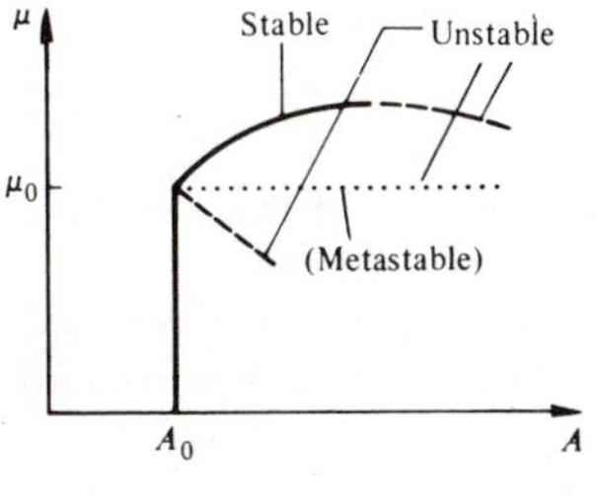

Figure 4.1 Alternative paths of a governing load or displacement factor μ.

†In a formal way Eqs. (4.2) and (4.3) may even be used to describe a barrier initiation, simply by replacing the defined γ with a higher value. Obviously, the essential information must come from elsewhere. We return to a similar formalism in Chap. 5.

The calculation of $\mathcal{G}$ was discussed in Sec. 3.2 on the assumption of a codirectional growth of the existing crack. If there is in-plane assymmetry with respect to that plane, that is, $K_{II} \neq 0$, this is not the real direction of growth, whether experimentally observed or theoretically predicted. The critical direction by Griffith analysis will be the one which maximizes the crack driving force under otherwise given conditions. We shall return briefly to this problem in Chap. 7.

Example 4.1 For the DCB specimen in Example 3.1 we have already derived the crack driving force

$$\mathcal{G} = \frac{P^2a^2}{BEI} = \frac{9}{4}\frac{u^2EI}{Ba^4}$$

in terms of force P or the displacement u. We want to know the action which starts the crack moving and the later development.

Assuming initiation from a precrack length $a = a_0$, the critical load or displacement are obtained by (4.2), i.e.,

$$P = P_0 = \frac{\sqrt{2\gamma BEI}}{a_0} \qquad \text{or} \qquad u = u_0 = \tfrac{2}{3}a_0^2\sqrt{\frac{2\gamma B}{EI}} = Z(a_0)P_0$$

What happens next depends on whether the load or the displacement is controlled, i.e., whether P or u can be identified here as the action parameter μ. In the first case, (4.2) has the solution

$$P = \frac{\sqrt{2\gamma BEI}}{a} \qquad \frac{dP}{da} < 0$$

which indicates an unstable crack, whereas in the last case

$$u = \frac{2a^2}{3}\sqrt{\frac{2\gamma B}{EI}} \qquad \frac{du}{da} > 0$$

implying stability. Equation (4.6) confirms the last result, in that

$$\left(\frac{\partial \mathcal{G}}{\partial A}\right)_\mu = \frac{1}{B}\left(\frac{\partial \mathcal{G}}{\partial a}\right)_u < 0$$

It is important to recognize that displacement control in this case (and often otherwise) is less fracture-prone than load control.† Any attempt by the crack front to "run off" would now be effectively countered by the increased flexibility of the two beams, so that for a given displacement u a reduced straining of the crack tip would ensue.

†The reader may verify that if the load $P = \sigma b$ were controlled action in Example 1.1, the crack would definitely have been unstable.

4.2 A GENERALIZED APPROACH CONSIDERING SEPARATION WORK

We retain the essential hypothesis on the growth of the crack, that is, $C = 2\gamma = \text{const}$, when the material and its environment are given. The content of this assumption will be discussed further with regard to real material behavior. In such a context, however, it is necessary to consider the plastification outside the fracture process zone, as would be a consequence in general of the concentrated straining. This means that the dissipation term D must be included in the energy balance (3.12) and (3.13), which then takes the form

$$\mathcal{G} - D \equiv W - \frac{d\Phi}{dA} - D = 2\gamma \tag{4.7}$$

when $dT/dA = 0$. Such a balance may also apply to the initiation of crack motion, as before, but we should note now that the dissipation term D generally depends on the history of separation and deformation, so that in the same configuration it takes on different values for a precrack (on virtual growth) and a propagated crack.

Condition (4.7) is illustrated in Fig. 4.2. By analytic simulation it would be possible to evaluate both $\mathcal{G}$ and D as dependent on the present load or displacement factor μ and the previous history. To continue (or initiate) motion the critical factor μ is that which makes the difference between the instantaneous functions $\mathcal{G}(\mu)$ and $D(\mu)$ equal to 2γ.

During local yielding $\mathcal{G}$ is essentially equal to the crack driving force of the fully elastic body. It has been recognized, furthermore, that $D(\mu)$ may be nearly as large as $\mathcal{G}$ (it even equals $\mathcal{G}$ if singularity requirements are not formally met at the crack tip). Thus it appears, on the basis of the generalized Griffith approach, that the critical action derived by purely elastic considerations may be far too small. Indeed, one should recognize the screening effect provided by the yield region at the crack tip: an amount of energy will be dissipated which would otherwise contribute to separation work in the process zone.

For the fracture parameter γ the situation is more diffuse than in the ideal world of Griffith. He considered γ to be related only to separation in regular lattices, which has an implication on pure cleavage fractures in single crystals. During real quasi-brittle fractures in polycrystals, on the other hand, γ should also include dissipation through nonhomogeneous slip within and between the grains,

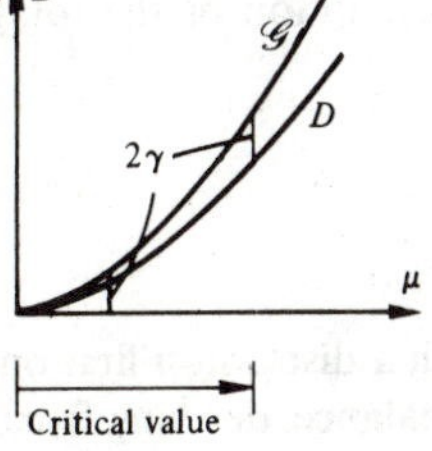

Figure 4.2 The screening effect of the dissipation term D. Increasing $D(\mu)$ will raise the critical value of the load or displacement factor μ if γ is a given property.

through some local void growth, and perhaps through viscous mechanisms such as diffusion and creep under certain conditions. This general picture may call for a coupling between γ and the separation mode, perhaps involving the crack velocity. Since mode I generally dominates, there is still a certain justification for taking γ to be a constant, as the crack moves slowly in a given material and environment. This applies for the continued crack motion, though not necessarily at the same magnitude for initiation.

Even if we accept conditions ensuring that γ will be a constant defining the process, it remains as an important problem to determine its value explicitly. Microscopic studies of the fracture process zone lead to nothing further than orders of magnitude. For application, the solution might go through some indirect calibration: knowing the critical load of some specimen by experiments, one could use this as the aim of an analytical approach wherein γ is adapted to conform with the result.

The fracture process zone was initially defined to be surrounded by a continuous medium and thereafter was assumed to be negligibly thick. However, this dimension may not always be disregarded, e.g., in some cases of ductile failure. It should further be recognized that a number of fracture events are not unconditionally predictable on the basis of a critical separation work. Since void growth is a dominant physical mechanism in ductile fractures, reasonably related to large local straining up to some critical level, it may be hard to identify an energy parameter which defines the process. Analytically, it may also require special constraints in the continuum representation or an explicit modeling of the process zone as such to anticipate a work of separation which is physically consistent and numerically well behaved. We are thus faced with two rather open concepts, 2γ and C, which by assumptions or by some calibration may possibly be coupled into an equation with the predictive potential. Efforts along this line have already been made (Sec. 9.6). Considerable interest is attached to the problem of describing stable and extensive crack growth, and some success here in terms of simple concepts is a welcome contribution. Whatever artifice contributes to the evaluation of C is less decisive than the final prediction. At this point, one should recall that Griffith's detailed handling of the elastic case may also be open to criticism. Neither of the two linearizations mentioned first in Sec. 3.3.1 seem to apply in the crack-tip region, whereas the success of the adopted formalism is unquestionable. A critical scalar measure, as provided by Griffith in the brittle case, is certainly the best one could hope for also in situations involving cracking and extensive yielding. However, in view of the complex nature of fracture it is doubtful whether any such one-parametric description of the total event can be fully realized.

4.3 LINEAR FRACTURE MECHANICS

In the spirit of the foregoing considerations we proceed with a discussion first on energetic premises. Equation (4.7) represents the governing balance, deriving from

(3.12) and (3.13) under the present quasi-static conditions. An extreme idealization may be to disregard the continuum dissipation D, which would lead to the treatment in Sec. 4.1. Given a certain value γ, such an analysis may be severely in error by predicting unrealistically small fracture loads (Fig. 4.2).

A better attempt should include D but in such a way that the simple results of elastic analysis will essentially be retained. One might reason as follows. If yielding is highly localized at the crack front considered, and if that front is far removed from the external surface and other cracks, the dissipation D should reasonably depend for a given material and environment only on the extent and mode of the crack motion. Specifically, if a single mode (I or III, in practice) dominates, one is motivated to look upon D as a given constant if crack initiation is considered and as some specific function of the traveled crack area (or the crack-tip translation) in the case of a continued small motion. The constant or function may be viewed as a material property in the given mode, available, in principle, from experiments which are guided by some underlying theoretical estimates.

On this basis (which does not exclude other means of approach to the same end), we shall review the content of so-called linear (elastic) fracture mechanics (LFM or LEFM). The name indicates that all necessary analysis can be formally referred to an elastic body since yielding has been localized to such an extent. We start by considering the first stage of cracking, which is the simplest case when D is taken to be a known constant. Next, slow crack motion is studied under conditions when D can be specified by the extent of the growth.

4.3.1 The Starting of Crack Motion

We assume a highly localized (or small-scale) yielding at the crack front. In addition to suggesting that D can be viewed as a constant of the material D_c for a crack to start moving, this also implies that $\mathcal{G}$ in (4.7) will be essentially the same crack driving force as derived for the purely elastic body in Sec. 3.2. Thus, we have the cracking condition

$$\mathcal{G} = \mathcal{G}_c \tag{4.8}$$

where $\mathcal{G}_c \equiv 2\gamma + D_c$ is the total dissipation or resistance associated with the starting event.† This implies that we shall be retracing the formalism discussed in Sec. 4.1, simply through substituting $\mathcal{G}_c$ for the surface-energy term 2γ. This modification of the elastic-body analysis was suggested independently by Irwin [5] and Orowan [188]. $\mathcal{G}$ will be a function of the load or displacement factor μ, the critical value of which is then determined by (4.8).

The dissipation D is generally the dominant right-hand term in (4.8) while the surface energy 2γ acts more in the way of a guiding signal. Their sum, $\mathcal{G}_c$ is the parameter of real concern in LFM and is also comparatively simple to determine by experiment. Such an experiment should comply with the assumptions of LFM,

†In the literature one also finds R_c used with the same meaning as $\mathcal{G}_c$.

so that D on cracking will not be influenced by small variations in the geometry of the test specimen. This is tantamount to requiring that typical specimen lengths be large compared with a characteristic size of the plastic zone. For a platelike specimen under opening action on a precrack through the thickness, that consideration will impose minimum requirements both on the crack length and the plate thickness. The last condition should be included to suppress the geometry effect whereby the state at the side surfaces ($\sigma_{33} \approx 0$, with x_3 normal to the plate) deviates widely from the state internally at the crack front ($\varepsilon_{33} \approx 0$, or approximately plane strain).

Assume that the experiment is in mode I, implying by (3.30) that

$$\mathscr{G} = \frac{\beta}{E} K_{\mathrm{I}}^2 = \mathscr{G}_c \tag{4.9}$$

and that the loading as cracking starts is recorded.† Using elastic analysis or expressions like those in Appendix C, we derive the corresponding stress intensity factor K_{I}, now called $K_{\mathrm{I}c}$ as it refers to the critical event. Equation (4.9) should be satisfied

$$\frac{\beta}{E} K_{\mathrm{I}c}^2 = \mathscr{G}_c \tag{4.10}$$

(where $\beta = 1 - \nu^2$, as for plane strain), from which $\mathscr{G}_c$ has been determined. However, since $\mathscr{G}_c$ is a material constant by assumption, so is $K_{\mathrm{I}c}$ by (4.10). This provides an equivalent form of Eq. (4.9); the substitution of $\mathscr{G}_c$ from (4.10) yields simply

$$K_{\mathrm{I}} = K_{\mathrm{I}c} \tag{4.11}$$

as our critical condition. The constant $K_{\mathrm{I}c}$ is the so-called *fracture toughness*, important as a design parameter but also as a means of *classifying* materials for given environment conditions (mostly temperatures): the larger the $K_{\mathrm{I}c}$ value the lower the probability that cracking will start under otherwise given conditions. Values of $K_{\mathrm{I}c}$ have been published and compiled in the literature (Sec. 7.4). Under favourable circumstances data on $K_{\mathrm{I}c}$ for a given case of interest may therefore be available from previous experiments.

We go back to pointing out that the interpretation of D_c as a typical constant depends on one mode being dominant. It also turns out that any inclusion of mode II will serve to bring the crack motion out of its original tangent plane, thus violating the assumption before (3.29) of a codirectional growth. However, if the action is mixed but still dominated by mode I, Eq. (3.29) should be considered a useful guide to extrapolations from mode I

$$K_{\mathrm{I}}^2 + K_{\mathrm{II}}^2 + \frac{K_{\mathrm{III}}^2}{1 - \nu} = K_{\mathrm{I}c}^2 \tag{4.12}$$

Likewise, if mode III dominates, one should refer to critical experiments in that mode, which effectively would replace the right-hand side of (4.12) by another

†We return in Chap. 7 to the special problems involved in defining the first cracking as related to measurements.

material constant $K_{\mathrm{IIIc}}^2/(1-\nu)$. We return later (Sec. 7.5) to a more complete discussion of nonopening modal contributions.

As we have noted, requirements on specimen thickness B and precrack length a must be satisfied for LFM to apply, e.g., Eqs. (4.9), (4.11), or (4.12). These requirements have been set down as†

$$B \geq 2.5\left(\frac{K_{\mathrm{Ic}}}{\sigma_Y}\right)^2 \qquad a \geq 2.5\left(\frac{K_{\mathrm{Ic}}}{\sigma_Y}\right)^2 \tag{4.13}$$

implying that the specimen dimensions should be *substantially* larger than the plastic-zone size. According to (2.40), the lengths differ by a factor of about 25. For a material of high toughness (often accompanied by low hardness), these requirements may impose prohibitively large dimensions in a testing situation and the need to look further than LFM to account for realistic dimensions as they occur in situ.

Example 4.2 A platelike tension specimen has a precrack of length $2a$ perpendicular to the tension axis and through the thickness; a is small compared with the lengths of the plate. Together with the thickness, a is large enough to permit the use of LFM. The fracture toughness is given as K_{Ic}. Determine the remote stress σ_∞ as cracking is started when a is given. Determine also the largest defect possible when σ_∞ is given.

Case 3 of Appendix C provides

$$K_{\mathrm{I}} = \sigma_\infty\sqrt{\pi a}$$

so that (4.11) will respectively give answers of

$$\sigma_\infty = \frac{K_{\mathrm{Ic}}}{\sqrt{\pi a}}$$

and
$$a = \frac{1}{\pi}\left(\frac{K_{\mathrm{Ic}}}{\sigma_\infty}\right)^2$$

For example, $K_{\mathrm{Ic}} = 40\ \mathrm{MPa\ m^{1/2}} \equiv 40\ \mathrm{MN\ m^{-3/2}}$ and $\sigma_\infty = 200\ \mathrm{MPa}$ would permit $a = 0.013$ m. Note the dimension of K_{I} and the great importance of the

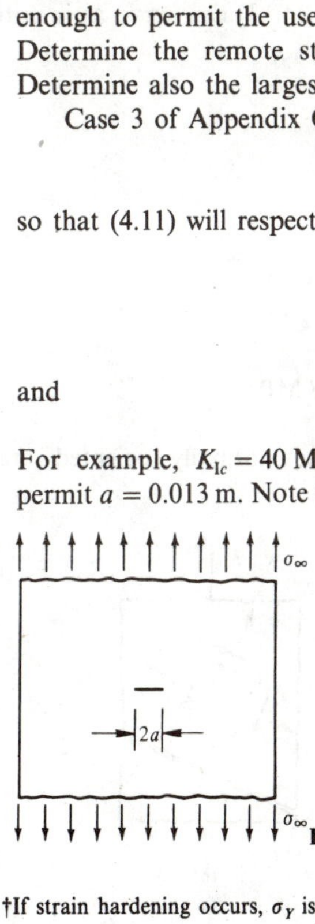

Example 4.2

†If strain hardening occurs, σ_Y is understood to be represented by its nominal value, $\sigma_{0.2}$.

fracture toughness for permissible defects. Comparing our results with the dimensional requirement (4.13*b*), we observe that σ_∞ should not exceed $\sigma_Y/\sqrt{2.5\pi} = 0.36\sigma_Y$ for LFM to apply in the present situation.

Example 4.3 Part (*a*) of the figure shows a long rectangular tension specimen in two projections. It contains a slanted, cracklike slit, size $2a$, which is assumed to be small compared with the cross-sectional lengths W and B. Using LFM, we want to estimate the remote tensile stress when cracking begins. Assume that $K_{Ic} = 50\text{ MPa m}^{1/2}$, $\nu = 0.3$, $a = 0.02$ m, and $\theta = 20°$.

We transform the remote state of stress onto the plane parallel to the slit (assuming that cracking will start somewhere inside the specimen) to obtain the normal and shear stress

$$\sigma'_\infty = \sigma_\infty \cos^2\theta \qquad \tau'_\infty = \sigma_\infty \sin\theta\cos\theta$$

indicated in part (*b*) of the figure. Referred to cases 3 and 19 in Appendix C, this gives the stress intensity factors

$$K_I = \sigma'_\infty\sqrt{\pi a} = \sigma_\infty\sqrt{\pi a}\cos^2\theta \qquad K_{III} = \tau'_\infty\sqrt{\pi a} = \sigma_\infty\sqrt{\pi a}\sin\theta\cos\theta$$

When K_I^2 dominates, as for the given θ, Eq. (4.12) can serve as a critical condition

$$K_I^2 + \frac{K_{III}^2}{1-\nu} = K_{Ic}^2$$

giving

$$\sigma_\infty = \frac{K_{Ic}/\sqrt{\pi a}}{\cos\theta\sqrt{1 + [\nu/(1-\nu)]\sin^2\theta}}$$

With the specified data this provides

$$\sigma_\infty = \frac{200}{0.94(1.025)} = 208\text{ MPa}$$

which is quite close to the value for the normally oriented crack, $\sigma_\infty = 200$ MPa.

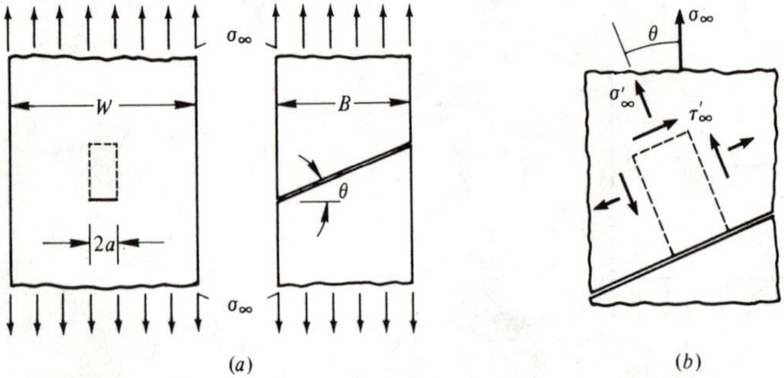

Example 4.3

4.3.2 Continued Quasi-Static Cracking

On the same introductory basis as that preceding Sec. 4.3.1, we state the condition for a further slow motion of the crack as

$$\mathcal{G}(\mu, A) = R(A - A_0) \tag{4.14}$$

The crack driving force should now depend on the crack area (explicitly and through some relation between the action parameter and the crack motion); the *resistance* R is assumed to depend uniquely on the crack area traveled

$$R = 2\gamma + D(A - A_0) = R(A - A_0)$$

where $D(A - A_0)$ is the continued dissipation function.

We have above adhered to the previous formulation in terms of a crack area. This one-parametric characterization anticipates, of course, constraints to the motion such as self-similarity, rotational symmetry, or plane motion. Only in the last case, when the crack length will be an equivalent parameter, can the function $R(A - A_0)$ or $R(a - a_0)$ be seen as an inherent property of the material.

It should also be noted that blunting of the crack through local yielding may be accompanied by some forward motion of the crack tip, which is no true reflection of fracture. The effect, illustrated in Fig. 4.6, can be anticipated by interpreting A_0 as the sum of the precrack area and the blunting projection. This could be of some importance in the more ductile cases of fracture.

Further, it would be natural to associate the limit as $A - A_0 \to 0$ with the discussion in Sec. 4.3.1, such that $D(0) = D_c$, $R(0) = \mathcal{G}_c\ (= \beta K_{Ic}^2/E)$. However, the fracture-toughness concept is tied traditionally to experimental techniques which may not always detect the event of true crack initiation. In such cases, when R increases monotonically, $R(0)$ will be somewhat overestimated by $\mathcal{G}_c$. We keep this reservation in mind, before returning to the experimental aspects in Chap. 7.

Since the equality (4.14) should remain valid as A increases, we must have

$$\left(\frac{\partial \mathcal{G}}{\partial \mu}\right)_A \frac{d\mu}{dA} + \left(\frac{\partial \mathcal{G}}{\partial A}\right)_\mu - \frac{dR}{dA} = 0 \tag{4.15}$$

such that

$$\frac{d\mu}{dA} = \frac{\{dR/dA - (\partial \mathcal{G}/\partial A)_\mu\}}{(\partial \mathcal{G}/\partial \mu)_A}$$

Therefore, the crack must be stable, $d\mu/dA > 0$, as long as

$$\left(\frac{\partial \mathcal{G}}{\partial A}\right)_\mu < \frac{dR}{dA} \tag{4.16}$$

whereas the transition to dynamic fracture is signaled by

$$\left(\frac{\partial \mathcal{G}}{\partial A}\right)_\mu = \frac{dR}{dA} \tag{4.17}$$

We have repeated the arguments running through Fig. 4.1 and Eqs. (4.4) to

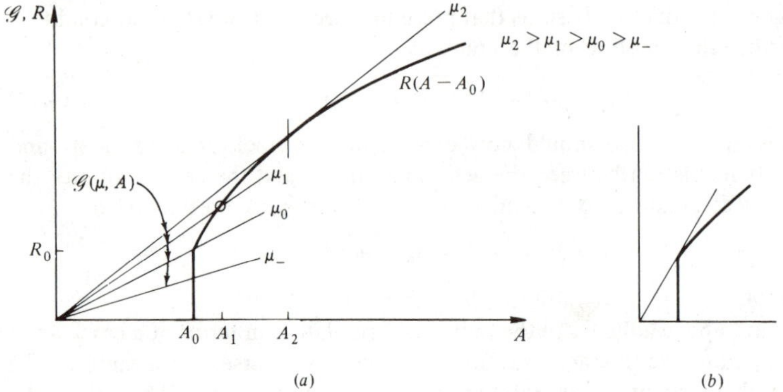

Figure 4.3 (*a*) The *R* curve and a family of rising $\mathscr{G}$ curves. Dynamic initiation is provoked by the smaller precrack (*b*).

(4.6), the only difference being that here the resistance term (R replacing 2γ) also depends on A.

Graphically the results can be interpreted as shown in Fig. 4.3. With the indicated functions $R(A - A_0)$ and $\mathscr{G}(\mu, A)$ we see that the load or displacement factor μ_- is subcritical (no crack initiation), while μ_0 starts the crack and μ_1 keeps it growing through the current crack area A_1. In the last case (4.14) alone is satisfied, whereas the pair of values μ_2 and A_2 satisfy (4.14) and (4.17) simultaneously, meaning that instability is imminent. It is clear, then, that the range $A_2 - A_0$ of stable growth depends both on the function $R(A - A_0)$ and on the geometry and external action (through $\mathscr{G}$). The farther to the left the point of tangency between R and a $\mathscr{G}$ curve the shorter the range of stable crack growth. Such cases are also fully conceivable, as in Fig. 4.3*b*, when the crack runs off immediately upon initiation, because then $(\partial \mathscr{G}/\partial A)_\mu$ is already greater than dR/dA.

Figures 4.3*a* and *b* differ only in the precrack area A_0. The trend toward loss of stable growth, shown here to accompany a reduction of A_0, has also found some experimental support.

The family of rising $\mathscr{G}$ curves in Fig. 4.3 is typical of many cases of load control. If, on the other hand, displacements are controlled (compare Example 4.1), the trend will usually be the opposite, as in Fig. 4.4. If R increases monotonically, we then find that the crack will remain stable for all lengths. Experiments designed to such conditions can be used to determine the R curve of a material. One then measures corresponding values of action and crack area during the stable growth, that is, $\mu(A)$, to yield $\mathscr{G}$ in terms of A by one of the methods of Sec. 3.2; whereupon $R(A - A_0)$ follows from (4.14).

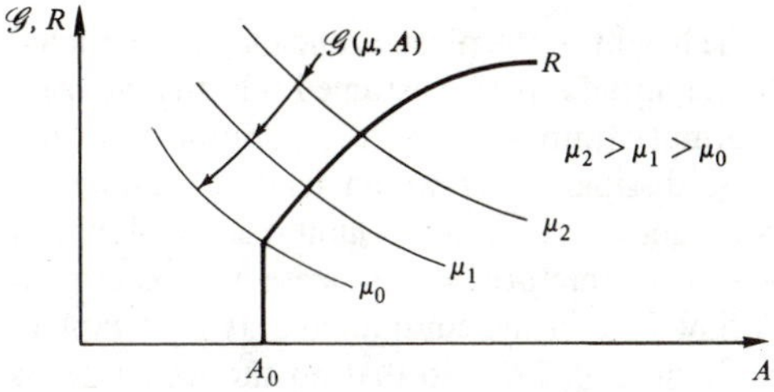

Figure 4.4 A stabilizing type of action. With falling $\mathscr{G}$ curves, tangency of $\mathscr{G}$ and R is ruled out or delayed.

4.3.3 Final Remarks

As a means of predicting the onset of fracture, LFM is the best-developed and most practicable part of fracture mechanics. Experimental methods are well established (compare Chap. 7), and material data are available both for classification and for making more direct estimates. Nevertheless, this theory has a severe limitation in the minimum size requirements for its validity. The possibility that we may have to go past these limits will be explored in the balance of this chapter, which may be seen as leading into or belonging to the area of nonlinear fracture mechanics.

In the present treatment of continued crack growth we have discussed the R-curve method, within the restriction of small-scale yielding and a limited amount of growth. Unconditionally, it would be difficult to accept the trace of $\mathscr{G}$ as a material property. An interesting length measure in this context will be that crack-tip displacement which corresponds to the transient, or rising part of the R curve. For the thick plates considered above, there is only one length with which this measure can be compared, namely the extent $c \approx 0.1(K_{Ic}/\sigma_Y)^2$ of the yield zone. Typically, the theory would then be expected to predict a length of stable growth up to the order of $(K_{Ic}/\sigma_Y)^2$, when $(\partial\mathscr{G}/\partial A)_\mu > 0$, as in Fig. 4.3. In the extreme of yield localization, this would totally exclude a stable crack growth, which seems to accord with observation and is also the conclusion we reached on a purely elastic basis in Eq. (4.6) and the comments below it.

4.4 QUASI-LINEAR FRACTURE MECHANICS

Section 2.3 discussed qualitities of the yielding at the crack front in a thick or thin plate. For comparatively large values of the thickness B it appeared that the total dissipation must be essentially governed by the interior conditions, i.e., the restriction $\varepsilon_{33} \approx 0$, which serves to suppress extensive yielding. "Comparatively large" can only refer to the ratio of B and the yield length $c \approx 0.1(K_{Ic}/\sigma_Y)^2$, as implemented in (4.13a). The constant $\mathscr{G}_c$ may therefore be viewed as a lower asymptote of the dissipation term $C + D$ with increasing B/c when the term is averaged over the thickness.

If B is small on the other hand, the height of the plastic zone may be restricted to the plate thickness in its order of magnitude. If it is assumed that the intensity of yielding (as expressed by the normal strain ε_{22}) before separation is rather insensitive to the thickness, then the dissipation D would tend to be linearly dependent on the thickness as the latter goes to zero. In this limit it is also doubtful that the separation should be viewed as fracture in the sense that energy is dissipated within a process zone. A flow field in the continuum may then explain the separation as related to local necking (Fig. 2.6*b*) so that no surface energy is activated.

Figure 4.5*a* summarizes the conjectures and empirical data. The solid curve indicates K_I-values typically measured at onset of cracking, notably when the assumption of localized yielding in the plane of the plate is satisfied. Curves 1 and 2 are the limiting cases mentioned above, curve 1 obviously representing the results within linear fracture mechanics and curve 2 being taken through the origin, as reasoned above. In an extended empirical form of the critical condition $\mathcal{G} = \mathcal{G}_c$ or

$$K_I = K_c \tag{4.18}$$

K_c^2 will represent the average dissipation $C + D$ over the thickness. The right-hand index I is omitted to emphasize the observation that cracking within a finite thickness is never restricted to mode I but instead is a mixture of modes I and III. This refers to more or less pronounced slanted fractures occurring at the plate surfaces. These shear effects may even penetrate a very thin plate (compare Fig. 2.6*b*) but appear also in thicker plates at *shear lips* at the surfaces. Such typical fractures are illustrated in Fig. 4.5*b*. Related to the difference in internal and external behavior is the *pop-in* phenomenon. It results when cracking is first

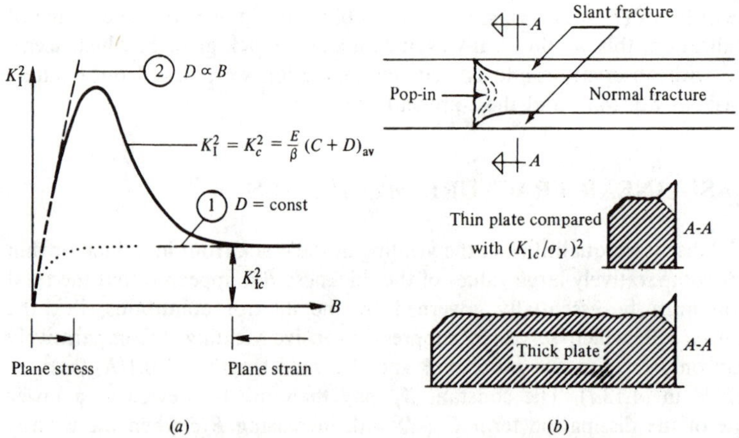

Figure 4.5 Transitional behavior from dominant plane stress and slant fracture to plane strain and normal fracture. Delayed shearing separation in the side ligaments is reflected by the peak in K_c and by pop-in.

provoked internally and is recorded as "thumbnail" cracking or "tunneling" as the crack moves (in one or more steps) inside while the external ligaments are still intact.† Not until the action corresponds to $K_\mathrm{I} = K_c$ will the new crack penetrate the plate. K_c should accordingly define as critical the start of a further crack motion which includes the whole thickness.‡ It has the formal appearance of a fracture parameter, but it will depend (at least) on the plate thickness in addition to the material properties. Experimental data on the dependence of K_c on the thickness are not readily available.

The inequalities (4.13) were necessary conditions for LFM to apply. The first served to suppress the relative importance of plate-side disturbances, so that plane strain was dominantly achieved at the crack tip through the thickness. The second is part of the precautions taken to ensure a sufficient yield localization in the plane of the plate, so that elastic analysis is essentially applicable. We now see how a *slight* relaxation of this last condition can be repaired in a very approximate manner. Reference is then made to the Irwin's study (as in Sec. 2.4), which led to the approximate small-scale yield lengths in platelike specimens

$$c = \frac{1}{\pi}\left(\frac{K_\mathrm{I}}{k\sigma_Y}\right)^2 \qquad \text{where } k = \begin{cases} 1 & \text{for thin plates} \\ \sqrt{3} & \text{for thick plates} \end{cases}$$

In that study $c/2$ also appeared as a translation of the local elastic field due to yielding. This displaced field might also have been caused by a crack of fictitious length $a + c/2$ in a fictitious, totally elastic body. Irwin suggested (rather than provided a conclusive argument) that real crack growth might be expected when K_I based upon this "equivalent" crack attained its critical value. This means that the K_I value to be inserted in (4.11) or (4.18) should be related to the crack length

$$a + \frac{c}{2} = a + \frac{1}{2\pi}\left(\frac{K_\mathrm{I}}{k\sigma_Y}\right)^2$$

instead of a. To illustrate, for a plane geometry as in Example 4.2, this would lead to the critical condition

$$\sigma_\infty = \begin{cases} \dfrac{K_{\mathrm{I}c}}{\sqrt{\pi[a + (1/6\pi)(K_{\mathrm{I}c}/\sigma_Y)^2]}} & \text{thick plate} \qquad (4.19a) \\[2ex] \dfrac{K_c}{\sqrt{\pi[a + (1/2\pi)(K_c/\sigma_Y)^2]}} & \text{thin plate} \qquad (4.19b) \end{cases}$$

giving a somewhat smaller value of σ_∞ than would be obtained without the length

†This behavior can be linked to the higher dissipative screening of the separation process at the surfaces, as implied above, or directly to the fact that stresses are higher in the interior plane-strain state (Chap. 2).

‡Operationally, K_c is usually interpreted as the value of K_I when unstable growth is about to start. While the two versions may be equivalent in practice, the last form is definitely tied to the dependence of K_c on the whole geometry, not only the thickness of the specimen.

correction. The tendency has been verified experimentally, and it also seems to agree reasonably well with some predictions from more basic assumptions.

Attention in (4.18) was focused on the starting of crack growth. An analog to the condition $K_I = K_{Ic}$ turned out to be $K_I = K_c$ more generally (though still within the small-scale yielding assumption), K_c depending here on the material *and* the thickness. A similar modification can be applied to the analysis of continued growth, in that the material resistance function $R(A - A_0)$ is replaced by a resistance $R(A - A_0, B)$. The latter must depend on the thickness B in such a way that the asymptote $R(A - A_0, \infty)$ coincides with the former $R(A - A_0)$. Diagrams corresponding to Figs. 4.3 and 4.4 then apply to one given thickness and have some experimental support. Indeed, this is the type of problem in which stable crack growth is of greatest concern, as the phenomenon is observed mostly in thin plates. We can reconcile the facts with the arguments in Sec. 4.3.3 by noting that the thickness has now entered as another characteristic length. It is quite possible that the comparatively large stable growth which may appear in thin plates is related to variations in the local modes of separation; compare the development of slant fracture (Fig. 4.5*b*, upper part). Finally, it may be noted that a crack-length correction of Irwin's type is being applied even in the context of continued growth to account for a slightly more extensive yielding than implied by (4.13*b*).

The results of Secs. 4.3 and 4.4 can be roughly summarized by the conditions

$$K_I = K_{Ic} \quad \text{or} \quad K_I = K_c \tag{4.20}$$

for conventional starting and

$$\mathcal{G} = \frac{\beta}{E} K_I^2 = R \tag{4.21}$$

for the total history of slow crack growth in mode I.† Although they have been discussed mainly on an energetic basis, starting from Griffith's theories, as they finally appear they are empirical relations apart from one essential theoretical observation, namely the condition that *yielding and the fracture process be so localized as to be governed by the single parameter* K_I. This leaves room for alternative arguments to the same final result (like the emphasized statement itself), which can therefore be viewed as being independent of physical considerations of the fracture process specifically. Thus, whether the local process is of a more or less brittle type is of no formal significance as far as the use of linear or quasi-linear fracture mechanics is concerned.

Example 4.4 The geometry and kinematics of the introductory exercise (Fig. 1.2) are reconsidered. The uniform thickness B assigned to the specimen is small enough for an overall plane-stress condition to be assumed, that is, $\eta = 1/(1 - \nu^2)$ according to (1.2). A consequence is the violation of the

†The alternative form $K_I = K_R$, where $K_R = \sqrt{ER/\beta}$ is the *crack growth toughness*, is obviously equivalent to (4.21).

thickness requirement (4.13*a*); however, (4.13*b*) is assumed to be well satisfied for any current value of the fracture (crack-growth) toughness, so that a quasi-linear approach can be invoked. Our aim is to study the initial and continued separation along the precrack line, under conditions when the separating force $P = \sigma bB$ is controlled [Part (*a*) of the figure]. The resistance function $R(a - a_0, B)$ is as indicated in part (*b*) of the figure

$$R = R_0\left[1 + f\left(\frac{\Delta a}{d}\right)\right] \qquad \begin{aligned} f(0) &= 0 \\ f'(0) &= \theta d \end{aligned} \tag{a}$$

where θ and d are measures of initial slope and transient length, respectively.† This resistance is to be compared with the separating action.

The substitution of

$$\frac{\eta Eu}{H} = \sigma = \frac{P}{bB}$$

from (1.1) in (1.4) gives the crack driving force in terms of the load P,

$$\mathscr{G} = w_0 H\left(\frac{P}{\sigma_0 bB}\right)^2 \tag{b}$$

where σ_0 is some reference stress and

$$w_0 \equiv \frac{\sigma_0^2}{2\eta E} \tag{c}$$

Further

$$\mathscr{G}_{,a} \equiv \left(\frac{\partial \mathscr{G}}{\partial a}\right)_P = -\frac{\partial \mathscr{G}}{\partial b} = \frac{2}{b}\mathscr{G} \tag{d}$$

Governing any stable crack growth are now Eqs. (*a*) to (*c*) and (4.14), i.e.,

$$\mathscr{G} = w_0 H\left(\frac{P}{\sigma_0 bB}\right)^2 = R_0(1 + f) \tag{e}$$

assuming that (4.16) together with Eq. (*d*) are satisfied; thus

$$\mathscr{G}_{,a} \le \frac{R_0 f'}{d} \qquad \text{or} \qquad \frac{2}{b}(1 + f) \le \frac{f'}{d} \tag{f}$$

The load $P = P_0$ and the subsidiary condition necessary to initiate such stable motion follow by the substitution of $\Delta a = 0$ or $b = b_0$ in Eqs. (*e*) and (*f*)

$$P_0 = \sigma_0 b_0 B\left(\frac{R_0}{w_0 H}\right)^{1/2} \qquad 2 < \theta b_0 \tag{g}$$

When P is a function of Δa or b as dictated by (*e*), the subsequent stable regime will terminate when P and Δa take such values that (*e*) and the equality (*f*) are simultaneously satisfied. Thus, the length $\Delta a = b_0 - b$ of stable growth

†Specifically, the function $f = \theta d[1 - \exp(-\Delta a/d)]$ has been used in later evaluations.

and the instability load $P = P_{max}$ are given successively through

$$1 + f = \frac{b}{2}\frac{f'}{d} \qquad \text{and} \qquad \frac{P_{max}}{P_0} = \frac{b}{b_0}\sqrt{1+f}$$

Part (c) of the figure illustrates the family of $\mathscr{G}$ curves as ordered by the load P. The chosen R curve has $\theta d = 2$. A combination of size and resistance is assumed such that $b_0/d = 5$, implying $\theta b_0 = 10$. The resulting trajectory of load versus crack length [part (d) of the figure] can be deduced graphically from part (c). In particular, the point of tangency between $\mathscr{G}$ and R signifies the end of stable growth when

$$\Delta a = 0.74d = 0.15b_0 \qquad P_{max} = 1.22P_0$$

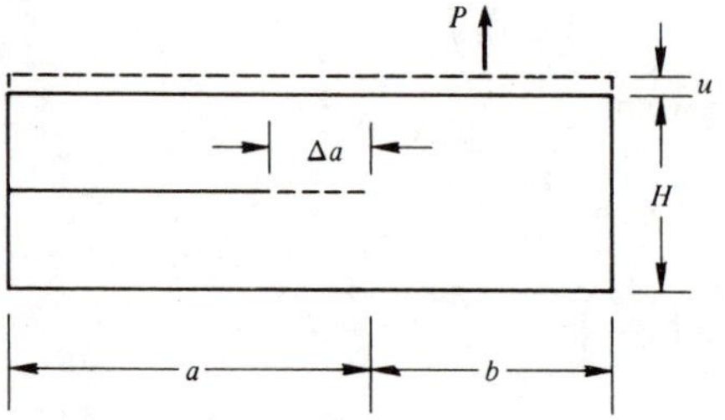

(a)

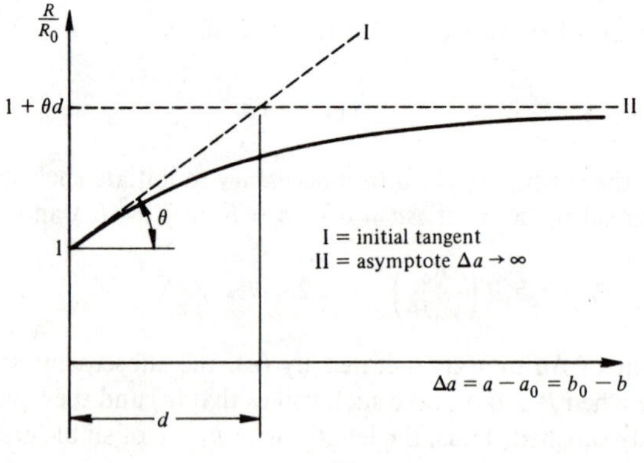

(b)

Example 4.4(a) and (b)

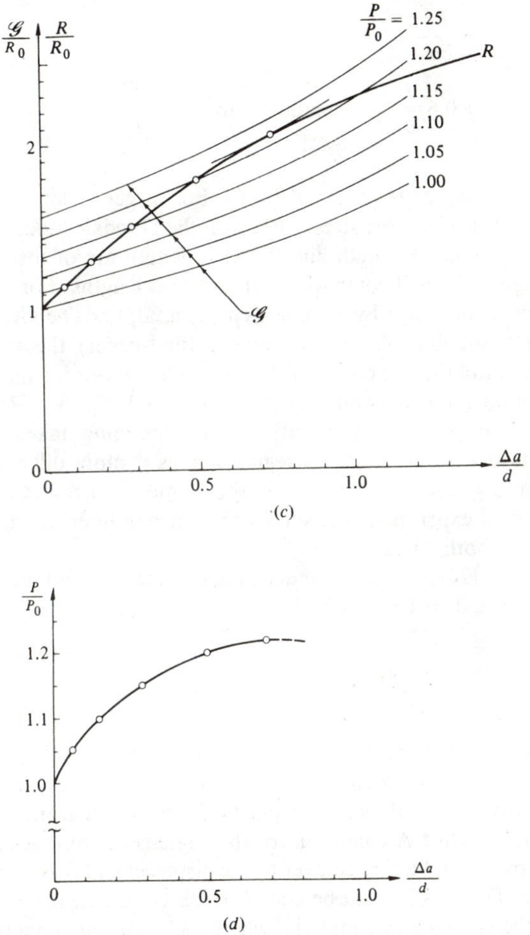

Example 4.4(c) and (d)

4.5 CRACK-OPENING DISPLACEMENT AS A CRITICAL QUANTITY

In Secs. 2.5, 2.6, and 2.8 the COD δ was considered in relation to an opening-mode loading. It represents a more or less well-defined relative displacement of the crack sides close to the front (Fig. 4.6). The reason for this displacement is yielding through the hinge mechanism in thick plates and through necking ahead of the crack in thin ones (see Fig. 2.6). The respective values of δ with ideal plasticity were

$$\delta \begin{cases} = \dfrac{K_{\mathrm{I}}^2}{E\sigma_Y} & \text{plane stress} \\ \approx 0.6\dfrac{K_{\mathrm{I}}^2}{E\sigma_Y} & \text{plane strain} \end{cases} \tag{4.22}$$

assuming such localized yielding as is governed by K_{I}. For larger yielding some analytical expressions exist for δ in plane stress, based on the Dugdale model; see, for example, (2.46), where the crack length and the yield length are of the same order of magnitude though still small compared with external lengths. For plane strain δ can be evaluated numerically by a finite-element analysis. The COD is uniquely defined within a small-displacement-gradient (first-order) theory for ideally plastic materials and can then be consistently derived by first-order analysis using elements which exhibit the proper singularity at the crack tip (Sec. 2.8.1). Within a second-order description, and/or with strain hardening taken into account, the gradual blunting in the crack-tip region makes it more difficult to identify a typical opening displacement. Of course, this is merely a reflection of the real situation as observed experimentally; some experience in interpreting the results will be called for in both cases.

Cottrell [200] and Wells [201] suggested independently that a condition for crack growth should be related to the COD; that is,

$$\delta = \delta_c \tag{4.23}$$

where δ_c is a typical constant for the material in its environment and for the thickness if the plate is quite thin. As briefly mentioned in Chap. 1, locally ductile fracture in metals is typically described by a large plastic deformation at the crack front which activates void growth.† A condition for the coalescence of crack and voids, so that the front moves, can be carried over to the displacement δ as derived for a homogeneous body. Thus, (4.23) can be considered a condition for locally ductile fracture, i.e., for the cracking in a material where inclusions are contained in a ductile matrix of sufficiently high temperature. This supplements the energy-based criteria and offers a way to bypass any assumption that yielding is localized at the crack tip. However, for the present it should be noted that (4.23)

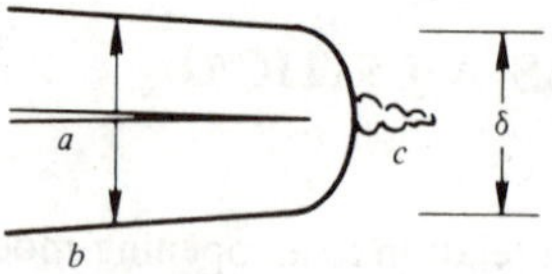

Figure 4.6 Blunting of a precrack tip, causing the COD δ and some forward motion in that same order of magnitude; a = precrack, b = deformed precrack, c = initiated crack to follow.

†The region of such critical void growth has a height and length of the order of δ in a plane situation. This exemplifies the fracture process zone when the local process is ductile.

is intended only to describe the *initiation* of crack growth (with the implied uncertainties in measurements), partly because the crack-tip geometry changes upon initiation, as discussed later.

During *local* yield, linear or quasi-linear fracture mechanics is also applicable and may serve to calibrate the constant δ_c. Introducing (4.20) in (4.22) and then (4.22) in (4.23) leads to

$$\delta_c = \begin{cases} \dfrac{K_c^2}{E\sigma_Y} & \text{very thin plate} \\[2ex] 0.6\,\dfrac{K_{Ic}^2}{E\sigma_Y} = \delta_{Ic} & \text{thick plate} \end{cases} \tag{4.24}$$

which determines δ_c if K_{Ic} or K_c is known, and conversely. The basic requirement here is that (4.20) and (4.23) be *identical* conditions when the yielding is localized, referring to a well-defined event of crack initiation.

For larger yielding δ_c according to (4.24) will continue to specify the material and the thickness, while expressions other than (4.22) apply to the left-hand side of (4.23). To illustrate, if we look at a thin plate as in Fig. 2.11*b*, the initiation condition will be

$$\frac{8\sigma_Y a}{\pi E} \ln\left[\left(\cos\frac{\pi\sigma_\infty}{2\sigma_Y}\right)^{-1}\right] = \frac{K_c^2}{E\sigma_Y} \tag{4.25}$$

referring to (2.46). This equation can be solved with respect to the critical value σ_∞. The result is displayed in Table 4.1 and Fig. 4.7, where σ_{lin} is the solution which would follow from using LFM blindly

$$\sigma_{\text{lin}} \equiv \frac{K_c}{\sqrt{\pi a}} \tag{4.26}$$

Table 4.1 Comparison of predictions by the Dugdale model and the Irwin correction

$\sigma_{\text{lin}}/\sigma_Y$	σ_∞/σ_Y By (4.25)	σ_∞/σ_Y By (4.19*b*)	$\sigma_{\text{lin}}/\sigma_Y$	σ_∞/σ_Y By (4.25)	σ_∞/σ_Y By (4.19*b*)
0.1	0.100	0.100	0.7	0.632	0.627
0.2	0.198	0.198	0.8	0.700	0.696
0.3	0.295	0.294	0.9	0.760	0.759
0.4	0.387	0.385	1.0	0.812	0.816
0.5	0.475	0.471	∞	1	$\sqrt{2}$
0.6	0.557	0.552			

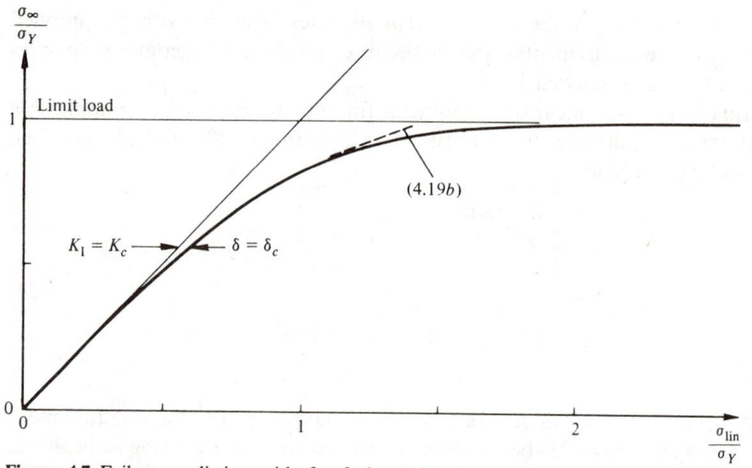

Figure 4.7 Failure prediction with $\delta = \delta_c$ by (4.25) (or with $J = J_c$) interpolating between the asymptotes $K_\mathrm{I} = K_c$ (brittle fracture) and $\sigma_\infty = \sigma_Y$ (plastic collapse).

Why this must be discarded as critical in the case of larger-scale yielding is obvious if we go to very high K_c values, i.e., tough materials where cracks do not grow easily. Regardless of such toughness, the plastic collapse mechanism (limiting plastic state of the given body) must exclude $\sigma_\infty > \sigma_Y$ as acceptable values. This condition is included in Fig. 4.7 (where the critical state is conservatively interpreted as failure).

Mainly to satisfy curiosity we have also included the result (4.19) by the Irwin correction in the table and figure. The agreement is almost incredible, except in the vicinity of the limiting stress $\sigma_\infty = \sigma_Y$. This final deviation should be gratefully accepted, as the Irwin correction must be totally insensitive to the far-field behavior which governs plastic collapse.

We return briefly to the problem of continued growth under large-scale yielding, with its implications for the δ value. A schematic picture of the fracture event is indicated in Fig. 4.8*a*. (Actually, the crack path is tortuous, and the walls are torn and irregular.) Upon initiation, when $\delta = \delta_c$, the crack emerges spear-shaped from the vaulted end of the precrack. The empirical form of δ at the *original* crack tip following this event is shown in Fig. 4.8*b*; δ increases as a result of further slow motion and may seem to approach a linear dependence on the forward displacement Δa after a transient start. This means that the backward opening angle *COA* in Fig. 4.8*a* rapidly reduces toward a constant value as growth proceeds. Another interesting parameter is the translated opening angle, formed between two lines from the current crack tip through opposite crack-side points at some specified distance behind the tip.† Even this angle seems to approach a constant value, and we are thus faced with two comparatively attractive candi-

†An equivalent measure is obviously the translated COD, the distance between the two points.

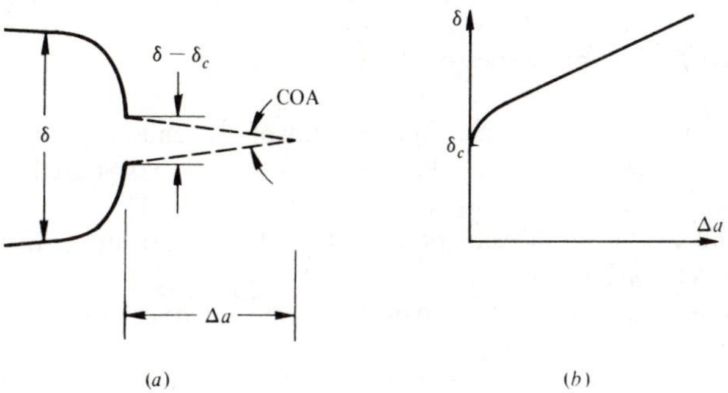

Figure 4.8 Schematic geometry (a) and recording (b) of continued crack growth.

dates for predicting the growth history. The first is the simpler to calculate and observe, but the second, the crack-tip opening angle (CTOA) may be the more fundamental quantity. Nevertheless, this is an area of fracture mechanics where much remains to be clarified by theory and experiment; a few elements of the discussion are recorded in Secs. 4.6 and 9.6.

Finally, it should be pointed out that the COD may also provide the criterion for crack initiation in mode III, analogous to mode I. In condition (4.23), that is, $\delta = \delta_c$, one can begin by inserting the solution (2.76) for small-scale yielding. When the result is required to be the same as would follow from linear fracture mechanics, i.e.,

$$K_{\mathrm{III}} = K_{\mathrm{III}c}$$

the dependence

$$\delta_c = \frac{2}{\pi} \frac{K_{\mathrm{III}c}^2}{G\tau_Y} \tag{4.27}$$

is obtained. For extensive yielding, e.g., with δ as given by (2.78), the equality of that displacement and δ_c by (4.27) would lead to a diagram very like that in Fig. 4.7 with τ, K_{III}, and $K_{\mathrm{III}c}$ replacing σ, K_{I}, and K_c, respectively. Even here a similar trend would follow by the Irwin modification

$$\tau_\infty = \frac{K_{\mathrm{III}c}}{\sqrt{\pi[a + (1/2\pi)(K_{\mathrm{III}c}/\tau_Y)^2]}}$$

[compare (4.19b)], except in the vicinity of the limit load.

There is a body of guidelines and experience related to the COD criterion. We return to some practical aspects in Chap. 7.

4.6 THE J INTEGRAL AS A CRITICAL QUANTITY

We have already noted general properties of the J integral:

1. It may be approximately invariant, i.e., independent of the chosen path of integration from one crack side to the other, before the crack has started to grow (Secs. 2.6 and 3.3.3).
2. Accordingly it should be a scalar measure of the state at the crack front because the path may be shrunk arbitrarily.
3. It can be experimentally determined from load-displacement curves of cracked specimens (Sec. 3.3.3).

Special physical connections have also been reviewed, such as its relation to the COD for ideal plasticity (Sec. 2.6), its relation to the stress-strain singularities for elasticity (Sec. 3.2.3) and strain-hardening plasticity (Sec. 2.8.2), its equality to the crack driving force in an elastic material (Sec. 3.2.3) and to the total dissipation during small-scale yielding in an elastoplastic material (Sec. 3.3.2). Experimental results may even indicate a more extended relationship with the COD.

Since this quantity obviously has an important bearing on crack-tip phenomena, it has been suggested [207] that J itself may be seen as critical in relation to cracking. Restricting our discussion to the opening mode, this gives a condition

$$J = J_c \tag{4.28}$$

for *initiation* of cracking, J_c being a typical property of material, environment, and plate thickness. Advantages of this choice are that the J integral as such is comparatively easy to evaluate or measure (not counting the simple cases of small-scale yielding, where it is in effect equivalent to $\mathcal{G}$, K_{I}, or δ). Quite as important are several supporting results from critical experiments. Disadvantages are that path independence is not generally ensured,† and a unique physical interpretation of J hardly exists. The severe limitations related to continued crack growth and certain geometry effects will be commented on later.

The results for small-scale yielding, ideal-plastic response

$$J = \begin{cases} \dfrac{1}{E} K_{\mathrm{I}}^2 = \sigma_Y \delta & \text{plane stress} \\[2ex] \dfrac{1-\nu^2}{E} K_{\mathrm{I}}^2 \approx 1.6 \sigma_Y \delta & \text{plane strain} \end{cases}$$

according to (3.37), (2.50), and (2.52) imply the relationships between the material

†One can easily verify this by evaluating J for alternative paths which cut through the yield zone of the Dugdale model.

parameters

$$J_c = \begin{cases} \dfrac{1}{E} K_c^2 = \sigma_Y \delta_c & \text{plane stress} \\[2ex] \dfrac{1-\nu^2}{E} K_{Ic}^2 = 1.6\sigma_Y \delta_{Ic} = J_{Ic} & \text{plane strain} \end{cases} \tag{4.29}$$

when predictions are required to be the same for well-defined initiation of such globally brittle fractures. As a result of the proportionality relation $J = \sigma_Y \delta$ even for extensive yielding in the Dugdale model, the prediction of Fig. 4.7 could follow just as well by the condition $J = J_c$ (replacing $\delta = \delta_c$).

Further insight into the nature of J for varying applied actions may be provided by the results in (3.41) to (3.44), U denoting the external work on the body before crack initiation. First, recalling that

$$J = \frac{U}{bB} \tag{3.43}$$

applies to the Fig. 1.2 situation regardless of material properties, we have here the J-P-u dependence shown in Fig. 4.9 if elastic–ideally plastic response and a fixed crack length are assumed. In this case there are two distinct regimes, elastic and postyield. If the fracture toughness of the material is so small that J_c [through (4.29)] does not exceed the corner value

$$J_0 = \frac{1}{2}\frac{P_0 u_0}{bB}$$

the response is essentially elastic until failure is predicted by the small-scale-yield version of (4.28), that is, $K_I = K_c$. This would be a globally brittle mode of failure

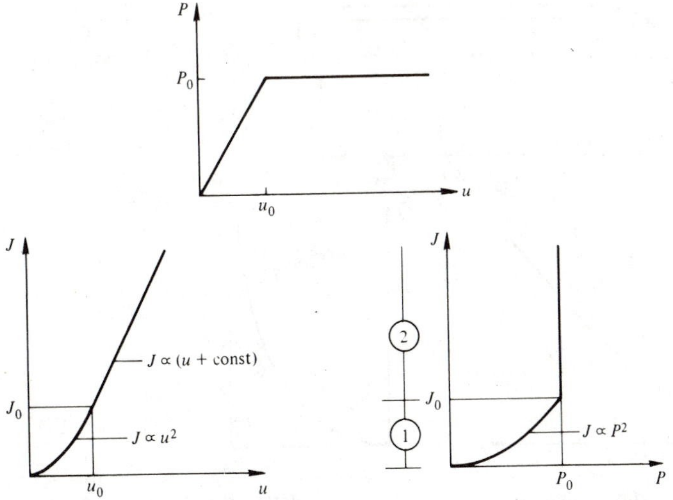

Figure 4.9 Elastic–ideally plastic behavior of the Fig. 1.2 test specimen before fracture.

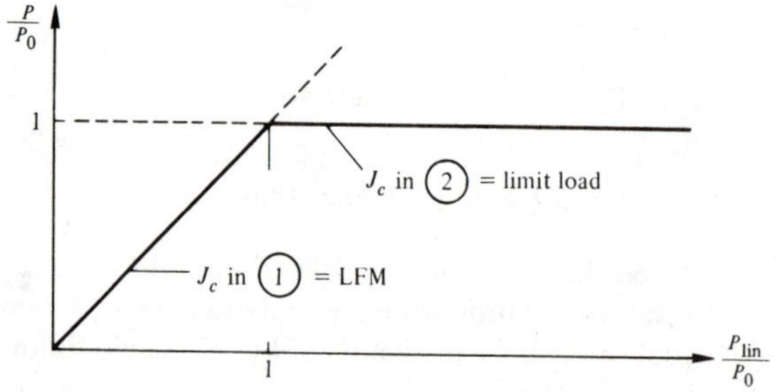

Figure 4.10 Failure prediction for the situation in Fig. 4.9: brittle fracture or plastic collapse; compare Fig. 4.7.

through cracking, predicting by linear fracture mechanics the critical load P_{lin}. If $J_c > J_0$, on the other hand, failure would be at the limit load $P = P_0$ by a plastic collapse mechanism which disregards cracking. The result is shown in Fig. 4.10.

The decoupling of cracking and plastic collapse derived above can hardly be met in real life, however. Instead, a plausible situation is one of some interaction, resulting in smoothening of the failure curves, as experienced in Fig. 4.7. This goes back to inhomogeneities of the stress field and/or nonlinearities of the stress-strain relation. Under such conditions the load-displacement curve is continuous, as indicated in Fig. 4.11*a*, final nonhardening being considered here for simplicity.

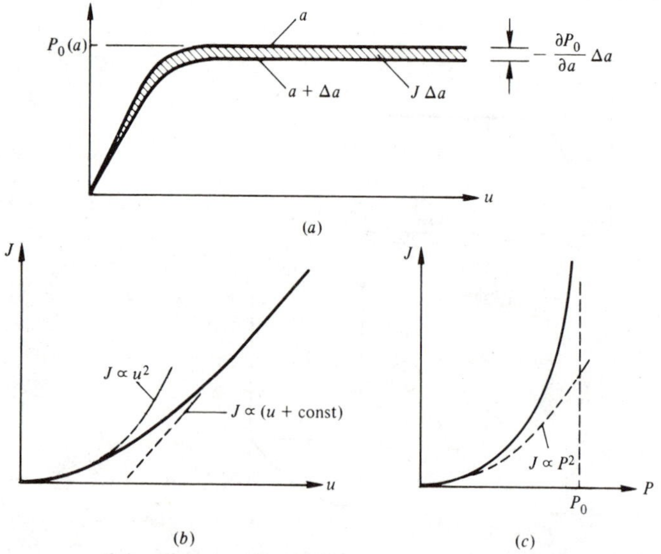

Figure 4.11 Elastoplastic behavior under final nonhardening conditions.

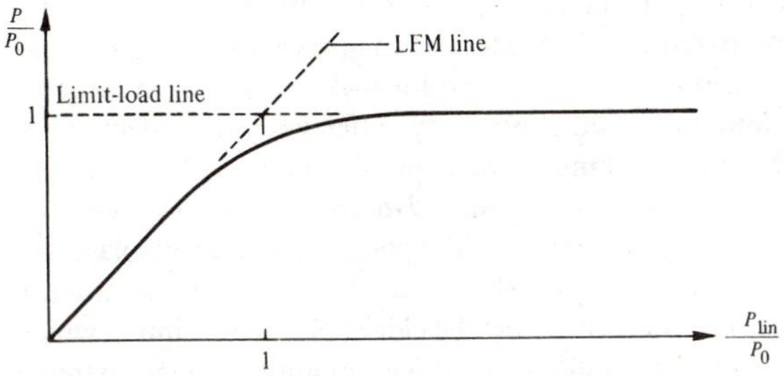

Figure 4.12 Failure prediction for the situation in Fig. 4.11: brittle fracture, plastic collapse, or transition; compare Fig. 4.7.

Proceeding from a version of the general formula (3.41),

$$J = -\int \left(\frac{\partial P}{\partial A}\right) du$$

we see a transition from $J \propto u^2$ in the initial elastic regime (left part of hatched area) to the linear part

$$J \to -u \frac{\partial P_0}{\partial A} \qquad \text{as } u \to \infty$$

for extensive yielding. Alternatively, J as a function of P must go to infinity as the limit load P_0 is approached. These conclusions, illustrated in Fig. 4.11*b* and *c*, can easily be pursued into a failure diagram looking like Fig. 4.12.

Our previous power-law hardening results [(2.103) and (2.104) with (2.109)] indicated the existence but not the extent of a J-dominated zone. This is a region where all relative distribution is fixed and the remaining one-parametric strength of the fields is given in terms of J. We have here an analog of the K-dominated zone met in linear elasticity, which may also be a region surrounding the yield zone if yielding is highly localized. In turn, J may dominate part of that yield zone, as just observed. At the very tip of the crack we find the fracture process zone, where such continuity concepts as K and J lose their meaning altogether. All the same, the hierarchy of zones is important in the sense that whatever happens *inside* a region of one-parametric dominance must be totally governed by that one parameter. Thus, we have seen in linear fracture mechanics that the yield extent and fracture event are both governed by the stress intensity factor provided yielding (and fracture process, by assumption) were so localized that they were surrounded by a K-dominated field. With larger yielding such K dominance breaks down,† but we can still look for some J dominance internally to whatever elastic regime remains and externally over the process zone. There may be cases,

†The outside elastic field then has no region where the first (singularity) term of the solution dominates.

however, where a possible J dominance is made fictitious by the fact that the J zone is not large enough compared with the extent of the process zone. This could certainly be the case if to the process zone we add a wake of unloading due to crack growth. Even crack-initiation predictions may suffer from this effect, since some motion is required for the cracking to be experimentally evident. Certain "J-phobic" geometries tend to set up such small J-dominated zones, typically those where single yield-strip mechanisms would operate under ideal plasticity conditions, e.g., case 1 of Table 2.1. We also recall that the one-parametric characterization of strains breaks down in the ideal-plasticity limit, which indicates that J may not perform well as a critical quantity under extreme low-hardening conditions.

The possible use of the J integral as a quantity governing stable crack growth has been a matter of some concern. As indicated above, a first condition for such use is that the crack motion be small compared with the original J zone, i.e.,

$$\Delta a \ll r_J \tag{4.30}$$

where r_J is some measure of this original J-zone size. However, with crack growth the surrounding strain distribution will also change, partly invalidating any assumption of proportional straining, which was the requirement for path independence of J. This is certainly the case very close to the crack tip; on the other hand, proportional straining under changing action and crack length can be approximately realized at distances $r > r_p$ from the crack tip which are large compared with the ratio $J/(dJ/da)$. This is a result due to Hutchinson and Paris [219], derived primarily for nonlinear elasticity but significant for elastoplastic response as well. The essential observation follows (recall the hierarchy of zones) that this r_p should be well contained within the original J zone for proportionality to dominate and J to reflect the event. This is tantamount to requiring

$$r_J \gg \frac{J}{dJ/da} \qquad \text{or} \qquad \frac{dJ}{J} \gg \frac{da}{r_J} \tag{4.31}$$

which is an inequality signifying that the proportional changes of strains due to incremental action are large compared with nonproportional changes due to increases in the crack length while the crack is stably growing.

For rather localized yielding, excluding possible J-phobic effects, one can estimate the orders of magnitude

$$O(r_J) = c \qquad O(J) = \sigma_Y \delta \propto \sigma_Y \frac{\sigma_Y}{E} c$$

with the yield-zone size c and COD δ as indicated by (2.39) or (2.40) and (4.22). Inserting this into inequality (4.31) gives

$$T \gg 1 \tag{4.32a}$$

where by definition

$$T = \frac{E}{\sigma_Y^2} \frac{dJ}{da} \tag{4.32b}$$

necessary for J characterization under the circumstances. In the literature [218] the symbol T for *tearing modulus* has been introduced as the right-hand group of (4.32*b*).

Similarly, for large-scale yielding the order of the J-zone size may be comparable to the uncracked ligament length b, requiring

$$b\frac{dJ}{da} \gg J \tag{4.33}$$

instead of (4.32).

Experiments must be invoked to bring magnitudes into inequalities (4.32) and (4.33) for various cases if the J integral is to be significantly related to stable crack growth. One may hope that the J integral as derived (correctly, see below!) for a given amount of crack advance will have a value which is uniquely related to the material and that it will allow for some important orders of crack motion. In such cases J-based diagrams of the same type as the Fig. 4.3 R curve can be applied to various materials under extensive yielding, $\mathcal{G}$ being then replaced by J; the resistance R will express the current J value during a stable growth. Then, we have

$$J = R \tag{4.34}$$

$$\left(\frac{\partial J}{\partial a}\right)_\mu < \frac{dR}{da} \tag{4.35}$$

by (4.14) and (4.16) as conditions governing stable crack growth. Note that the $R(a - a_0)$ curve above must be identical to the one of Sec. 4.3.2 for a given material, because such a parameter, by assumption, should not reflect different conditions (small- or large-scale yielding) for undertaking the analysis.†

Experience with such experiments is still limited and somewhat mixed. Path and geometry independence in J has been reported for smaller amounts of crack growth in several cases (up to the order of 5 percent or even more), whereas some geometry dependencies observed even from zero motion may reflect the J-phobic effect discussed above. In general, the J integral seems to predict cases of quite limited crack growth; specifically this may involve the important capacity for estimating initial cracking stability by satisfying (4.34) and (4.35). Conversely, instability is predicted by the inequality opposite that in (4.35).

The J expression (2.48) for a given path remains valid during crack growth, by definition, and then continues to be the basic way of computing the integral. The task of devising other means to estimate J, equivalent of (3.42) to (3.44), is not always trivial. We illustrate this point by considering a record of load versus

†Condition (4.31) having been derived on the assumption of *either* plane stress or plane strain, there is no rational parallel in the J formalism of the thickness-dependent R curves for thinner plates. That empiricism included a transition tending from plane strain and normal fracture toward plane stress and slant fracture with the growth of the crack, whereas the bulk of the present applications may reasonably be in plane strain.

displacement and ligament (or crack) length as depicted in Fig. 4.13, assuming the validity of (3.43) and (3.44):

$$J = \frac{\eta}{Bb_0} \int_0^{u_1} P \, du \qquad \eta = 1, 2 \tag{4.36}$$

for *initial* cracking. Dominantly proportional straining is achieved with the satisfaction of (4.30) and (4.31), which implies that the problem can be formally addressed by considering nonlinear elastic response. For this hypothetical material, w in (2.48) is the strain energy, which does not depend on a specific path toward the current state P_1, u_1, b_1. A simple way to identify the current J value is therefore to assume a history of virgin loading in a specimen precracked to the ligament length b_1, the result being

$$J = \frac{\eta}{Bb_1} \int_0^{u_1} P(u, b = b_1) \, du \tag{4.37}$$

This integral is the area below the dotted loading curve in Fig. 4.13, which should not be confused with that below the actual trajectory for simultaneous loading and cracking, drawn solid.

Example 4.5 The striplike specimen of Fig. 1.2 was further considered in Example 3.4 and a comment to (3.43). For this specimen, having the thickness B, we assume that a record of load, displacement, and length of a stably growing crack has been established experimentally, e.g., Fig. 4.13. The aim is to extract from this information values $J = R$ which correspond to the set of observed crack lengths or ligament sizes.

The use of (4.37) in the present situation is awkward, in that additional specimens, differing in precrack size, would be required to provide the virgin loading curves. Instead we replace the linear elastic-strain energy Φ in

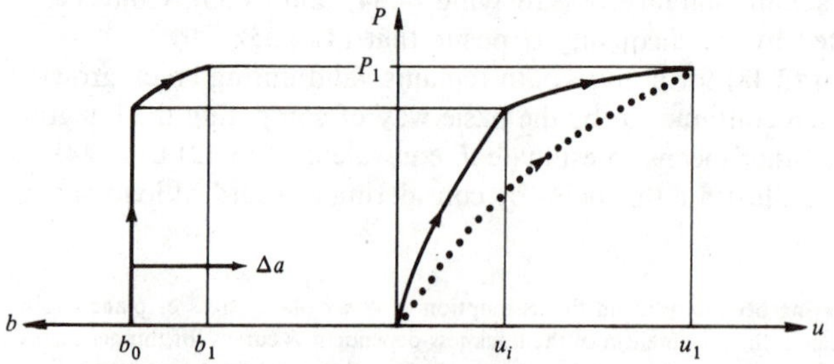

Figure 4.13 Load path with the b_0 uncracked ligament size and subsequent crack growth (*solid curve*) and with the b_1 uncracked ligament size (*dotted curve*). Displacement u_i at actual onset of cracking.

Example 3.4 by

$$\int_0^{\varepsilon_1} \sigma \, d\varepsilon = \int_0^{u_1} \frac{P}{Bb} \frac{du}{H}$$

to yield

$$J = \frac{1}{B} \int_0^{u_1} \frac{P}{b} \, du$$

in terms of corresponding values P and b. This integral can easily be derived from the record.

Evaluation techniques for the same practical end have been devised for other types of specimen, e.g., those of bending character. Appendix H may be consulted for references.

Example 4.6 The evaluation in Example 4.4 will be appropriate at stress levels well below general yield. If σ_0 therein is identified as the initial yield stress σ_Y, a requirement should be that $R = R_0(1+f)$ be small compared with $w_0 H$. This condition, once satisfied, may be violated if for example the in-plane specimen dimensions are sufficiently reduced. An assessment of crack growth under such large-scale yielding conditions will presently be attempted for the same type of specimen. We intend to describe a limited amount of stable growth, when the local process is essentially plane-strain normal fracture, even when overall conditions in the specimen are closer to plane stress. The tool chosen for this investigation is the J integral. Unlike the procedure in Example 4.5, which was aimed at the experimental determination of an R curve through the record of load versus displacement and crack length, we start here by assuming a given R curve in the linearized form

$$R = R_0(1 + \theta \, \Delta a) \qquad \theta = \text{const}$$
$$R_0 \equiv J_c \tag{a}$$

to serve as a basis for predicting the behavior when the load is increased. A rough approximation of the tension-bar constitutive law when the strain may considerably exceed the elastic limit ε_Y is taken as

$$\frac{\varepsilon}{\varepsilon_Y} = \left(\frac{\sigma}{\sigma_Y}\right)^m \qquad \varepsilon_Y = \frac{\sigma_Y}{E}$$

[compare (2.96) and Fig. 2.27]. Going through (2.95), we adapt this to the present overall conditions, i.e., zero strain along the strip and zero stress normal to it, to obtain

$$\frac{\varepsilon}{\varepsilon_Y} = \left(\frac{\sqrt{3}}{2}\right)^{m+1} \left(\frac{\sigma}{\sigma_Y}\right)^m$$

in the vertical components ε and σ.

Using, say, (2.48) and the arguments in Example 3.4, we derive

$$J = w_0 H \left(\frac{P}{\sigma_Y b B}\right)^{m+1} \tag{b}$$

where

$$w_0 = \left(\frac{\sqrt{3}}{2}\right)^{m+1} \frac{m}{m+1} \frac{\sigma_Y^2}{E} \tag{c}$$

and

$$J_{,a} \equiv \left(\frac{\partial J}{\partial a}\right)_P = -\frac{\partial J}{\partial b} = \frac{m+1}{b} J \tag{d}$$

Obviously, this parallels similar expressions in Example 4.4.

Stable crack growth is assumed to be governed by Eqs. (*a*) to (*c*) and (4.34), i.e.,

$$J = w_0 H \left(\frac{P}{\sigma_Y b B}\right)^{m+1} = R_0(1 + \theta\, \Delta a) \tag{e}$$

with the added condition that (4.35) be satisfied,

$$J_{,a} \le R_0 \theta \qquad \text{or} \qquad \frac{m+1}{b}(1 + \theta\, \Delta a) \le \theta \tag{f}$$

But, this description by J (or by nonlinear elasticity, in effect) is meaningfully applicable to the crack growth only if conditions (4.30) and (4.31) are satisfied. In the latter context the requirement by (4.33) is

$$\theta b \gg 1 + \theta\, \Delta a \tag{g}$$

which may well include condition (*f*) for the cracking to be stable.

To initiate such crack growth it is necessary that

$$P = \sigma_Y b_0 B \left(\frac{R_0}{w_0 H}\right)^{1/(m+1)} \equiv P_0 \qquad m + 1 < \theta b_0 \tag{h}$$

Stability is eventually lost when Eqs. (*f*) and (*e*) are simultaneously satisfied; thus

$$1 + \theta\, \Delta a = \frac{\theta b}{m+1} \qquad P = P_0 \frac{b}{b_0}(1 + \theta\, \Delta a)^{1/(m+1)}$$

giving

$$\Delta a = b_0 - b = \frac{b_0}{m+2}\left(1 - \frac{m+1}{\theta b_0}\right) \tag{i}$$

$$P = P_0 \frac{m+1}{\theta b_0}\left(\frac{1 + \theta b_0}{m+2}\right)^{(m+2)/(m+1)} \tag{j}$$

Whether this load pertains to a total instability or just the internal jump (pop-in) of the crack will depend also on the specimen thickness through the relative strength of the side ligaments.

Illustrative parametric values may be taken as $m = 6$ and $\theta b_0 = 20$, for which the terminal length and load by Eqs. (*i*) and (*j*) are $\Delta a = 0.08b_0$, $P = 1.06P_0$. Such a prediction may, of course, suffer in realism by the assumptions made, e.g., by the linearization of the resistance curve.

PROBLEMS

4.1 Assume some Griffith behavior of a material during crack growth, i.e.,

$$\mathcal{G} = 2\gamma = \text{const}$$

With load control, assess the stability of cracking for the cases considered in Probs. 2.2, 2.3, 3.1, and 3.2.

Answer: Problems 2.2 and 3.1 are unstable, 2.3 stable, and 3.2 metastable.

4.2 The flexibility of a testing machine is represented by linear springs as in the figure, each having a compliance equal to $Z_t/2$. The relative displacement u somewhere away from the tested specimen is assumed to be controlled. With Griffith behavior of the double cantilever beam, determine the maximum compliance Z_t of the machine for which the crack will be stable.

Answer:
$$Z_t = \frac{4a^3}{3EI} \qquad I = \frac{Bh^3}{12}$$

follows by (3.25), (4.6), and Example 3.1.

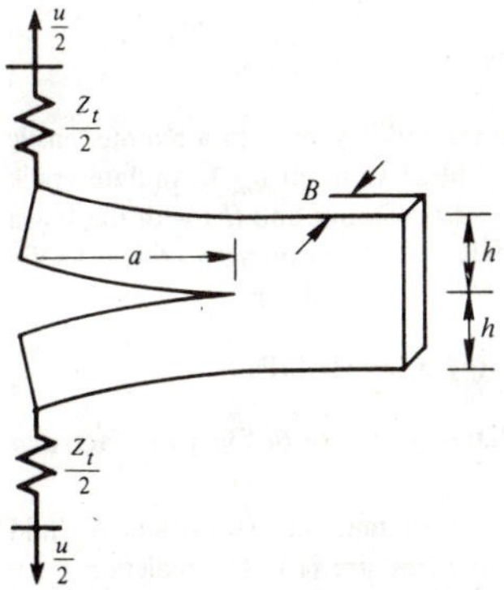

Problem 4.2

4.3 A platelike tension bar may have a central transverse crack through the thickness. The remote tension is $\sigma_\infty = 300$ MPa, the width is $2W = 300$ mm, and the fracture toughness is $K_{Ic} = 100$ MPa m$^{1/2}$. Use LFM to verify the critical crack length $2a = 67$ mm.

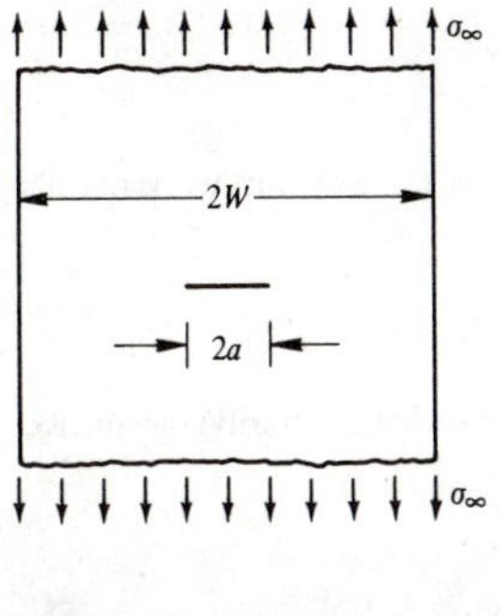

Problem 4.3

4.4 A circular cylindrical bar simultaneously carries an axial tensile force S and a torsional moment T. A surface precrack has been formed in the cross-sectional plane, the depth a being small compared with the diameter D and the length of the crack. Use LFM to verify that $S = 4.0$ MN will induce further cracking when $T/S = D/8$, $D = 100$ mm, $a = 2$ mm, $K_{Ic} = 50$ MPa m$^{1/2}$, and $\nu = 0.2$. *Hint:* Equations (C.7) and (C.8) with $d/D \approx 1$ apply.

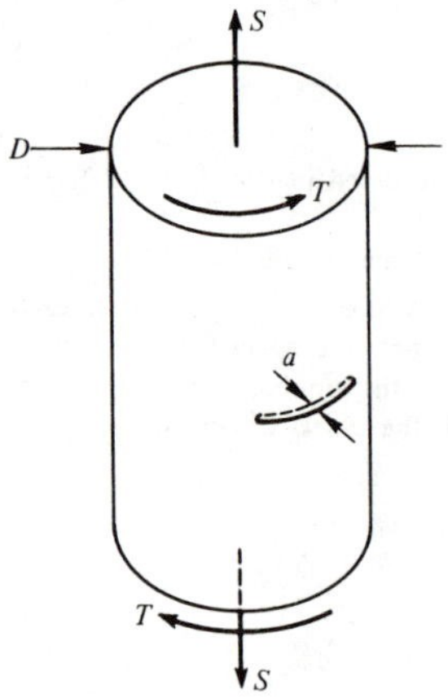

Problem 4.4

4.5 A thin plate has a small central precrack through the thickness and is subject to a remote tensile stress σ_∞ normal to the crack (as in Prob. 4.3, $W \to \infty$). The critical value of σ_∞ to initiate crack motion is sought, on the basis of quasi-linear fracture mechanics (a) without and (b) with the Irwin correction for yield length; (c) COD evaluation and the Dugdale model, assuming a critical COD. With the data $a = 10$ mm, $K_c = 100$ MPa m$^{1/2}$, and $\sigma_Y = 1000$ MPa, verify the results

$$(a)\ \sigma_\infty = 565 \text{ MPa} \qquad (b)\ \sigma_\infty = 524 \text{ MPa} \qquad (c)\ \sigma_\infty = 528 \text{ MPa}$$

Verify also that a is smaller than $2.5(K_c/\sigma_Y)^2$, which substantiates [compare (4.13b)] the fact that (a) is an invalid solution.

4.6 The following has a bearing on Prob. 2.7. For a total range of material parameters K_{Ic} and σ_Y the failure load P is sought. Asymptotic cases are the brittle-fracture (LFM) prediction

$$P \equiv P_{\text{lin}} = K_{Ic} \frac{B\sqrt{a}}{6.82}$$

if K_{Ic} is small, and the total-yield (limit-load) prediction

$$P \equiv P_0 = 0.257 B a \sigma_Y$$

if K_{Ic} is large, as indicated by dashed lines in the figure. The student may further verify the coordinates of the following points:

Point 1. The limit of brittle-fracture predictions according to (4.13b)
Point 2. The intercept of the above asymptotes
Point 3. As derived by the Irwin correction for yield length, given (somewhat arbitrarily) the abscissa of point 2

With the two asymptotes and points 1 and 3 of departure and transit, the solid line has been estimated as a reasonable approximation of a failure diagram. Compare Fig. 4.12.

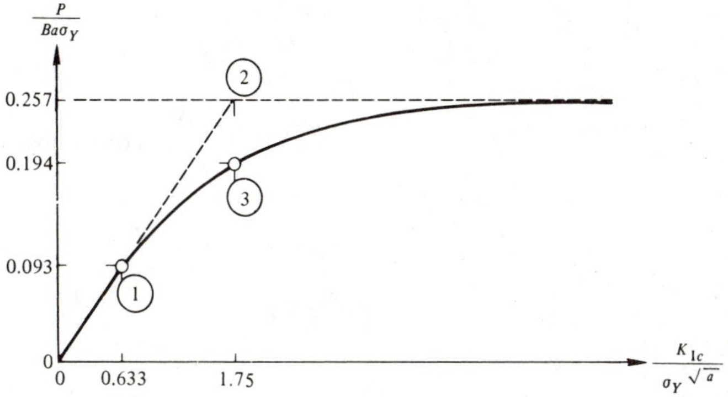

Problem 4.6

4.7 Referring to Prob. 3.3, verify the critical J-integral value $J_{Ic} = 0.125\ \mathrm{MN\,m^{-1}}$ for crack initiation, assuming that initial crack motion is observed at the load-displacement value $u = 1.25\ \mathrm{mm}$. (But real life may not be as simple as it sounds.) Deduce from this the fracture toughness value $K_{Ic} = 170\ \mathrm{MPa\,m^{1/2}}$, given $E = 210\ \mathrm{GPa}$ and $\nu = 0.3$.

CHAPTER

FIVE

DYNAMIC CRACK GROWTH

The motion of a crack leads to sudden unloading (or negative loading) in the surface traversed by the crack front. This will emit stress waves, which may return upon reflection in characteristic surfaces to influence the further motion of the crack. In the quasi-static analyses of earlier chapters the development of the crack was assumed to be so slow that the dynamic interference was terminated and damped out. This motivated the neglect of all particle accelerations in the body. A different situation must be faced when the crack velocity is large, i.e., comparable to wave velocities, as anticipated in Chap. 1.

5.1 DYNAMIC ENERGY BALANCE

During a dynamic event the total energy balance (3.9) must be satisfied. Most analyses have been restricted by the same assumption as made in static LFM, i.e., yielding so localized at the crack tip that the continuum dissipation D can be superposed on the separation work C into a total resistance R, typical of the given material. This amounts to directing the analysis formally at an elastic body, such that

$$\mathscr{G} \equiv W - \frac{dT}{dA} - \frac{d\Phi}{dA} = R \tag{5.1}$$

is satisfied. It has been suggested (and partly verified by experiment) that R of the given material may be taken as depending uniquely on the crack-tip velocity $\dot{a}$ in a first approximation. A monotonic increase (Fig. 5.1) is often met. An anomaly may govern the initiation, nevertheless, in the sense that an initiation barrier admits that the resistance $R = R_0$ toward a first motion exceeds the limiting value $R(0)$ defined by letting $\dot{a}$ approach zero from above. The left-hand side of (5.1) is assembled as the crack driving force $\mathscr{G}$ in a notation similar to that used in static LFM.

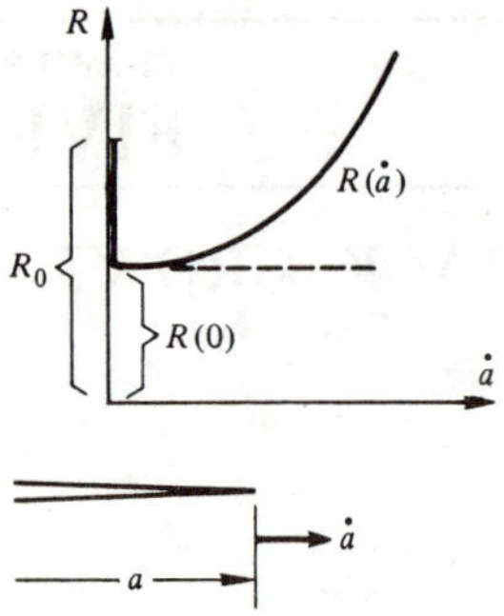

Figure 5.1 Types of dynamic resistance curves: a constant (*dashed*) or monotonic function of $\dot{a}$, including an initiation barrier $R_0/R(0) > 1$.

5.2 A UNIAXIAL MODEL

The analysis of dynamic crack growth is quite complex. This chapter can only quote a few typical results for two-dimensional cases, but relevant aspects of the problem will be related to a simple uniaxial model. It can be visualized as a straight, homogeneous bar under central action due to some active loading and to reactions within a length of complete attachment to a rigid substrate (Fig. 5.2). It is assumed that the attachment can be broken, so that a cracklike surface of debonding will move along the bar. We aim at finding the velocity $\dot{a}$ of the crack front and pertinent stresses and particle velocities.

Typical solutions of the equation of motion [(A.7) of Appendix A] will be mentioned first. With particle velocities $\dot{u}_1 \equiv \dot{u} = \partial u/\partial t$ and normal stresses $\sigma_{11} \equiv \sigma$ along the axis $x_1 \equiv x$, and with a vanishing component b_1 of the body force, the equation is

$$\frac{\partial \sigma}{\partial x} = \rho \frac{\partial^2 u}{\partial t^2} \tag{5.2}$$

which can be combined with Hooke's law,

$$\sigma = E\varepsilon = E\frac{\partial u}{\partial x} \tag{5.3}$$

into

$$c_0^2 \frac{\partial^2 u}{\partial x^2} = \frac{\partial^2 u}{\partial t^2} \qquad c_0 \equiv \sqrt{\frac{E}{\rho}} \tag{5.4}$$

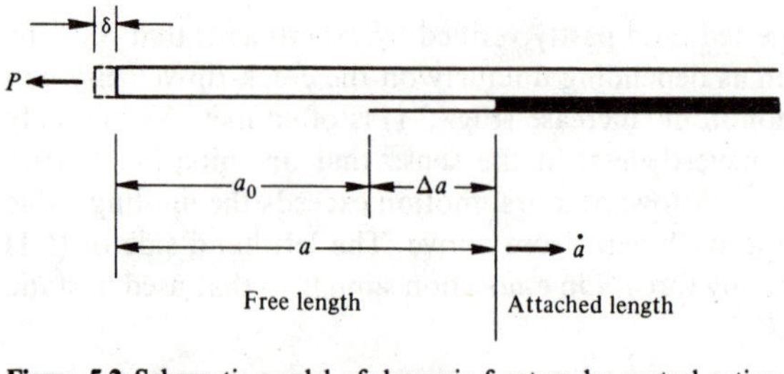

Figure 5.2 Schematic model of dynamic fracture by central action.

[If the bar were restricted from undergoing lateral displacements, the modulus E would have to be replaced by $E(1-\nu)/(1-\nu-2\nu^2)$, such that c_0 should be changed into a constant c_1, defined later.] Equation (5.4) has solutions of the form

$$u = f(\xi) = f(x \mp c_0 t) \tag{5.5}$$

because

$$\frac{\partial^2 u}{\partial x^2} = \frac{\partial^2 f}{\partial \xi^2} \qquad \frac{\partial^2 u}{\partial t^2} = c_0^2 \frac{\partial^2 f}{\partial \xi^2}$$

We can interpret these from the position of an observer moving with the speed c_0 in the positive (upper sign) or negative (lower sign) x direction. The observer will then be currently situated at $x = \pm c_0 t + \text{const}$ and will always see the same displacement value $u = f(\text{const})$. This means that (5.5) represents signals through pulses or waves traveling in the positive or negative x direction with a constant speed c_0 and a constant profile. Corresponding to (5.5) is the longitudinal stress wave

$$\sigma = E\frac{\partial f}{\partial \xi} \equiv Eh(\xi) = Eh(x \mp c_0 t) \tag{5.6}$$

also moving with the speed c_0, and the particle velocity

$$\dot{u} = \mp c_0 \frac{\partial f}{\partial \xi} = \mp c_0 h(x \mp c_0 t) = \mp \frac{c_0}{E}\sigma \tag{5.7}$$

the signs still referring to waves in the positive and negative x directions, respectively.

Another way of expressing the particle velocity, which applies to the vicinity of the crack front, will appear if we consider the separation event in detail. In Fig. 5.3 point Q is an identified material particle of the bar. When the crack front has passed the original position of Q by the length $da = \dot{a}\,dt$, Q will have been displaced backward through $du = -\varepsilon\, da$ because the detached length has attained the axial strain ε. This corresponds to the particle velocity

$$\dot{u} = -\varepsilon \dot{a} = -\frac{\dot{a}}{E}\sigma_t \tag{5.8}$$

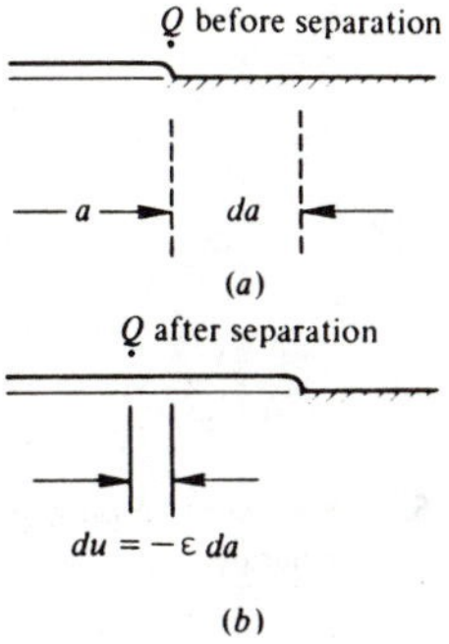

Figure 5.3 Position of a material point Q (a) before and (b) after separation.

where σ_t is the total stress to the left of the crack front. When the initial crack growth is considered, σ_t will be the sum of the original stress σ_0 before cracking ($\sigma_0 = P/S$ for the loading in Fig. 5.2 when S is the cross-sectional area of the bar) and an emitted stress wave σ_e due to the separation; i.e.,

$$\sigma_t = \sigma_0 + \sigma_e \tag{5.9}$$

the last term being related to the particle velocity through (5.7)

$$\dot{u} = \frac{c_0}{E}\sigma_e \tag{5.10}$$

with the proper sign of (5.7) representing the backward wave. (Recall that the bar in front of the crack is assumed to be totally fixed.)

Equations (5.8) to (5.10) can be solved with respect to the three unknowns σ_e, σ_t, and $\dot{u}$, the result being

$$\sigma_e = -\frac{\alpha}{1+\alpha}\sigma_0 \qquad \sigma_t = \frac{1}{1+\alpha}\sigma_0 \qquad \dot{u} = -\frac{\dot{a}}{1+\alpha}\frac{\sigma_0}{E} \tag{5.11}$$

in terms of the relative crack speed

$$\alpha = \frac{\dot{a}}{c_0} \tag{5.12}$$

However, the crack speed $\dot{a}$ (or α) remains an unknown in (5.11). To determine this essential value we still have at our disposal the energy balance (5.1)

$$\mathcal{G} \equiv W - \frac{1}{B}\frac{dT}{da} - \frac{1}{B}\frac{d\Phi}{da} = R(\dot{a}) \tag{5.13}$$

where the contact width B has been introduced. We further accept the resistance function R described in Fig. 5.1, even in the present case, where the separation may develop between different materials.

The signal that the crack is growing comes with the emitted stress wave. Hence, if the situation close to initiation is considered, the external work W per unit of crack area should be taken as zero. The increment dT of kinetic energy during the first motion corresponds to the length $(\alpha^{-1} + 1)\,da$ between the emitted wavefront and the crack front having particle velocities $\dot{u}$ according to (5.11); see

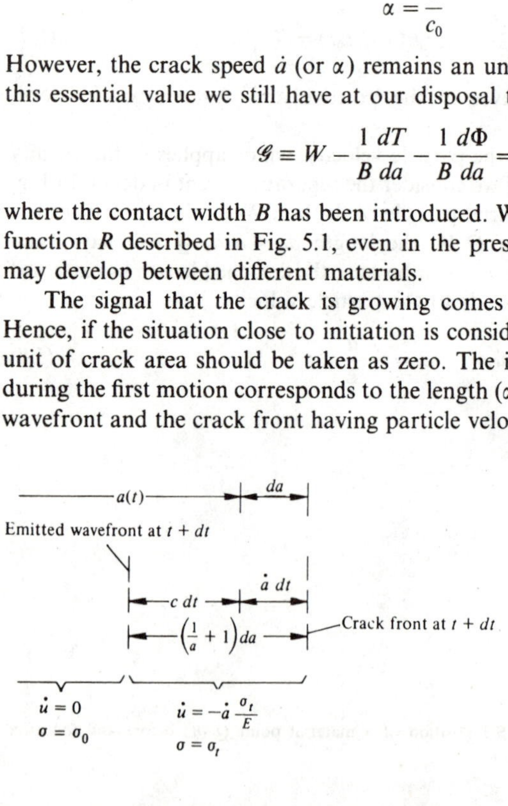

Figure 5.4 Local bookkeeping of wavefronts generated by separation.

Fig. 5.4. This means that

$$dT = \tfrac{1}{2}\rho S\left(\frac{1}{\alpha}+1\right)da\,\frac{\dot{a}^2}{(1+\alpha)^2}\frac{\sigma_0^2}{E^2} = \frac{\alpha}{1+\alpha}\frac{\sigma_0^2}{2E}\rho S\,da$$

S being the cross-sectional area. The increment $d\Phi$ of elastic energy corresponds to the total length $(\alpha^{-1}+1)\,da$ carrying the stress σ_t; originally its part $c_0\,dt = da/\alpha$ to the left of the front carried σ_0 and its part $\dot{a}\,dt = da$ to the right was free of stress. Again, this means that

$$d\Phi = \tfrac{1}{2}S\left(\frac{1}{\alpha}+1\right)da\,\frac{1}{(1+\alpha)^2}\frac{\sigma_0^2}{E} - \tfrac{1}{2}S\frac{da}{\alpha}\frac{\sigma_0^2}{E} = -\frac{1}{1+\alpha}\frac{\sigma_0^2}{2E}S\,da$$

whereupon (5.13) provides

$$\mathscr{G} = \mathscr{G}_0 g(\dot{a}) = R(\dot{a}) \tag{5.14}$$

with

$$\mathscr{G}_0 = \frac{\sigma_0^2}{2E}\frac{S}{B} \tag{5.15}$$

a "static" crack driving force, and

$$g(\dot{a}) = \frac{1-\alpha}{1+\alpha} = \frac{c_0-\dot{a}}{c_0+\dot{a}} \tag{5.16}$$

a dynamic correction, which interpolates between unity for $\dot{a}=0$ and zero for $\dot{a}=c_0$. The fact that the crack driving force vanishes in the last case must imply that the signal velocity c_0 is an upper bound of the crack velocity $\dot{a}$ under the circumstances. The function $g(\dot{a})$ according to (5.16) is indicated by curve 1 in Fig. 5.5.

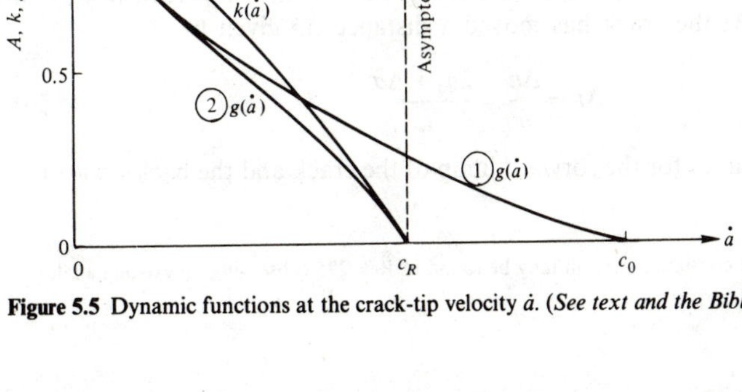

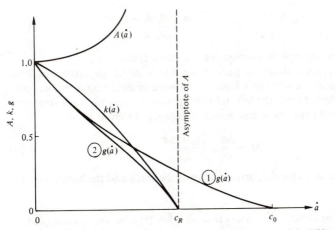

Figure 5.5 Dynamic functions at the crack-tip velocity $\dot{a}$. (*See text and the Bibliography for reference.*)

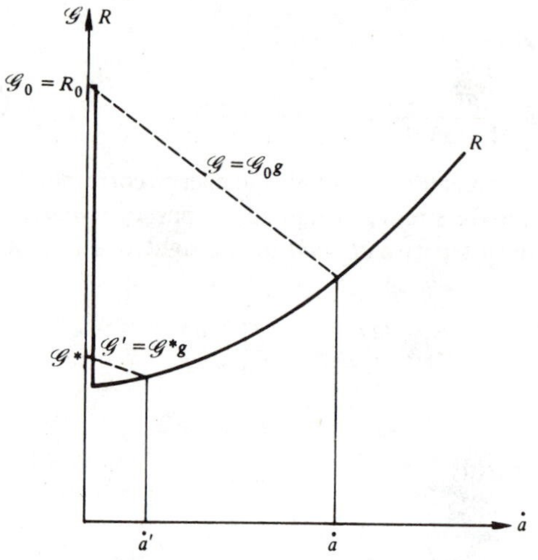

Figure 5.6 Graphical determination of first velocity $\dot{a}$, and $\dot{a}'$ after incidence of reflected waves.

For cracking to start, any existing initiation barrier $R_0/R(0) > 1$ must be overcome. This leads to the initiation condition (compare Fig. 5.1)

$$\mathcal{G}_0 = R_0 \tag{5.17}$$

so that the crack speed immediately to follow is given by (5.14) through

$$\mathcal{G} = R_0 g(\dot{a}) = R(\dot{a}) \tag{5.18}$$

A graphical interpretation of this equation is shown in Fig. 5.6, the desired solution being the speed $\dot{a}$ with corresponding values $\mathcal{G} = R(\dot{a})$ of crack driving force and resistance, respectively. In particular, if $R(\dot{a}) = R = \text{const}$ (the dashed line in Fig. 5.1), it follows from (5.18) that

$$R_0 \frac{1-\alpha}{1+\alpha} = R \qquad \alpha = \frac{\dot{a}}{c_0} = \frac{R_0/R - 1}{R_0/R + 1} \tag{5.19}$$

and we have here an explicit solution for the crack speed.

If the progressively detached part of the bar is under no further external action, the results above will apply from the moment of initial crack motion until the first wave reflected from the left end impinges on the moving crack front.† In this time interval Δt the crack has moved a distance Δa given by

$$\Delta t = \frac{\Delta a}{\dot{a}} = \frac{2a_0 + \Delta a}{c_0} \tag{5.20}$$

which equates the times for the forward jump of the crack and the back-and-forth

†More detail on the continued motion may be found in Ref. 275 (observing an existing analogy) and in Ref. 276.

displacement of the wavefront. Here a_0 denotes the initial free length of the bar. From (5.20) we find

$$\Delta a = \frac{2\alpha}{1-\alpha} a_0 = a_0 \frac{1+\alpha}{1-\alpha} - a_0 \tag{5.21}$$

The message brought by the reflected wave is essential for the further development. If a constant load P were prescribed to act at the left end (Fig. 5.2), it can be shown that the crack speed would suddenly increase and new stress waves would be emitted. The history of incident reflected waves would go on to approach a steady state, in which case the crack speed would be constant and all emitted and reflected stresses would add to zero. We note in this case that

$$\dot{u} = -\frac{\dot{a}}{E}\sigma_0 \tag{5.22}$$

with σ_t in (5.8) now being equal to σ_0 and

$$dT = \tfrac{1}{2}\rho S\, da\, \dot{a}^2 \frac{\sigma_0^2}{E^2} \qquad d\Phi = \tfrac{1}{2} S\, da \frac{\sigma_0^2}{E^2} \qquad W\, da = P\frac{\sigma_0}{E} da$$

expressing the kinetic and elastic energy of the detached length da and the external work due to stretching of that length. With the energy balance (5.13) we then arrive at

$$\mathscr{G}_0(1-\alpha^2) = R(\dot{a}) \tag{5.23}$$

from which the stationary value $\dot{a}$ can be determined.

An even more interesting case may occur when the displacement δ, instead of the load P, is kept constant after the initiation of cracking. The left-end reflection will then superimpose a forward-moving stress wave on the wave first emitted from the crack front, so that the resulting particle velocity becomes zero. By (5.7) and (5.10) the reflected stress must have the same size σ_e and sign as the emitted one. When this reflected stress impinges on the crack front, we have a total stress equal to

$$\sigma_t = \sigma_0 + 2\sigma_e + \sigma_e' \tag{5.24}$$

replacing (5.9), and with the generated *new* stress pulse σ_e' a particle velocity

$$\dot{u}' = \frac{c_0}{E}\sigma_e' \tag{5.25}$$

The results in Eqs. (5.14) to (5.16) can therefore be carried over when σ_0 and $\dot{a}$ are replaced by $\sigma_0 + 2\sigma_e = \sigma_0(1-\alpha)/(1+\alpha)$ according to (5.11*a*) and the new velocity $\dot{a}' = \alpha' c_0$. This provides the new crack driving force and its resistance

$$\mathscr{G}' = \mathscr{G}^* \frac{1-\alpha'}{1+\alpha'} = \mathscr{G}^* g(\dot{a}') = R(\dot{a}') \tag{5.26}$$

with

$$\mathscr{G}^* = \mathscr{G}_0 \left(\frac{1-\alpha}{1+\alpha}\right)^2 = R(\dot{a})\frac{1-\alpha}{1+\alpha}$$

yielding the new velocity $\dot{a}'$ (Fig. 5.6). $R(\dot{a})$ has been introduced according to (5.14), and a condition for $\dot{a}' > 0$ to exist is seen to be

$$\mathscr{G}^* = R(\dot{a})\frac{1-\alpha}{1+\alpha} > R(0)$$

Therefore, if the resistance is so dependent on the crack speed that

$$R(\dot{a}) < \frac{1+\alpha}{1-\alpha}R(0) \tag{5.27}$$

cracking in the present case must stop when the reflected wave impinges. Then Δa by (5.21) defines the total jump of the crack front. This phenomenon of crack arrest is important in itself and, as we have seen, strongly affected by the boundary conditions. Intuitively it may appear quite reasonable for a condition like $\delta = \text{const}$ to have a more retarding effect than $P = \text{const}$. Further, it has been pointed out that the crack front cannot distinguish between the two cases until the message has been imparted by a reflected wave.

5.3 TWO-DIMENSIONAL CASES

Some features of the one-dimensional analysis apply in a more general context. Specifically, this holds for the energy balance and the significance of reflected waves. Thus, for two-dimensional plane or antiplane cases, Freund [238] and Eshelby [237] have shown that the form in (5.14) remains valid before the incidence of such waves. $\mathscr{G}_0$ should then be replaced by a crack driving force $\mathscr{G}(a)$ derivable by static considerations; in general this coincides with the familiar static driving force under the given boundary conditions only in the first stages of crack motion. The waves generated are far more complex than in the uniaxial case; in the plane-strain problem they can be broken into *dilatational* (or irrotational) waves traveling with a speed of

$$c_1 = \sqrt{\frac{E}{\rho}\frac{1-\nu}{1-\nu-2\nu^2}} \tag{5.28}$$

(see Appendix F) and *rotational* (or equivoluminal) waves traveling with a speed of

$$c_2 = \sqrt{\frac{G}{\rho}} = \sqrt{\frac{E}{\rho}\frac{1}{2(1+\nu)}} \tag{5.29}$$

An example of the first type was mentioned following (5.4). The so-called *Rayleigh waves* may also appear near free surfaces, with particles moving in elliptical paths and a speed of propagation c_R somewhat smaller than c_2

$$c_R = \begin{cases} 0.88c_2 & \text{for } \nu = 0 \\ 0.92c_2 & \text{for } \nu = 0.3 \\ 0.95c_2 & \text{for } \nu = 0.5 \end{cases}$$

Plane strain applies here, but plane stress would give similar values. The dilatational velocity c_1 is the largest, going to infinity as ν approaches $\frac{1}{2}$. To give an idea of magnitudes we record the following values for *steel*

$$c_1 \approx 6000 \text{ m s}^{-1} \qquad c_0 \approx 5000 \text{ m s}^{-1} \qquad c_2 \approx 3200 \text{ m s}^{-1} \qquad c_R \approx 3000 \text{ m s}^{-1}$$

The Rayleigh wave has a particular bearing on the crack problem, where the generation of free surfaces is the typical event. It turns out that c_R will be a theoretical upper bound of the velocity of an isolated crack tip for mode I under elastic response, corresponding to c_0 of the preceding example. In both cases a limiting condition is the fact that a positive energy of separation cannot arrive at a crack tip moving supercritically, $\dot{a} > c_0$ or $\dot{a} > c_R$. The dynamic correction $g(\dot{a})$ in the present two-dimensional case must therefore be an interpolation between unity for $\dot{a} = 0$ and zero for $\dot{a} = c_R$. This function was determined [238, 247] in plane strain as curve 2 of Fig. 5.5. A fair approximation of this result is the linear one

$$g(\dot{a}) \approx 1 - \frac{\dot{a}}{c_R} \tag{5.30}$$

to be introduced in (5.31), the energy balance valid before incidence of reflected waves,

$$\mathcal{G} = \mathcal{G}(a)g(\dot{a}) = R(\dot{a}) \tag{5.31}$$

From this again we can determine the starting velocities $\dot{a}$, just the same as with (5.17) and (5.18) or Fig. 5.6. Two typical cases are illustrated in Fig. 5.7. Fig. 5.7*a* gives the speed $\dot{a}_0$ suddenly attained if an initiation barrier has been overcome, while Fig. 5.7*b* indicates the gradual increase of velocity which might follow a preceding growth which was stable up to $a = a_0$.

The crack velocity influences the stress field at the crack tip. In the radial coordinate from the tip there is a singularity of the same kind ($r^{-1/2}$) as in the static

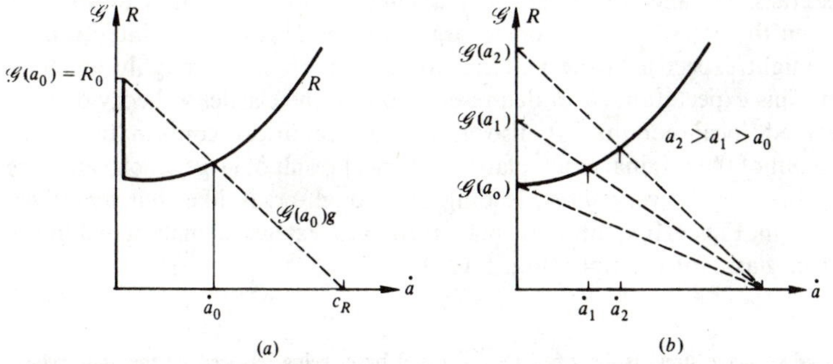

Figure 5.7 Graphical determination of the first crack velocities: $\dot{a}_0$ with an initiation barrier or the starting function $\dot{a}(a)$.

case, while in the circumferential variation the velocity $\dot{a}$ enters as a governing parameter. For mode I

$$\sigma_{ij}(\dot{a}, \theta, r \rightarrow 0) = \frac{K_{\mathrm{I}} f_{ij}(\dot{a}, \theta)}{\sqrt{r}} \tag{5.32}$$

with similar relations applying in modes II and III. The *dynamic* stress intensity factor K_{I} turns out to determine the crack driving force by an expression similar to (4.9), i.e.,

$$\mathscr{G} = \frac{\beta}{E} K_{\mathrm{I}}^2 A(\dot{a}) \tag{5.33}$$

Assuming plane strain ($\beta = 1 - \nu^2$) the function $A(\dot{a})$ is defined as

$$A(\dot{a}) = \frac{\delta_1(\delta_2^2 - 1)}{(1 + \delta_2^2)^2 - 4\delta_1\delta_2} \frac{1+\nu}{\beta} \qquad \delta_1 = \sqrt{1 - \left(\frac{\dot{a}}{c_1}\right)^2} \qquad \delta_2 = \sqrt{1 - \left(\frac{\dot{a}}{c_2}\right)^2} \tag{5.34}$$

as illustrated in Fig. 5.5.† Introducing further the static relationship

$$\mathscr{G}(a) = \frac{\beta}{E} K_{\mathrm{I}}^2(a) \tag{5.35}$$

where $K_{\mathrm{I}}(a)$ is a stress intensity factor corresponding to $\mathscr{G}(a)$, and comparing the first of (5.31) and (5.33), we also see the dependence

$$\frac{K_{\mathrm{I}}}{K_{\mathrm{I}}(a)} = \sqrt{\frac{g(\dot{a})}{A(\dot{a})}} \equiv k(\dot{a}) \tag{5.36}$$

The function k is shown in Fig. 5.5; however, such a unique relation between the dynamic and the static stress intensity factor is subject to the same restrictive condition as (5.31), i.e., that no reflected waves have yet engaged the crack tip. The vanishing of k when $\dot{a} = c_R$ again indicates that the Rayleigh-wave velocity is an upper bound of $\dot{a}$.

Properties of the distribution functions $f_{ij}(\dot{a}, \theta)$ are reviewed in Figs. 5.8 and 5.9. Figure 5.8 gives the circumferential stress σ_θ as scaled to its value on the extended crack line and shows that its maximum value is displaced to abscissas $\theta_m > 0$ when the crack speed becomes large enough. This result is interesting in that one might expect a brittle material to develop cracking roughly in the θ_m direction. This expectation (with due reservation on the elasticity theory defining such detailed local conditions) also has some empirical confirmation since *branching* out of the original crack plane is a typical result of large crack velocities. The branches may be very short, leading to a rough crack like that tentatively reproduced in Fig. 5.10*a*, or they may turn into extensive macrocracking as observed in glass, for example (Fig. 5.10*b*).

†The case of plane stress is described by $\beta = 1$ and by c_1 being replaced by the *plate velocity* $c_p = \sqrt{E/(1 - \nu^2)\rho}$. The denominator of $A(\dot{a})$ taken to zero is the implicit expression of the Rayleigh-wave velocity; compare $\dot{a} = c_R$ in Fig. 5.5.

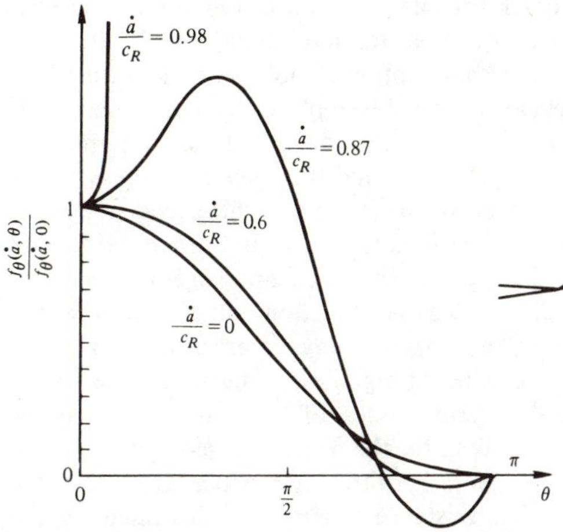

Figure 5.8 The circumferential singular stress at the crack tip scaled to its value on the crack line. (*Based on results in Refs. 227 and 235.*)

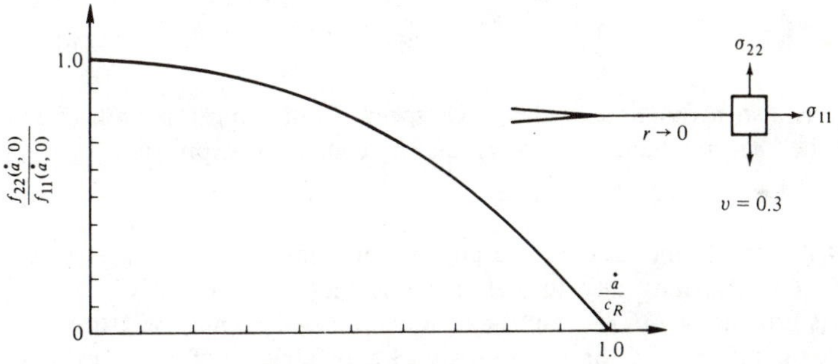

Figure 5.9 The transverse singular stress at the crack tip scaled to the longitudinal stress. [Eq. (F.15).]

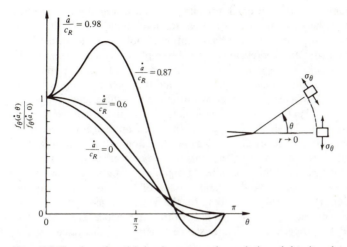

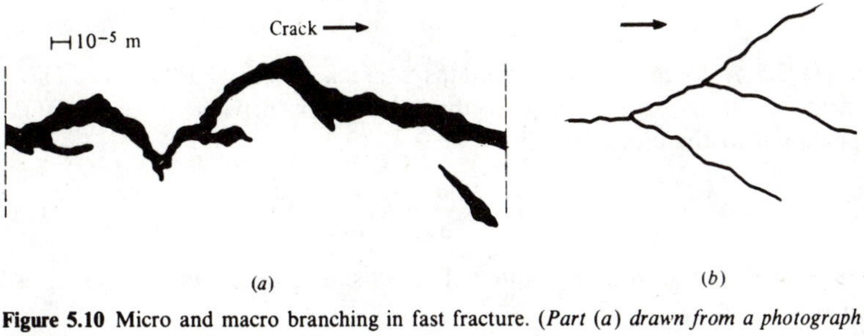

Figure 5.10 Micro and macro branching in fast fracture. (*Part (a) drawn from a photograph in Ref. 299.*)

Another interesting effect of crack velocity is shown in Fig. 5.9 by the ratio $f_{22}(\dot{a}, 0)/f_{11}(\dot{a}, 0)$ of the singular stresses normal to, and collinear with, the crack direction. As $\sigma_{33} = \nu(\sigma_{11} + \sigma_{22})$ under the assumptions, it follows that a rather high level of spherical (hydrostatic) stress at $\dot{a} = 0$ will be replaced by a more deviatoric state when $\dot{a}$ increases. This should favor a higher degree of yielding and may therefore indicate that the plastic dissipation around the crack tip increases with velocity. A similar effect is also to be expected from the branching tendency, when more than one crack path may be occasionally active and when bridges of unbroken material may be left in the wake of the crack tip. Together, this may partly explain the rising tendency of the resistance function, indicated in Fig. 5.1. It remains a disturbing fact, however, that this rise may empirically take place at lower crack velocities than those producing significant effects by the above arguments. Viscous processes and certain modes of microcracking may be additional mechanisms necessary to account for the fact. A consequence seems to be that velocities $\dot{a}$ exceeding $c_R/2$ are met rather rarely in practice. Linking of the maximum speed to properties of the resistance function is illustrated by the following example.

Another way of expressing the energy balance during dynamic cracking should be pointed out. By analogy with the static relation (4.10) we can define a *dynamic fracture toughness* K_{ID} through the identity

$$R(\dot{a}) \equiv \frac{\beta}{E} K_{ID} A(\dot{a}) \tag{5.37}$$

whereby K_{ID} is also a function of the crack speed. Combining this with (5.33) reveals that the energy balance $\mathcal{G} = R(\dot{a})$ has the equivalent expression

$$K_I = K_{ID} \tag{5.38}$$

similar to (4.11). While the latter was a critical condition for cracking to start, (5.38) is relation governing any further dynamic motion. Obviously, no new information is provided by (5.38), but the formulation as such may be attractive. In particular, the failure of $\mathcal{G}$ or K_I to satisfy (5.1) or (5.38) if they drop below the possible range of the right-hand function would imply crack arrest.

For some fundamentals related to Sec. 5.2 Appendix F may be consulted. Experimental aspects are considered in Sec. 7.3. Appendix H includes a wider review.

Example 5.1 A large specimen contains a crack of initial length a_0. The geometry is assumed to be such that the crack driving force $\mathcal{G}(a)$ is proportional to the crack length

$$\mathcal{G}(a) = \mathcal{G}_0 \frac{a}{a_0} = R_0 \frac{a}{a_0} \tag{a}$$

where $\mathcal{G}_0 = R_0$ is the initiation value. The resistance of the moving crack is given by

$$R(\dot{a}) = R(0)\left[1 + \left(\frac{\dot{a}}{v}\right)^n\right] \tag{b}$$

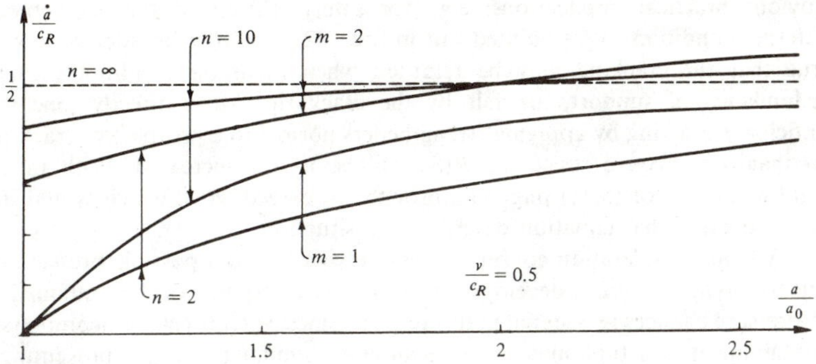

Example 5.1

where v is a reference velocity and $n > 1$ is a dimensionless exponent, both characterizing the material. The ratio

$$m = \frac{R_0}{R(0)} \tag{c}$$

signifies the strength of a possible initiation barrier. A relationship between the crack length a and the crack velocity $\dot{a}$ is sought, insofar as (5.30) and (5.31) govern the propagation, i.e.,

$$\mathscr{G}(a)\left(1 - \frac{\dot{a}}{c_R}\right) = R(\dot{a}) \tag{d}$$

The introduction of Eqs. (*a*) to (*c*) in (*d*) gives

$$\frac{a}{a_0}\left(1 - \frac{\dot{a}}{c_R}\right) = \frac{1}{m}\left[1 + \left(\frac{\dot{a}}{v}\right)^n\right] \tag{e}$$

which is easily solved for a/a_0 as dependent on m, n, v/c_R, and $\dot{a}/c_R$. Typical results are shown in the figure, assuming $v/c_R = 0.5$. It appears that the velocity will increase smoothly from zero if there is no initiation barrier, $m = 1$, while the crack starts impulsively if $m > 1$. Further, it is evident that $\dot{a} \approx v$ may act as an effective upper bound of the crack speed if the resistance function $R(\dot{a})$ has a very large gradient above this argument value; i.e., if the exponent n is large. It may serve no purpose to continue the curves farther than shown in the figure because reflected waves may invalidate such results. Finally, the assumption that R depends only on $\dot{a}$ excludes (in particular, when $m = 1$) the consideration of R-curve effects like those discussed in Sec. 4.3.2.

5.4 CRACK ARREST OR TOTAL FAILURE?

It turns out that a crack can move rapidly indeed; one might like to ask what possibilities there are of suppressing such a development. The problem has obvious practical implications, e.g., for safety. Effects of the geometry and external conditions were pointed out in Sec. 5.2; even in a broader context it is true that the cracking may be retarded when favorable conditions, such as redundancy of support, are felt by the crack tip. More directly, one might anticipate cracking by applying strengtheners normal to a prospective crack path, so that the effective resistance $R(\dot{a})$ will be locally increased. With this, the right-hand side of (5.31) might be brought to exceed $\mathscr{G}(a)$, which would imply crack arrest if that equation described the situation.

A final consideration concerns pressure vessels and a possible pressure drop due to leakage if a crack develops. The mechanical response of the contained fluid to pressure changes is essential in this respect. Since water is nearly incompressible, containing it in a tube may produce an effect similar to that of preventing the transverse deformation of the elastic bar [compare the comment following (5.4)] while ν is allowed to approach $\frac{1}{2}$ to simulate incompressibility. Even if the signal velocity in the water remains finite, contrary to the limit of c_1 by (5.28), it may still be high enough for the pressure drop within the tube to run ahead of the crack tip. With this, there will also be an immediate drop in the crack driving force, such that a very limited amount of cracking is produced. Quite different may be the consequences of initial cracking if the content is a gas. Here, the signal velocity could be smaller than the velocity of cracking, which means that the crack tip would currently experience the total pressure. Rapid and extensive—even catastrophic—fracture could be the consequence.

PROBLEMS

5.1 For the central debond problem illustrated schematically in Fig. 5.2, assume that the end displacement δ is first increased slowly to its critical strength δ_0 for cracking to start. With δ kept at this value, assume that the crack then jumps through the debond length Δa until arrest. δ_0 and Δa are measured; otherwise, the relevant geometric and material data are known. Assume that the arrest is effected by the first reflected wave, after an essentially elastic response. Deduce from this information the crack velocity $\dot{a}$ and the corresponding dynamic resistance R which has been operative during the jump.

Answer: Equations (5.15) and (5.17) give the initial resistance

$$R_0 = \frac{E}{2}\left(\frac{\delta_0}{a_0}\right)^2 \frac{S}{B}$$

With $\alpha \equiv \dot{a}/c_0$, (5.21) and (5.18) next provide the values sought

$$\alpha = \frac{\Delta a/a_0}{2 + \Delta a/a_0} \qquad R = \frac{R_0}{1 + \Delta a/a_0}$$

in terms of known right-hand quantities.

Comment: If not directly applicable, this result also pertains qualitatively to the double-cantilever crack-arrest experiment illustrated in Fig. 7.8. Varying R_0 (by artificially blunting the crack tip, for example), or a_0 will produce varying jumps Δa and thus a set of correlated values $(\dot{a}, R)$.†

†In the double-cantilever-beam test these are averages in some sense.

Each member of the set is as an experimental point in a diagram of R versus $\dot{a}$ and may serve in an attempt to establish the resistance function $R(\dot{a})$. Under proper conditions (small-scale yielding) a single curve may reasonably fit the experiments, as anticipated in Figs. 5.1, 5.6, and 5.7.

5.2 An elastic bar is centrally loaded by a force at one end and by adhesive joining to a rigid support extending from the other. The resistance to initial debond at the end of the joint is R_0, and the further dynamic resistance is taken as

$$R(\dot{a}) = R(0)\left[1 + \left(\frac{2\dot{a}}{c_0}\right)^n\right]$$

where c_0 is the longitudinal-stress wave speed and n a material parameter. Debond is effected by slowly increasing the load.

(*a*) Assume an initiation barrier $R_0/R(0) = 6$ and determine the first velocity $\dot{a}$ of the crack front running along the joint.

(*b*) The actively loaded end of the bar is kept stationary after the above crack initiation. Will crack arrest be produced by the first incidence of reflected waves?

(*c*) When the initial free length of the bar is a_0, determine the total crack jump Δa.

Answer: (*a*) $\dot{a} = c_0/2$, (*b*)–(*c*) yes, $\Delta a = 2a_0$.

5.3 An illustration of the energy transfer during dynamic, essentially elastic cracking can be referred to the Prob. 5.2; $\dot{a} = c_0/2$ is the crack velocity. In the figure the energy content is displayed versus traveled crack length Δa. Φ is the total strain energy, having the original value $\Phi_0 = (\sigma_0^2/2E)Sa_0$. T is the total kinetic energy, and $RB\,\Delta a$ is the total work expended on separation. Verify by inspection that the slopes (with $W = 0$) satisfy the energy balance (5.13). Verify also the coordinates, observing that the discontinuities at $\Delta a = a_0/2$ reflect the stage when the first emitted wave has just traveled the length of the freely suspended bar.

Comment: A peculiarity of the present problem is the fact that the kinetic energy goes back to zero at the instant of crack arrest. Generally this is not the case, as can be recorded by oscillations occurring after the crack has stopped. Figure 7.8 illustrates such an effect.

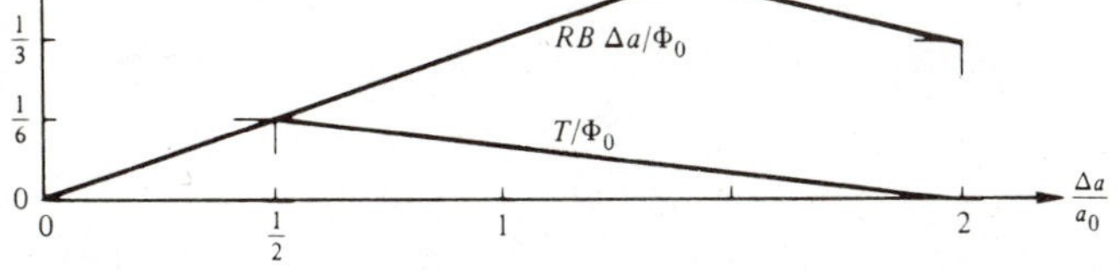

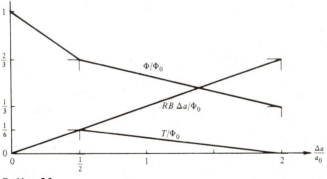

Problem 5.3

CHAPTER

SIX

FATIGUE

As observed in Chap. 1, slow crack growth may also take place under a pulsating (as opposed to a monotonically increasing) load. Such fatigue fractures may develop under load maxima which are well below the critical values for monotonic action because the repeated cycles have a cumulative damaging effect on the material.

Fatigue fractures may start from macroscopic precracks like those already considered but may also occur in originally uncracked bodies. Viewed microscopically, the fracture is often initiated in slip zones adjacent to the outer surface or to internal inhomogeneities (voids or inclusions) in the material. The mechanism will then be a cumulative relative displacement of strata between slip planes, like cards in a deck, until small cracklike intrusions have been produced in the slip directions. Initiation may also occur in the grain boundaries (particularly at triple points, where more than two grains meet). Among initiated microcracks, one or a few may develop further, provoked by the stress concentration (often aided by a corrosive environment) at the crack fronts. The direction of propagation may change, from the slip orientation mentioned above (stage I) to an average cracking normal to the maximum tension direction (stage II). The growth may proceed in a transcrystalline mode, either by progressive microplastic straining, which causes a typical striation pattern in the fracture faces, or by cleavage. The last form is favored by low temperatures or the presence of brittle inclusions. The growth may also be of intercrystalline type, as when the bonding is deficient or the environment aggressive, or it may involve the coalescence of voids within or between the grains (high temperatures and stresses). Finally, a dynamic crack propagation may complete a total failure.

The duration, expressed by the number N of load cycles, of the initiation and stage I period is generally not small compared with the duration of the macrocrack growth phase, stage II. At present, unfortunately only the last part seems open for detailed analyses within the frame of fracture mechanics. It may be noted, however, that such important cracks as those emanating from weld defects spend all or most of their lifetime in stage II. This is the type of crack growth which will be discussed below.

6.1 EMPIRICAL RELATIONS DESCRIBING CRACK GROWTH

In the spirit of LFM, we shall primarily assume stress levels so small that the singular part of the elastic solution governs the fracture process. Thus, the stress intensity factor is again invoked to define the local action uniquely. Following the cyclic variation of the far stress field, this factor is assumed to vary with a range

$$\Delta K = K_{\max} - K_{\min} \tag{6.1}$$

between the extreme values $K_{\max}$ and $K_{\min}$. It will be understood that growth occurs in a single mode, dominantly mode I since the body of applications and experimental results is found here. The pertinent index of K is therefore deleted to simplify the notation. For mode I, specifically, we shall also temporarily assume that $K_{\min} \geq 0$.

The crack size is assumed to be described by a single length a, such that the current intensity of growth is uniquely defined by the increment per cycle or the (smoothened) crack rate da/dN. This will be the essential dependent variable to consider.

With these assumptions it is reasonable to consider a functional relationship of the form

$$\frac{da}{dN} = f(\Delta K, K_{\max}, K_c, \Delta K_t, E, \nu, \sigma_Y, \sigma_U, \varepsilon_d, m, \kappa_i) \tag{6.2}$$

including such independent variables as the following:

1. Those which define the action locally, ΔK and $K_{\max}$ (or $K_{\min}$)
2. Those which define the material macroscopically, i.e., Young's modulus E, Poisson's ratio ν, the yield stress σ_Y (or $\sigma_{0.2}$), the other plasticity parameters σ_U (nominal ultimate stress), ε_d (ductility), and m (hardening exponent), the fracture toughness K_c, and a threshold value ΔK_t, discussed below
3. Possibly a set of microscopic parameters κ_i ($i = 1, 2, \ldots$)

The threshold value ΔK_t signifies that not every existing (initiated) crack will expand, so that a critical condition must be met for da/dN to exist. Further, it is obvious that an upper bound for $K_{\max}$ must be the static parameter K_c.

By dimensional considerations only, (6.2) can be restricted to

$$\frac{da}{dN} = \left(\frac{\Delta K}{E}\right)^2 F\left[\lambda, \frac{K_c}{\Delta K}, \frac{\Delta K_t}{\Delta K}, \frac{\sigma_Y}{E}, \frac{\sigma_U}{E}, m, \nu, \varepsilon_d, (\beta\kappa)_i\right] \tag{6.3}$$

where

$$\lambda = \frac{K_{\max}}{\Delta K} \tag{6.4}$$

and where F is a function of the dimensionless parameters; β_i is the necessary multiplier of κ_i, for example, $(E/\Delta K)^2$ if κ_i is a length. By physical reasoning one can attribute certain properties to the function F. While taking $K_{\max}$ toward K_c

should make F unbounded,† it is probable that the dependence of F on K_{max} is small when loads are well below that critical level. This is true because the *hysteresis*, i.e., cyclic dissipation of energy, should be primarily related to the deterioration and should be expected to be governed by the *range* ΔK of stress intensity. Accordingly, the threshold value ΔK_t can be interpreted directly as that minimum of ΔK which is required to make the crack grow at all.

The above considerations are summarized in Fig. 6.1. Reasonably interpreted, the diagram may also include alternating mode I loading, that is, $K_{min} < 0$, which signifies that the crack closes during part of the load cycle. We can then assume that the *opening* action during the cycle works the way of ΔK above. This means that ΔK should be interpreted as K_{max} in all relations (above and below) when $K_{min} < 0$ and $K_{max} > 0$, as illustrated in the figure.‡

Farther than this it is hard to get by general reasoning. To specify F one must rely on additional restrictive hypotheses and (primarily) on experimental data. A particularly simple hypothesis will be followed by a brief review of empirical relations.

Considering first a case of repeated tensile loading, $\lambda = 1$, the hypothesis is that the crack growth is linearly related to the COD δ. If δ during loading exceeds a threshold value δ_t, a permanent step Δa of the crack is assumed to remain upon unloading; thus [312]

$$\Delta a \propto \delta_{max} - \delta_t \tag{6.5}$$

The factor of proportionality should be small, empirically of the order of 1 percent. By (4.22) we also have

$$\delta \propto \frac{K^2}{E\sigma_Y} \tag{6.6}$$

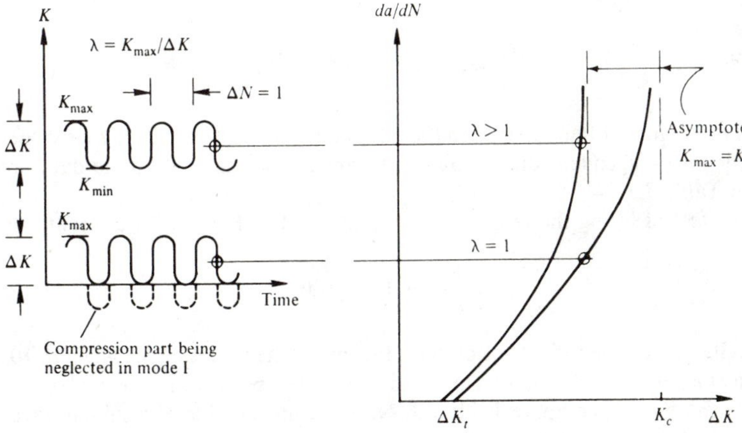

Figure 6.1 Estimate of the crack rate da/dN, dependent on the range ΔK and level of stress intensity.

†It is safe to identify this case with fast fracture.

‡A possible modification of this statement is implied by the later footnote referring to Elber.

for ideal plasticity, such that (6.5) and (6.6) can be summarized as

$$\frac{da}{dN} = \frac{A}{E\sigma_Y}[(\Delta K)^2 - (\Delta K_t)^2] \tag{6.7}$$

where ΔK_t is the threshold value corresponding to δ_t and A is a material parameter.

For small to moderate rates of growth (up to $da/dN \approx 1\ \mu$m) it has been found that (6.7) agrees tolerably well with observations for several metals if A is taken equal to $k\sigma_Y/E$, k being a number in the interval from 2 to 4. This gives the very simple function

$$F(1, \ldots) = k\left[1 - \left(\frac{\Delta K_t}{\Delta K}\right)^2\right] \tag{6.8}$$

in (6.3), to show that under the circumstances neither the fracture toughness, macroplasticity, nor microscopic parameters are explicit in the relation. Except when ΔK is very small, we may further neglect the last term in (6.8),† so that the propagation law for repeated tension simplifies to

$$\frac{da}{dN} = k\left(\frac{\Delta K}{E}\right)^2 \tag{6.9}$$

However, many experimental results are not accounted for by the above hypothesis. It is customary to attempt a description by a relation [307] like

$$\frac{da}{dN} = C_1(\Delta K)^n \tag{6.10}$$

(i.e., "Paris' law") or the generalization into

$$\frac{da}{dN} = C_1[(\Delta K)^n - (\Delta K\ )^n] \tag{6.11}$$

where n is a number often referred to the interval from 2 to 4 or even higher values and C_1 is an n-dependent dimensional constant of the material. Some data are given in Table 7.3.

The effect of lifting the range of K, so that $\lambda > 1$ in Fig. 6.1, or alternatively

$$R \equiv \frac{K_{min}}{K_{max}} \equiv 1 - \frac{1}{\lambda} > 0 \tag{6.12}$$

is normally to increase the rate of growth. For moderate values of ΔK when (6.10) is applicable, this effect can be empirically included by assuming n and C_1 to depend weakly on λ; compare Table 7.3. Near the threshold level a similar effect can be included by making ΔK_t somewhat λ-dependent. For instance,

†Indications are $\Delta K_t \approx 5$ MPa m$^{1/2}$ for some steels and $\Delta K_t \approx 2$ MPa m$^{1/2}$ for some aluminums.

$\Delta K_t \approx 4\,\mathrm{MPa\,m^{1/2}}$ for $\lambda = 2$ (or $R = \frac{1}{2}$), while $\Delta K_t \approx 6\,\mathrm{MPa\,m^{1/2}}$ for $\lambda = 1$ (or $R = 0$) in some mild steels.†

For larger values of ΔK an increase of λ may bring $K_{\max} = \lambda\,\Delta K$ up to the vicinity of K_c, which would cause an increase in the rate of growth. Modifications of (6.10) to comply with this behavior have been suggested within the frame

$$\frac{da}{dN} = C_2 \frac{f(\lambda)(\Delta K)^n}{[1 - (\lambda\,\Delta K/K_c)^p]^q} \tag{6.13}$$

where, for instance, $f = \lambda$, $p = q = 1$ [315] or $f = 1$, $p = 2$, $q = n/4$. By using (6.13) one includes the asymptote $da/dN \to \infty$ as $K_{\max} \to K_c$.

Some of the material parameters listed in (6.2) and (6.3) are contained in the dimensional constants C_1 and C_2. Clearly, the many alternative relations (6.7) to (6.11) and (6.13) illustrate the high degree of uncertainty involved in describing crack growth due to cyclic loading. Some of the discrepancies, however, should be related to varying physical conditions. When the action is limited to give $da/dN < 1\,\mu\mathrm{m}$, say, the striation mechanism may occur in many metals and $2 \le n < 3$ would be representative. In the same metals the void-growth mechanism may be activated by increasing the stresses and the temperature, which would call for a higher n if (6.10) were to apply. With this in mind, we cannot reasonably expect one continuous function to describe da/dN in all varieties.

Looking back at the result (6.3) by dimensional analysis, we see that only the presence of parameters κ_i can explain a nonquadratic dependence on ΔK of normal rates (when $K_{\max}$, K_c, and ΔK_t are not essential). It may also appear reasonable that the relations describing a "new" phenomenon like fatigue cannot be based only on macroparameters used in a previous context. At larger velocities, however, it may be sufficient to anticipate effects from the asymptote $K_{\max} = K_c$ or a possible violation of the assumption of small-scale yielding when apparently nonquadratic dependence on ΔK will be explained.

6.2 FATIGUE-LIFE CALCULATIONS FOR A GIVEN LOAD AMPLITUDE

We seek the relation between crack length a and the number of cycles N for a given loading range, for example $\Delta\sigma_\infty$. On this basis it will be possible to estimate the

†Extrapolating this into the alternating regime, we find that it may be an oversimplified prescription to transform an alternating K history into one which repeats in tension ($\lambda = 1$). The whole real history appears to influence ΔK_t. It is possible, however, that this variation in ΔK_t is more or less fictitious, or rather associated with an effect pointed out by Elber [322]. He found that under alternating load the crack might close *before* K passed through zero from above and open again with a corresponding delay. This behavior must be related to plastic deformation at the crack tip. Accordingly, one might correctly assume ΔK to be somewhat smaller than $K_{\max}$ under alternating load. This has the same effect as increasing ΔK_t and may therefore explain some of the λ dependence which has been attributed to ΔK_t.

relation between N and $\Delta\sigma_\infty$ when fracture occurs. A precrack is assumed to exist, of length a_0, which is sufficient for the crack to propagate by fatigue. This progress is taken to be described by (6.10), assuming the influence of ΔK_t to be small and observing that the form (6.13), for example, may improve the rate prediction only in the final stages of life.

In principle, the procedure is very simple. Integrating (6.10) gives

$$N - N_0 = \int_{a_0}^{a} \frac{da}{C_1(\Delta K)^n} \tag{6.14}$$

which is the $N(a)$ relationship sought, N_0 being the number of cycles that might have gone into producing the a_0 crack. Final fracture occurs for that crack length which follows from satisfying the condition

$$K_{\max} = K_c \tag{6.15}$$

Inserting this critical length $a = a_c$ as the upper integration limit in (6.14), we arrive at $N_c = N(a_c)$ the fatigue life sought. Generally, these operations can be carried out by numerical techniques. The relations (6.14) and (6.15) also apply, of course, if a displacement range is given or if, in general, ΔK is some other smooth function of a.

To be specific and analytic, we shall now assume a far-field stress loading σ_∞ with range $\Delta\sigma_\infty$, producing the current value and range of the stress intensity factor

$$K = \sigma_\infty\sqrt{\pi a}\, f(a) \qquad \Delta K = \Delta\sigma_\infty\sqrt{\pi a}\, f(a) \tag{6.16}$$

respectively. $f(a)$ is some geometry-dependent function (compare Appendix C). As an approximation, which often gives only a minor error in the life prediction, we further replace the function $f(a)$ by its initial value $f(a_0)$ during the integration. This implies

$$\Delta K = \Delta K_0 \sqrt{\frac{a}{a_0}} \qquad \text{where} \qquad \Delta K_0 = \Delta\sigma_\infty\sqrt{\pi a_0}\, f(a_0)$$

to be inserted in (6.14). The result will be

$$N - N_0 = \begin{cases} \dfrac{a_0}{C_1(\Delta K_0)^n}\dfrac{2}{n-2}\left[1 - \left(\dfrac{a_0}{a}\right)^{n/2-1}\right] & n > 2 \qquad (6.17) \\[2ex] \dfrac{a_0}{C_1(\Delta K_0)^n}\ln\dfrac{a}{a_0} & n = 2 \qquad (6.18) \end{cases}$$

Final fracture takes over when

$$K_{\max} = \sigma_{\infty\max}\sqrt{\pi a}\, f(a) = K_c \tag{6.19}$$

The solution of (6.19), $a = a_c$, can then be introduced in (6.17) or (6.18) to give the fatigue life N_c. In particular, if $a_c \gg a_0$, (6.17) yields the following relation

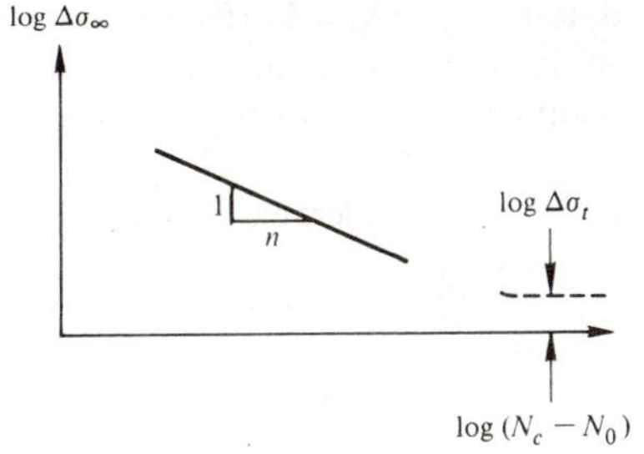

Figure 6.2 A prediction of the fatigue life $N_c - N_0$ in stage II growth.

between N_c and $\Delta\sigma_\infty$

$$(\Delta\sigma_\infty)^n(N_c - N_0) = \frac{a_0}{C_1[\sqrt{\pi a_0}f(a_0)]^n(n/2 - 1)} \tag{6.20}$$

as indicated in Fig. 6.2.

It might be interesting to investigate a possible relationship between the last results and empirical fatigue data such as usually presented in S-N or Wöhler diagrams. First, it should be noted that the assumption of small-scale yielding has excluded from our discussion most fractures which occur after only a few cycles, say, $N < 10^3$, that is, "low-cycle" fatigue. In the regime of longer life, $N > 10^5$, or "high-cycle" fatigue, good qualitative agreement between the present results and measurements has been reported. In particular, the predicted slope of the curve in Fig. 6.2 is a satisfactory estimate. Attention should however be called to the fact that the prepropagation life N_0 is unknown in the present results, limiting application and verification to cases where N_0 is small. This happens when flaws are initially present, as mentioned at the beginning of the chapter.

We know that the S-N diagram may also have an asymptote $\Delta\sigma_\infty > 0$ for $N \to \infty$, that is, a *fatigue limit*. This minimum of $\Delta\sigma_\infty$ necessary to produce fatigue failure is closely related to the threshold effect discussed above. Indeed, if the integration has been based on (6.11) instead of (6.10), the fatigue limit would have appeared as the stress-range value

$$\Delta\sigma_\infty = \frac{\Delta K_t}{\sqrt{\pi a_0}f(a_0)} \equiv \Delta\sigma_t$$

referring to Fig. 6.2.

Example 6.1† A small, approximately circular crack has been discovered internally in a tension member in the plane normal to the tension axis. The member is subjected to a repeated stress σ_∞ far from the crack, such that

$$0 \leq \sigma_\infty \leq \Delta\sigma_\infty = \text{const}$$

during one load cycle. The material is a high-strength steel having a fracture

†Adapted from Ref. 377.

toughness $K_{Ic} = 60\ \text{MPa m}^{1/2}$, a threshold intensity $\Delta K_t = 5\ \text{MPa m}^{1/2}$, and parameters $n = 4.0$ and $C_1 = 10^{-12}\ \text{MN}^{-4}\ \text{m}^7$ for use in (6.10) and (6.11). The crack radius is originally $a_0 = 2$ mm, and yielding is assumed to be so localized that the above linear theory applies.†

We first determine the stress range which can be tolerated without the crack growing. This follows from the condition

$$\Delta K(a_0) \le \Delta K_t$$

and upon substitution from (C.1)

$$\frac{2}{\pi}\,\Delta\sigma_\infty\sqrt{\pi a_0} \le \Delta K_t$$

yielding

$$\Delta\sigma_\infty \le \frac{5}{\sqrt{4(0.002)/\pi}} = 99\ \text{MPa}$$

We next ask for the highest permissible stress range $\Delta\sigma_\infty$ if total fracture is not to occur within $N = 10^4$ cycles. Equation (6.10) is assumed to hold approximately from the starting of crack motion, which can be verified as a final step in the evaluation.

Total fracture may occur when

$$\frac{2}{\pi}\,\Delta\sigma_\infty\sqrt{\pi a} = K_{Ic}$$

that is, when the radius equals

$$a = a_c = \frac{\pi}{4}\left(\frac{K_{Ic}}{\Delta\sigma_\infty}\right)^2 = \frac{2827}{(\Delta\sigma_\infty)^2}$$

Equation (6.10) further provides

$$\frac{da}{dN} = C_1\left(\frac{2}{\pi}\,\Delta\sigma_\infty\sqrt{\pi a}\right)^4$$

and by integration

$$\int_{a_0}^{a_c} a^{-2}\,da = \left(\sqrt{\frac{4}{\pi}}\,\Delta\sigma_\infty\right)^4 C_1 N$$

or

$$\frac{1}{a_0} - \frac{1}{a_c} = \frac{1}{a_0} - \frac{(\Delta\sigma_\infty)^2}{2827} = \left(\frac{4}{\pi}\right)^2 \times 10^{-12} \times 10^4(\Delta\sigma_\infty)^4$$

that is,

$$(\Delta\sigma_\infty)^4 + 2.18 \times 10^4(\Delta\sigma_\infty)^2 - 3.08 \times 10^{10} = 0$$

by which

$$\Delta\sigma_\infty = 406\ \text{MPa}$$

†It can be argued that the plastic-zone size with a cyclic range ΔK is smaller, in a sense, than the size if ΔK were applied by a monotonic loading (compare Fig. 6.5). This supports the suggestion that the rate equations governed by ΔK are applicable even if inequalities like (4.13) (with ΔK replacing K_{Ic}) are not satisfied.

This corresponds to $a_c = 17$ mm $\approx 8a_0$, and we note that the result could not change much if $1/a_c = 0$ were introduced in the above calculations. Since $(\Delta K_t/\Delta K)^n \leq (99/406)^4 = 0.003$, it was also permissible to predict the propagation with (6.10) instead of (6.11) as a basis.

6.3 EFFECTS OF CHANGING THE LOAD SPECTRUM

A structure is not often loaded as simply as assumed in Sec. 6.2. Compound load spectra (Fig. 6.3) are more usual. Formally, one might then consider using, say, (6.14) to find the length $\Delta a(i)$ of propagation through the given $N(i)$ cycles within each of the consecutive sequences $i = 1, 2, \ldots$ with a fixed load range $\Delta P(i)$ [or similarly, if this is the case, a fixed displacement range $\Delta u(i)$] and to add the contributions successively. If the sequences are few compared with the total number of cycles and not too dissimilar, this may be an acceptable strategy. However, each change in itself has a large impact on the state at the crack tip, which is not accounted for in the linear superposition. For instance, a large amplitude in tension will leave residual compressive stresses at the crack tip, which in turn may reduce the rate of growth if the succeeding amplitudes are small; this is case 1 in Fig. 6.4. In this way even a total crack arrest may be achieved. On

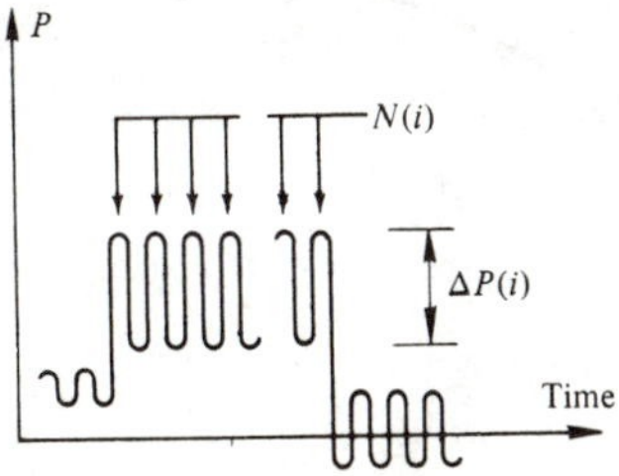

Figure 6.3 A varying load spectrum.

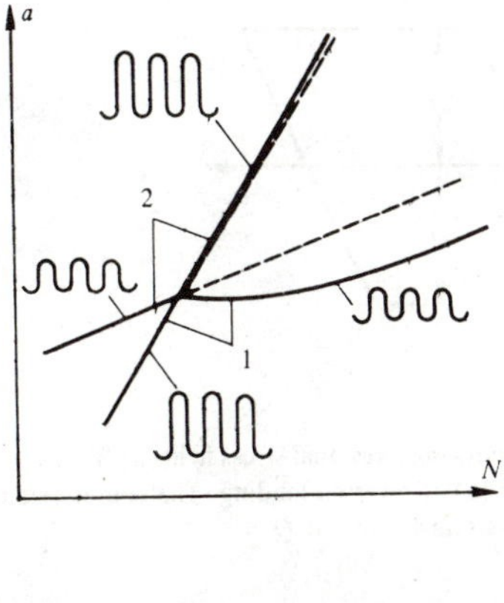

Figure 6.4 Effects of changing the load amplitude from large to small (path 1) or from small to large (path 2).

the other hand, a sequence of large amplitudes after a sequence of smaller ones produces a rate of growth hardly any larger than might be expected for stationary conditions; this is case 2 (Fig. 6.4). The disturbances in the rate of growth are of transient type, each vanishing with time (but this may take thousands of cycles to achieve). In the meantime, they have influenced the total damage; besides, frequent or large changes will lead to a cumulation of aftereffects. This is a complex product of a heterogeneous load history, which cannot be treated by simple superposition.

The reason for the retardation in the type 1 transition is that residual stresses remain at the crack tip after the previous sequence. Figure 6.5 indicates the stresses in a thin plate after a first opening loading (Fig. 6.5*a*), the stresses added due to load removal (Fig. 6.5*b*), and the residual stresses after the loading and unloading (Fig. 6.5*c*). A compression zone has been formed at the crack tip, which is not easily penetrated by the crack tip if the later amplitudes are small. With regard to the unloading step (Fig. 6.5*b*) it may be noted that this is tantamount to superposing the effects of a reverse load on the virgin material after *doubling* the yield stress. Compare the stress-strain history in a typical point Q (Fig. 6.5*d*).

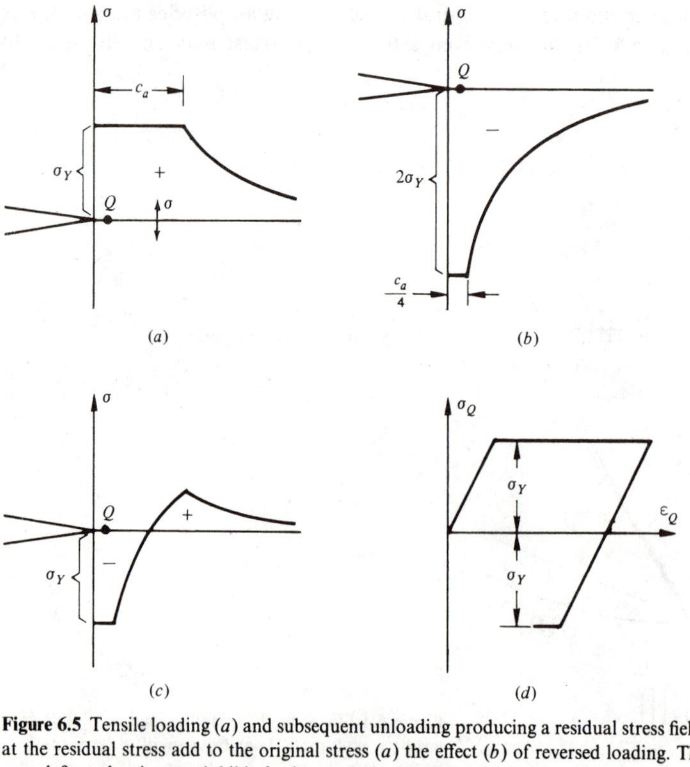

Figure 6.5 Tensile loading (*a*) and subsequent unloading producing a residual stress field (*c*). To arrive at the residual stress add to the original stress (*a*) the effect (*b*) of reversed loading. The compression zone left at the tip may inhibit further growth. (*d*) Stress-strain history at Q.

6.4 EFFECTS OF ENVIRONMENT

Given time, the environment will strongly affect the state at the crack tip. It is even possible for a crack to grow under a *constant* action if conditions are chemically aggressive enough (so called *stress-corrosion cracking*). When cyclic loading is added, the combination (corrosion fatigue) may yield a higher rate of growth than the two would give by simple superposition.

The previous discussion and the data in Chap. 7 are valid for normally dry air, the usual reference. Merely an increase in humidity will decrease fatigue life. Conditions in seawater are even less favorable, as observed by a shorter fatigue-initiation period, a high rate of growth, and a lower (possibly even vanishing) fatigue limit.

One can seek to improve the corrosion resistance in several ways, e.g., by initial heat treatment or by cathodic protection during operation. It is possible that in the first case the normal crack-growth data may suffer by the treatment, and in the second hydrogen production may increase brittleness, counteracting the desired effect. This is to exemplify that problems related to corrosion fatigue are not easily handled.

Temperature (see also Sec. 7.4) is another important aspect of the environment. A moderate increase, e.g., from 0 to 100°C, may improve the fatigue properties to a certain extent, whereas very high temperatures may be clearly unfavorable. In these cases, a crack may grow again under a constant action (creep fracture, related to creep behavior of the material); the addition of load cycling leads to an intensified total effect. At high temperatures the cracking may be the effect of microvoids growing and coalescing along grain interfaces, through a joint mechanism of diffusion and plastic deformation.

To return to the phenomenon of stress-corrosion cracking, we mention that the cause is assumed to be electrochemical reactions interacting with plastic slip, thus tending to break down layers which would otherwise protect the crack surfaces. Hydrogen embrittlement may participate in the process. Special combinations of material and environment promote this detrimental effect, e.g., aluminum alloys and seawater or copper alloys and ammonia. In small-scale yielding the necessary plastification can be expressed by a minimum level of the stress intensity factor required for growth. Figure 6.6 illustrates the development of a

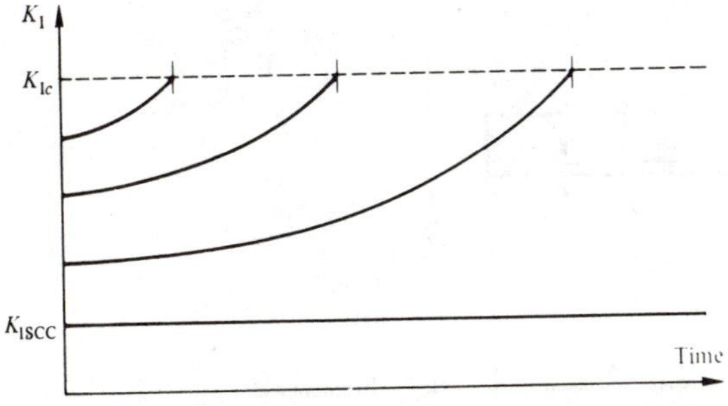

Figure 6.6 Crack growth by stress corrosion when $K_I > K_{ISCC}$. Terminal fracture when $K_I = K_{Ic}$.

crack during stress corrosion; as time goes on, the crack length and hence K_I will increase, until final fracture occurs when the condition $K_I = K_{Ic}$ is satisfied. The minimum value of the stress intensity factor to start the crack growing is K_{ISCC} (for stress-corrosion cracking). This environment and material parameter designates a threshold effect, much as did ΔK_t in the previous discussion.

In metals science (but not in the context of ceramics) the notion of fatigue is normally associated only with cyclic loading. The somewhat wider interpretation is therefore implied in the present section. Readers who want to pursue these subjects will find further information in the references for Chaps. 6 and 8.

PROBLEMS

6.1 A large plate has a small internal crack through the thickness. The crack has the initial length $2a_0$ and is normal to a remote tension σ_∞, which pulsates between the two levels σ_{max} and σ_{min}. Using the propagation law in Eqs. (6.3) and (6.8)

$$\frac{da}{dN} = k\left[\left(\frac{\Delta K}{E}\right)^2 - \left(\frac{\Delta K_t}{E}\right)^2\right]$$

verify the fatigue-life estimation

$$N_c = \frac{E^2}{k\pi(\Delta\sigma)^2} \ln \frac{(K_c/\sigma_{max})^2 - (\Delta K_t/\Delta\sigma)^2}{\pi a_0 - (\Delta K_t/\Delta\sigma)^2}$$

$$\Delta\sigma \equiv \sigma_{max} - \sigma_{min}$$

6.2 A DCB specimen (Examples 3.1 and 3.3) is subject to loads P pulsating between 3 and 5 kN. The initial crack length is $a_0 = 100$ mm, and we further specify $B = 10$ mm, $h = 20$ mm. The crack is assumed to propagate according to (6.10), with values of the constants

$$n = 3 \qquad C_1 = 10^{-10}\ \mathrm{MN}^{-3}\,\mathrm{m}^{11/2}$$

If the fracture toughness is taken as $K_{Ic} = 75\ \mathrm{MPa\ m}^{1/2}$, verify the critical crack length $a_c = 122.5$ mm and the fatigue life $N_c = 11{,}300$.

6.3 A circular pipe has the wall thickness B and a longitudinal crack of length L and depth a, as shown. A net internal pressure p pulsating between the values $p_0/2$ and p_0 may effect crack growth. The evaluation assumes $L \gg a$, such that the (relatively stationary) crack ends $y = \pm L/2$ do not retard the crack growth near $y = 0$ significantly. The initial crack depth is $a = a_0 = 5$ mm, and we further assume $B = 25$ mm, $R = 500\ \mathrm{mm} \gg B$.

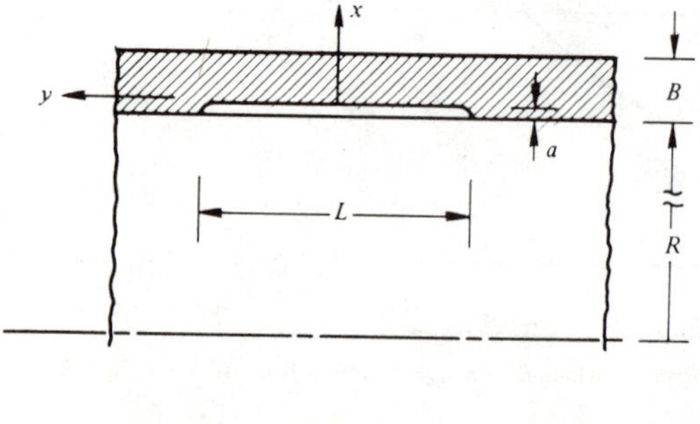

Problem 6.3

(*a*) Determine the maximum of p_0 for which the crack remains at rest when the threshold stress intensity range equals $\Delta K_t = 4\,\text{MPa m}^{1/2}$.

(*b*) Under a substantially higher range of load the crack is assumed to propagate according to (6.10). Determine from this the local life, i.e., the number N of pressure cycles before the crack penetrates the wall, if $n = 3$, $C_{\text{I}} = 10^{-12}\,\text{MN}^{-3}\,\text{m}^{11/2}$, $K_{\text{I}c} = 40\,\text{MPa m}^{1/2}$, and $p_0 = 10\,\text{MPa}$. Assume $K_{\text{I}} \approx 1.4\sigma_\phi\sqrt{\pi a}$ [compare (C.5); $\sigma_\infty = \sigma_\phi$ being the undisturbed circumferential stress] for simplicity.

Answer: (*a*) $p_0 = 2.3\,\text{MPa}$, (*b*) $a_c = 6.5\,\text{mm}$, $N_c = 2.3 \times 10^5$.

CHAPTER

SEVEN

EXPERIMENTAL METHODS AND RESULTS

Experiments are necessary to guide and confirm theoretical efforts and to provide material data for the practical applications of fracture mechanics. The latter class should be easily repeatable, i.e., designed along simple lines and leading to well-defined results, with preferably a broad field of application. In meeting such elementary requirements fracture mechanics is faced with severe problems, which have only been partially solved. The problems concern such items as (1) producing a test specimen having a well-defined crack with (2) a well-defined stress field in (3) an undistorted surrounding material; (4) measuring crack initiation and crack length; (5) recording a critical load (or displacement) which moves the crack; (6) translating the recordings into material data; and (7) assuming that items 4 to 6 apply to quasi-static conditions, developing testing procedures which apply to dynamic (impulsive) loading and/or rapid crack growth.

Since no obvious or unique way of proceeding exists for each testing situation, it is essential that experiments in fracture mechanics be based on commonly accepted norms as far as general routines are concerned. Such norms include the well-known American Society for Testing and Materials, Standards ASTM E399 for dealing with initial crack growth within the frame of linear fracture mechanics. We consider some of its main features below. Reference will also be made to testing procedures which may apply in the nonlinear regime of crack growth. Methods which concern dynamic loading or dynamic crack growth are briefly mentioned, and a few material data are presented in a separate section. Finally, some experimental results, with related theories, are given for cases of asymmetric crack growth. More detailed information may be sought in the Bibliography.

7.1 DETERMINATION OF STATIC FRACTURE TOUGHNESS

The aim is to determine the fracture toughness K_{Ic} for a given material (compare Sec. 4.3). To ensure that the results will be nearly independent of the geometry of the test specimen, minimum requirements on thickness and precrack length have been proposed as

$$B \geq 2.5\left(\frac{K_{\mathrm{I}c}}{\sigma_Y}\right)^2 \qquad a \geq 2.5\left(\frac{K_{\mathrm{I}c}}{\sigma_Y}\right)^2 \tag{4.13}$$

The numbers 2.5 are empirically based and may lead to large dimensions—often prohibitively so—as discussed later. Awkward in practice is the fact that $K_{\mathrm{I}c}$ is not known before the experiment, so that one first has to estimate a value to indicate which dimensions may satisfy the above inequalities. If the value which results from the experiment does not meet the requirements after all, that test must be considered invalid for predicting $K_{\mathrm{I}c}$ and a new test run on a larger specimen.

The specimen types, three-point loaded bending and compact-tension (CT), usually favoured in $K_{\mathrm{I}c}$ testing are shown in Fig. 7.1. They are typical (especially the second) in that comparatively small volumes are involved. When the load has been recorded, the respective stress intensity factors can be determined by formulas (C.9) and (C.10) in Appendix C. If cracked sheets are to be investigated, local buckling may be a practical problem and overall tensioned specimens will be preferred, e.g., cases 3 to 5 of Appendix C.

Processing the precrack requires a special technique. We want a crack with the front normal to the specimen sides; it must be sharp (not as when produced by sawing) and surrounded by material which has experienced a minimum of permanent deformation or damage. This is normally achieved by allowing a fatigue crack to develop, starting from a so-called chevron notch, as in Fig. 7.2, and controlled by load cycling in such a way that $K_{\mathrm{I,max}}$ is sufficiently small compared with $K_{\mathrm{I}c}$. The chosen type of notch favors symmetric fatigue cracking and forestalls the trend toward a thumbnail shape (Sec. 4.4). The precrack should also meet certain detailed requirements for final geometry. These are precautions

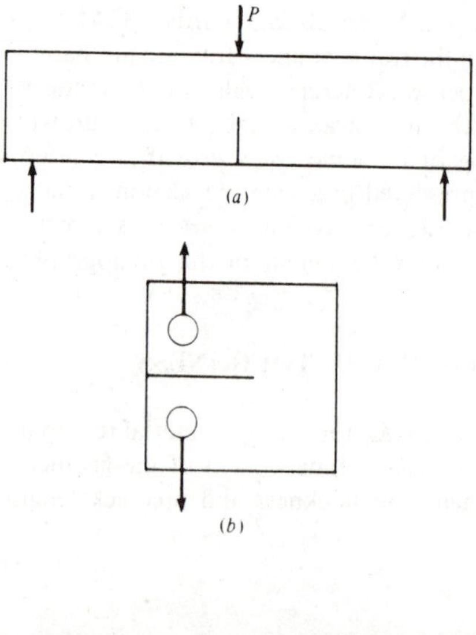

Figure 7.1 Two important types of specimen for testing fracture toughness: (*a*) three-point bending; (*b*) compact tension.

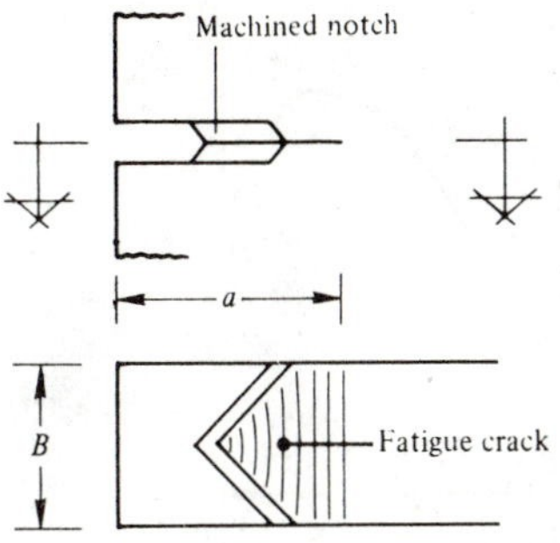

Figure 7.2 Ways of achieving a cracklike notch.

for a well-defined crack to be initiated in minimally disturbed surroundings, whereas the thickness requirement contributes to cultivating a state of plane strain (except close to the side surfaces) near the crack front.

While items 1 to 3 may thus appear to be well provided for, recording a critical load is even more difficult. The problem is to identify an initial crack growth. Within the ASTM standard under discussion, it is assumed that the load and the crack-mouth-opening displacement V are recorded simultaneously, V by means of a so-called clip gage carrying strain gages (Fig. 7.3). Three types of results can be envisaged (Fig. 7.4). Case 1 is quite simple; the almost linear response suggests a dominantly elastic behavior with a fixed crack area before the breaking point and a dynamic crack growth until total fracture. Critical with regard to initiation is the load P_c, which coincides with the maximum load in the present case. The trial value of the fracture toughness is the K_I value corresponding to P_c. Case 2 reflects the phenomenon of pop-in, or internal growth (Sec. 4.4); even here the peak load value P_c is assumed to be critical. More diffuse is the situation in the shown case 3, however. The nonlinearity of that curve might be due to plastic behavior with a fixed crack, a decreasing stiffness following stable growth, or (usually the case) some combination of the two patterns. The nominal critical load P_c is determined by the offset secant method shown, with an additional condition (also used for case 2) to invalidate any test if the maximum load exceeds P_c by more than 10 percent. This last restriction is to make sure that fracture will be

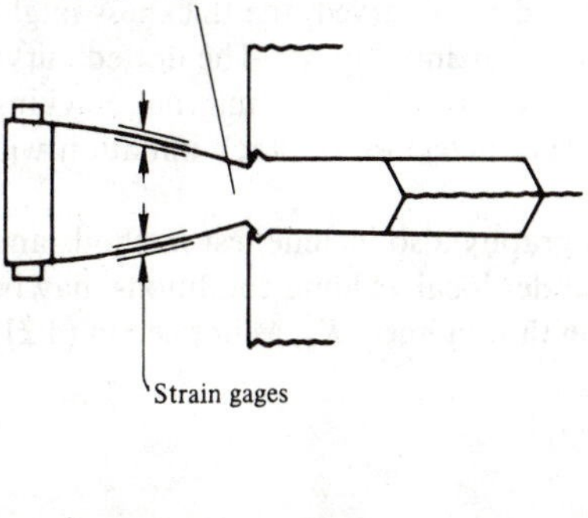

Figure 7.3 Measurement of the crack mouth opening.

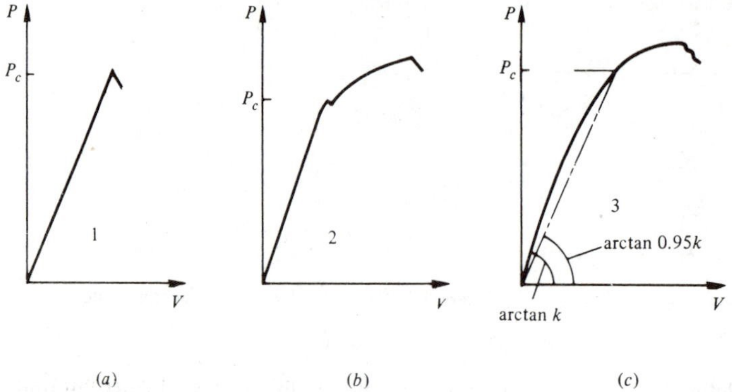

Figure 7.4 Typical results of a static fracture test.

dominantly involved in the nonlinearity. On the other hand, the secant intercept may include some stable crack growth (up to about 2 percent of the precrack length if the nonlinearity is mainly a result of cracking) even below P_c. The last observation may indicate a certain specimen dependence of the corresponding toughness value K_{Ic}; however, the method is simple and has much practical experience behind it.

Data on the fracture toughness K_{Ic} in Sec. 7.4 reflect the fact that the fracture toughness is often highly *temperature-dependent*, increasing with temperature. It also appears that as far as metals are concerned, LFM may be limited in application to high strength and/or low temperatures. In such cases the quotient K_{Ic}/σ_Y is comparatively small, so that realistic cracks and plate thicknesses satisfy condition (4.13). On the other hand, low-strength steels and some aluminums, for example, may be far too ductile in this context.

The size requirements (4.13) are aimed at promoting a state of local plane strain with strictly contained yielding and at limiting the included crack growth to the order of this plane-strain plastic-zone size. The last observation implies a simple means of ensuring that the recording will not be too far from actual initiation. If the thickness requirement (4.13*a*) is not satisfied, the result may still have a relevance as the fracture toughness K_c referring to *that particular* thickness (Sec. 4.4). Further, if the actual initiation could be observed, the thickness might be considerably reduced with the result still approximating K_{Ic}. The dotted curve in Fig. 4.5*a* indicates such an improved prediction, related to internal cracking (prelude to pop-in). Techniques for the sharper detection of crack initiation will be mentioned in Sec. 7.2.

The references for Chap. 7 in the Bibliography also include test methods and results for the *R* curve. The determination under local yielding conditions may be in terms of the resistance *R* or the crack-growth toughness K_R, as implied in (4.21) with the footnote.

7.2 DETERMINATION OF A CRITICAL CRACK-OPENING DISPLACEMENT OR CRITICAL *J* INTEGRAL

A load-displacement diagram with pronounced curvature, as in case 3 above, might well lead to trial values of a fracture toughness which do not comply with (4.13), thus indicating a problem within the regime of nonlinear fracture mechanics. Such trial values [particularly if $a < 2.5(K_{Ic}/\sigma_Y)^2$] could not represent material properties, and one would have to look for other quantities which might be more fundamental. Two possibilities are the COD δ and the J integral. The COD has an obvious physical interpretation but may be difficult to determine with accuracy.† For the J integral the opposite may be true even if high skill and special testing facilities are required. Opinion is divided over which of the two should have preference, although they are closely related quantities. $J = \sigma_Y \delta$ has been derived for very thin plates within the Dugdale assumption, and a similar relationship, $J = M\sigma_Y\delta$ where usually $1 < M < 2$, appears to apply for thicker plates. Within the uncertain identification of δ one might even conceive of some way of defining it quantitatively in terms of J, with σ_Y as the nominal stress ($\sigma_{0.2}$) in cases of moderate hardening. Then, the J integral could be considered an alternative way to calculate or measure the simpler physical concept δ, notably when, as here, crack initiation is the issue and not continued crack growth.

The COD can be recorded more directly in cases of extensive yielding, invoking the hinge mechanism mentioned in Sec. 2.3 and shown in Fig. 7.5. If the crack-face displacements can be explained as essentially resulting from rotations of the specimen parts about an axis situated a distance b/n ahead of the crack tip, it follows that

$$\delta = V\frac{b/n}{a + b/n} \tag{7.1}$$

This rigid-body description is directly applicable for loading in the vicinity of the

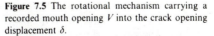

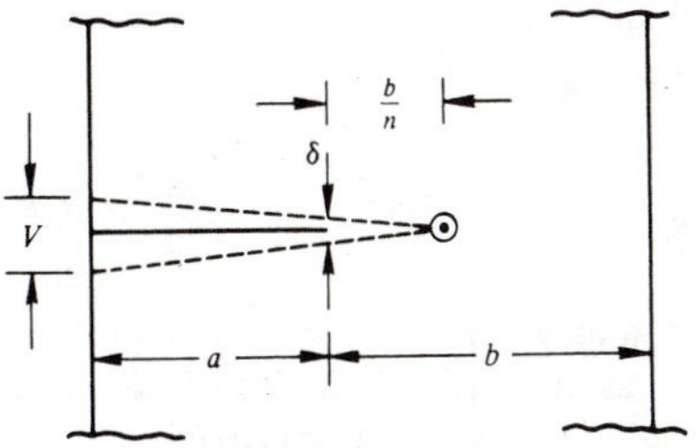

Figure 7.5 The rotational mechanism carrying a recorded mouth opening V into the crack opening displacement δ.

†This is due in part to the smoothly curved crack front (Sec. 4.5). A certain arbitrariness is therefore involved in the COD concept, even if it can be uniquely identified within the analysis on a first-order assumption.

plastic limit, b denoting the uncracked ligament width and n being a number close to (or somewhat above) 2 for usual specimen geometries. A realistic content of (7.1) may otherwise appear only if n is looked upon as a variable during the loading; then we are not much better off than if δ were to be measured directly. Such measurements can be undertaken, however, going through direct observation of specimens which have been broken open after cooling or, more recently, using silicon rubber to produce replicas of the crack. From such measurements empirical forms of the function n have also been suggested.

Traditionally, the critical value of δ has often been referred to the original crack tip and the maximum load. Before this recording some stable crack growth will normally have taken place, geometrically as indicated in Fig. 4.8 and with such measured effects as in Fig. 7.4, case 3. Conditions during crack motion will depend on the shape and dimensions of the test specimen, which implies that this recording *cannot* represent a material property. Insofar as the actual separation can be observed, however, there are strong indications that the δ value at *initiation* of crack growth is a constant for the material in its environment. Again, a certain minimum thickness will be necessary to provide well-defined plane-strain conditions, as discussed below. The corresponding critical value δ_{Ic} can be related to the fracture toughness K_{Ic} through (4.24*b*) if the latter quantity is also referred to actual crack initiation.

Some detailed guidelines for the testing situation can be found in the British Standards BS 5762:1979 [382]; the discussion for Chap. 7 in the Bibliography contains further relevant information.

In Sec. 4.6 arguments for seeing the J integral as a critical quantity describing initiation were reviewed, as well as certain aspects of its behavior with increasing action or crack length. In combination with finite-element methods, the defining equation (2.48) remains a significant tool for evaluating J in a given testing situation, but the alternative methods noted in Sec. 3.3.3 are more directly related to the simple recordings and are preferred when applicable. The use of (3.41*a*), stating that $J \approx -(\Delta U/\Delta A)_u$ with reference in Fig. 3.6, may seem attractive in itself, but the process of finding the difference in external work U for neighboring crack sizes is often error-sensitive. If possible, further specialized relations like (3.43) and (3.44) should be used because of their simplicity. In general they can be summed up as variations on the theme

$$J = \eta \frac{U}{bB} \tag{7.2}$$

where the coefficient η ranges between 1 and somewhat in excess of 2. Even this form is limited to cases of deep precracks, as in Fig. 7.6, and/or large-scale yielding, whereas a right-hand term may be added to account approximately for other effects, e.g., elastic effects. Guidelines for the testing situation have been prepared in the ASTM Standard E813-81 [387]; see the Bibliography also for related information.

So that the initiation measurements can be considered valid to provide

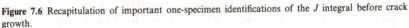

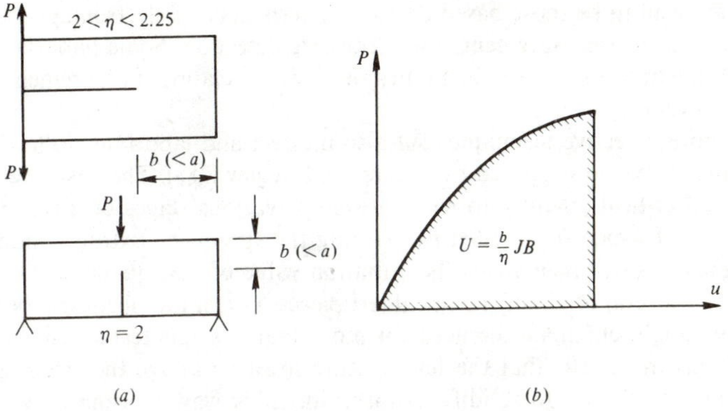

Figure 7.6 Recapitulation of important one-specimen identifications of the J integral before crack growth.

material properties δ_{Ic} or J_{Ic} the necessary thickness has been proposed as

$$B \geq m\delta_{Ic} \approx \frac{m}{M}\frac{J_{Ic}}{\sigma_Y} \qquad 1 < M < 2 \tag{7.3a}$$

[observe (4.29)], with m in the interval from 25 to 100. To compare these thicknesses with condition (4.13a) we substitute $J_{Ic} \approx K_{Ic}^2/E$; then (7.3$a$) takes the form

$$B \geq \frac{m}{M}\frac{K_{Ic}^2}{E\sigma_Y} \tag{7.3b}$$

With normal ratios of E and σ_Y, for example, about 1000 for steels, this is a considerably weaker requirement than (4.13a). The last condition was to make B large compared with the plastic zone size, whereas (7.3) asserts only that B should be large compared with the COD. The discrepancy may be surprising, as the minimum of B observed to eliminate a thickness dependence should have no relation to the underlying theoretical estimates. The explanation can be found in the different phenomena being directly observed; whereas (7.3) is clearly aimed at actual initiation, (4.13a) is to suppress variance due to a certain amount of crack growth. With the same reference, true initiation in both cases, we should have the same requirement regardless of the quantity measured. However, a thickness as small as given by (7.3) may still be a matter of some uncertainty.

The complex nature of ductile crack initiation makes it difficult to record directly. Tunneling behavior would make a simple optical method inapplicable (except in cases of material transparency), and on the microscale it may be difficult even to define the event since damage in front of the crack is a gradual process (voids growing and interacting, etc.). The present and prospectively most successful methods may seem to be those based on electric-resistance of electromagnetic effects, acoustic-emission measurement, and ultrasonic detection. Some published recordings well illustrate the severe difficulties one has in relating measurements to the initiation event.

Perhaps a more objective technique, but also indirect and laborious, follows the multispecimen principle suggested by Landes and Begley [388]. The basic idea is for the assumed critical quantity to be measured at various stages of traveled crack length, each of which is recorded by opening the specimen. Extrapolation back to zero actual growth then yields the initiation value of that quantity. This implies the need of several originally identical test pieces. Use of the silicon-rubber replica technique might eliminate the need for more than a single test specimen. A weakness of this method is that the temperature must be above the freezing point for the rubber solution to solidify. Another possible way of using only a single specimen is based on an idea as the following [418]. If small unloadings and reloadings (so that the deformation interval is essentially elastic) are performed at different stages during the crack growth, it may be possible to deduce the current effective crack length from the slope of the unloading-reloading curve. Sophisticated measuring facilities are needed, however, and both favorable and unfavorable experiences have been reported.

7.3 DETERMINATION OF DYNAMIC FRACTURE TOUGHNESS

More than referring to stress singularities the headline expression is now understood to encompass whatever data may characterize the resistance toward initial separation under a rapid loading or toward continued rapid crack growth. The field of application and research has attracted considerable interest but is intrinsically so complex that analysis, measurement, and even terminology may suffer from lack of total insight. For fractures involving small-scale yielding, such that the stress intensity factor is a governing parameter, we must distinguish between two critical values:

K_{Id} (which could as well have been denoted K_{Ic}) is the initiation value when the stress intensity factor has been increased with a rate $\dot{K}_I$ sufficiently high to influence the fracture resistance of the material. K_{Id} is then a function of $\dot{K}_I$ and relates to a crack which is still *stationary*.

K_{ID} is the value of the stress intensity factor needed to keep the crack in *continued motion*. In a first approximation, K_{ID} may be viewed as a function only of the crack velocity $\dot{a}$.

In more ductile fractures, which cannot be described in terms of K_I, less strictly defined energy measures have been used. $\Delta E/A$ below is taken to mean some average loss of mechanical energy per unit of crack area A after total dynamic fracture. Further, even in the dynamic case the COD seems to be a parameter which can be used to characterize the separation event.

Several experimental techniques are in use to provide critical data; we mention here Charpy or drop-weight tests with special equipment, crack-propagation studies with analytical guidance, and the use of shadow optics (method of caustics). The first is well known in routine testing situations. The notched test piece† is hit by a hammer, and the total energy loss ΔE_t can be measured by the difference in hammer elevation before and after breakthrough. Without correction, however, this energy loss is not suitable for characterization of the actual fracture, since it includes (damped out) kinetic energy and further dissipation by yielding away from the crack tip. Special techniques may make it possible to extract essentials from ΔE_t in such a way that a characteristic total fracture energy $\Delta E/A$ is estimated. Still, it is difficult to find how this energy is related to the fracture velocity and even to see how this velocity might be defined.

It is easier to formulate the problem in brittle cases, where the stress intensity factor is characteristic. In the instrumented impact test, transducers are attached to the hammer and/or the test piece to convey instantaneous values of local strains, while a clip gage may record the crack-mouth opening displacement. From simple quasi-static interpretations the data lead to a relationship between $K_\mathrm{I} = K_{\mathrm{I}d}$ and $\dot{K}_\mathrm{I}$ for fracture to initiate. The results indicate that $K_{\mathrm{I}d}$ may depend markedly on $\dot{K}_\mathrm{I}$ (Sec. 7.4) but in different ways for the various materials. For low to moderate strength steels it is generally observed that $K_{\mathrm{I}d}$ is *reduced* when $\dot{K}$ increases within normal rates, whereas $K_{\mathrm{I}d}$ may increase in many high-strength metals. The latter trend, which seems to apply for all metals if the velocity becomes very high, has been connected with adiabatic heating at the crack tip.

Error sources in the determination of $K_{\mathrm{I}d}$ are probable difficulties in identifying the crack initiation and the fact that quasi-static relations, e.g., between force or crack opening and the stress intensity factor, are in use even in a dynamic case. Since the first stress waves and oscillations may influence the initiation in a way not easily recorded, the results derived as indicated above should be viewed with some caution. The use of more refined measuring techniques may reduce such uncertainties, however [425, 441].

Typical recordings from an instrumented impact test are illustrated in Fig. 7.7.

The determination of $K_{\mathrm{I}D}$ may go through accurate measurements of the crack velocities, which are used next as input values in numerical simulation programs to provide the corresponding $K_{\mathrm{I}D}$ function. Wedge loading on a double cantilever or a compact-tension specimen has been frequently used, sometimes in combination with an optical method. In these cases the cracking is usually started from a blunted crack tip; this serves as a controlled initiation barrier which builds up energy to drive the crack dynamically once it has started. During the crack motion the mouth opening is fixed by the wedge contact; this has a retarding effect which may eventually arrest the crack [267, 431].

The case above is rather analogous to one in Sec. 5.2, where an end

†In the instrumented impact test the specimen is generally of the three-point bend type, as in Fig. 7.1*a*. Notches should be cracklike, produced, for example, by controlled fatigue (Sec. 7.1).

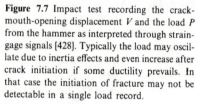

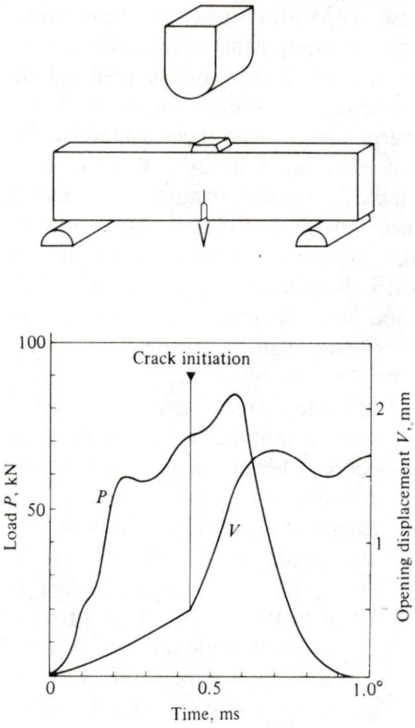

Figure 7.7 Impact test recording the crack-mouth-opening displacement V and the load P from the hammer as interpreted through strain-gage signals [428]. Typically the load may oscillate due to inertia effects and even increase after crack initiation if some ductility prevails. In that case the initiation of fracture may not be detectable in a single load record.

displacement δ was kept fixed at its value δ_0 to start the motion and the crack velocity $\dot{a} = \alpha c_0$ and jump Δa were related to the initiation barrier R_0/R through (5.18) and (5.21). Manipulating the variables, one might deduce the velocity and the operative resistance here in terms of measured crack jump and (through δ_0) initial resistance, as in Prob. 5.1. Indeed, analyses related to the double-cantilever specimens have disclosed reference diagrams which directly correspond to the latter results. This admits the approximate assessment of a resistance curve $R(\dot{a})$ or $K_{ID}(\dot{a})$ (see the comment to Prob. 5.1) [432, 433].

The use of shadow optics (method of caustics) is very convenient for transparent materials, e.g., Plexiglass. When a cracked plate is loaded in mode I, the stress concentration will have a double effect by reducing the thickness and changing the refractive index. If the plate is then illuminated by a light beam from one side, a nearly circular shadow will be cast on a screen on the other; this shadow emanates from the crack tip and has a diameter which is simply related to the stress intensity factor, both in K_{Id} and K_{ID} measurements. By making use of only the thickness-reduction effects the method can also be applied in reflection to opaque materials [438, 439].

For materials with favorable optical properties photoelastic methods are also being successfully applied in recording K_{Id} or K_{ID} values.

Typical recordings from a fast fracture and arrest DCB test are illustrated in Fig. 7.8.

7.4 SOME MATERIAL DATA RELATED TO MODE I CRACKING

The data in this section are not intended as a source for practical uses but as an indication of typical values or trends. The somewhat arbitrary content has been

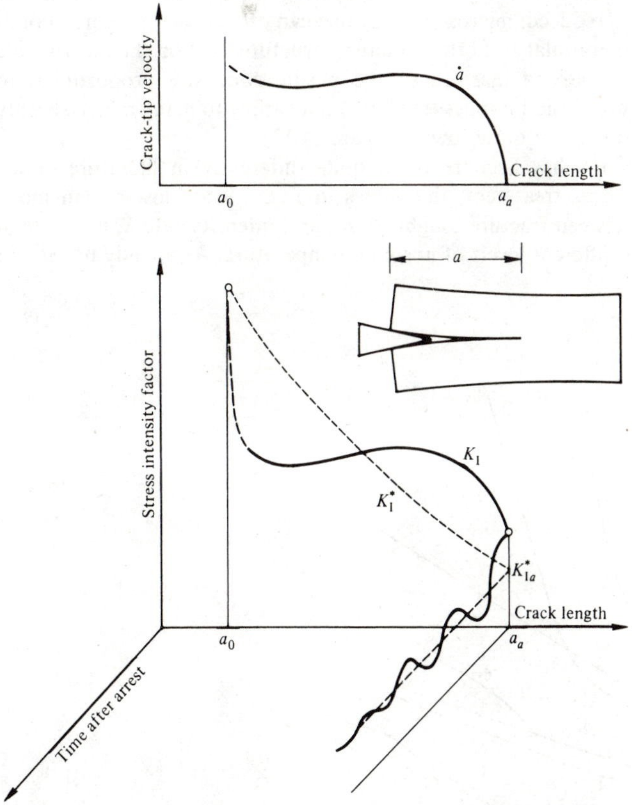

Figure 7.8 Dynamic fracture in a DCB specimen (schematic) originating from a blunted crack, as by Kalthoff et al. [440]. The crack velocity $\dot{a}$ and the dynamic stress intensity factor K_I as functions of crack length a. Both $\dot{a}$ and K_I can be measured — by high-speed photography and the method of caustics, for example. Keeping the wedge stationary stops the crack at length a_a, after which the stress intensity factor oscillates toward K_{Ia}^*, which corresponds statically to the wedge-opened condition.

determined by easy availability and by the intention of illustrating fully some dependencies on temperature and loading rate. In a real situation the material in question is often a given quantity, and it may not be easy to find representative fracture-mechanical data. The task was made easier by a review article [464] containing directions to the published literature systematically organized in terms of material composition. Specific data may be available through this guide. It is essential for full information to consult the sources directly, also because of the importance of material processing in addition to the chemical composition.

Figures 7.9 and 7.10 display some result for the steels and alloys described in Table 7.1; the weight percentages of the steels are listed in Table 7.2.

Figures 7.9 and 7.10 show that the fracture toughness K_{Ic} is strikingly more temperature-dependent than the yield stress σ_Y. The toughness increases with Θ while the strength is reduced; microscopically one can often observe a corresponding transition from granular to fibrous fracture structure, and on the macro scale there corresponds a strong increase of the plastic zone size proportional to $(K_{Ic}/\sigma_Y)^2$. The last conclusion implies that LFM may apply to many materials only when the temperatures are quite low; compare (4.13).

K_{Ic} for material 8 has been reported quite differently in literature, much depending on the heat treatment; the values in Fig. 7.9 are lower than most.

The relation between fracture toughness K_{Id} and intensity rate $\dot{K}_I$ is shown in Fig. 7.11a to d for different levels of absolute temperature. As already noted, the

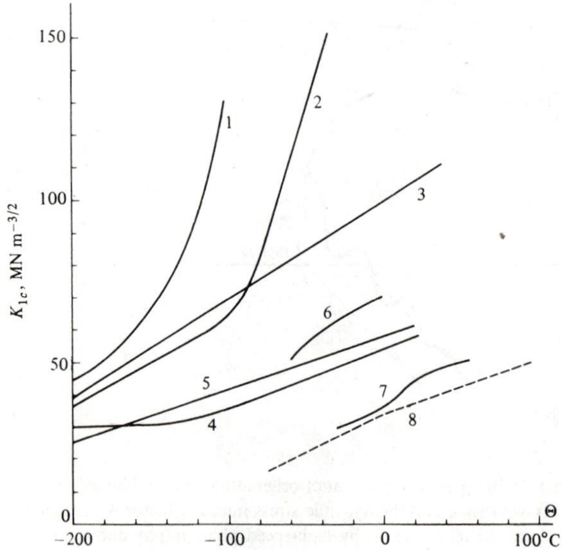

Figure 7.9 Fracture toughness as function of temperature Θ for metals listed in Table 7.1.

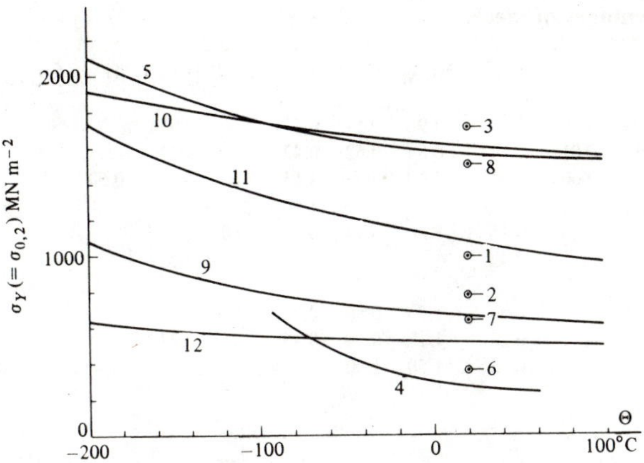

Figure 7.10 Yield stress, partly as function of temperature, for metals listed in Table 7.1.

Table 7.1 Description of steels and alloys

No.	Type	Description	Ref.
1	HY 130	Structural, high-toughness	455
2	A 517-F	Structural, pressure-vessel quality	455
3	18Ni(250)	Structural, maraging, for high strength but above average toughness	455
4	BS 4360/43A	Structural, low-strength	456
5	0.4C, 9Ni, 4Co	Special-purpose	456
6	SIS 140727	Cast iron	38
7	Ferritic cast iron	Cast iron	38
8	AISI 4340	Structural, high-strength	38
9	QT 35	Special-purpose	456
10	0.4%CNiCrMo	Special-purpose	456
11	IMI 680	Titanium alloy†	456
12	DTD 502A	Aluminum alloy‡	456

†2.25Al, 0.30Fe, 4.3Mo, 0.26Si, 10.8Sn.
‡4.15Cu, 0.75Mn, 0.4Mg, 0.2Fe, 0.7Si, 0.1Zn, 0.03Ti.

loading rate affects the various metals quite differently. Materials 10 and 11 are clearly more suited than 9 and 12 for use under dynamic conditions. For material 4 the intensity rates seemed to be rather unimportant; that is, $K_{\mathrm{I}d} \approx K_{\mathrm{I}c}$ within the related span of $\dot{K}_{\mathrm{I}}$. Similarly, the yield stress seemed to be only weakly rate-dependent for material 10, whereas otherwise it seemed to increase somewhat with $\dot{K}$ [456].

More scattered data (for only one temperature) are included in Table 7.3.

Table 7.2 Weight percentages of steels

No.	C	Mn	P	S	Si	Ni	Cr	Mo	Co	V	Ti	Cu
1	0.12	0.79	0.004	0.005	0.35	4.96	0.57	0.41	....	0.056		
2	0.17	0.89	0.015	0.015	0.19	0.84	0.52	0.42	....	0.04	....	0.30
3	0.003	0.002	0.001	0.004	0.003	17.10	....	4.65	7.60	....	0.50	
4	0.17	0.83	....	....	0.08							
5	0.40	0.14	....	....	0.15	8.80	0.13	0.15	3.80			
6	3.7	0.45	....	....	2.4							
7	3.8	0.48	....	....	2.4							
8	0.40	0.70	....	....	0.30	1.75	0.80	0.25				
9	0.18	1.20	....	....	0.20	1.10	0.90	0.45	....	0.07		
10	0.40	0.50	....	....	0.25	1.50	1.00					

Most of these results apply to aluminum alloys and to fatigue data related to

$$\frac{da}{dN} = C_1(\Delta K_1)^n \tag{6.10}$$

The fact that C_1 and n may take on different values as the ratio λ varies reflects the influence of the mean stress level during the cycle. Figure 7.12 illustrates how data for n (here 2.30) and C_1 (here 3.48×10^{-9}, or 7.1×10^{-11} in SI units) can be deduced from a number of test results. The techniques for recording such crack growth may vary from simple inspection of the side surfaces to electric-resistance measurements or deduction from an observed crack mouth opening. It is worth noting that the fatigue parameters C_1 and n are far less temperature-dependent than the fracture toughness. Note also that (6.10) is a linear relation between $\log(da/dN)$ and $\log \Delta K$ (Fig. 7.12).

Figure 7.13*a* and *b* shows temperature dependencies of K_{Ic} and Charpy energy (ΔE_t in Sec. 7.3) for two medium-strength steels. Apparently, the transition from low- to high-energy, granular to fibrous fracture is recorded in roughly the same way in the two parameters, though with some relative displacement along the temperature axis. Usually, an upper plateau value of K_{Ic} cannot be established, because the toughness becomes so great (and the strength so small) that the requirements (4.13) are not satisfied with a test specimen of reasonable size. In this respect Fig. 7.13*a* is exceptional. Attempts have been made to correlate the Charpy energy and the K_{Ic} temperature curves empirically, e.g. [462].

We repeat that the data in this section are meant to be illustrations, presented with no attempt at completeness or practical applicability. With regard to factors which may influence the fracture strength, we again emphasize the importance of earlier thermal or mechanical processing of the metal. Heating may influence grain size and the structure of included particles, both important for the fracture strength, and a mechanical process may produce anisotropies such that the material is strengthened in some directions and weakened in others. The consequence of some types of grain-boundary precipitation is insufficient bonding resulting in intercrystalline fractures.

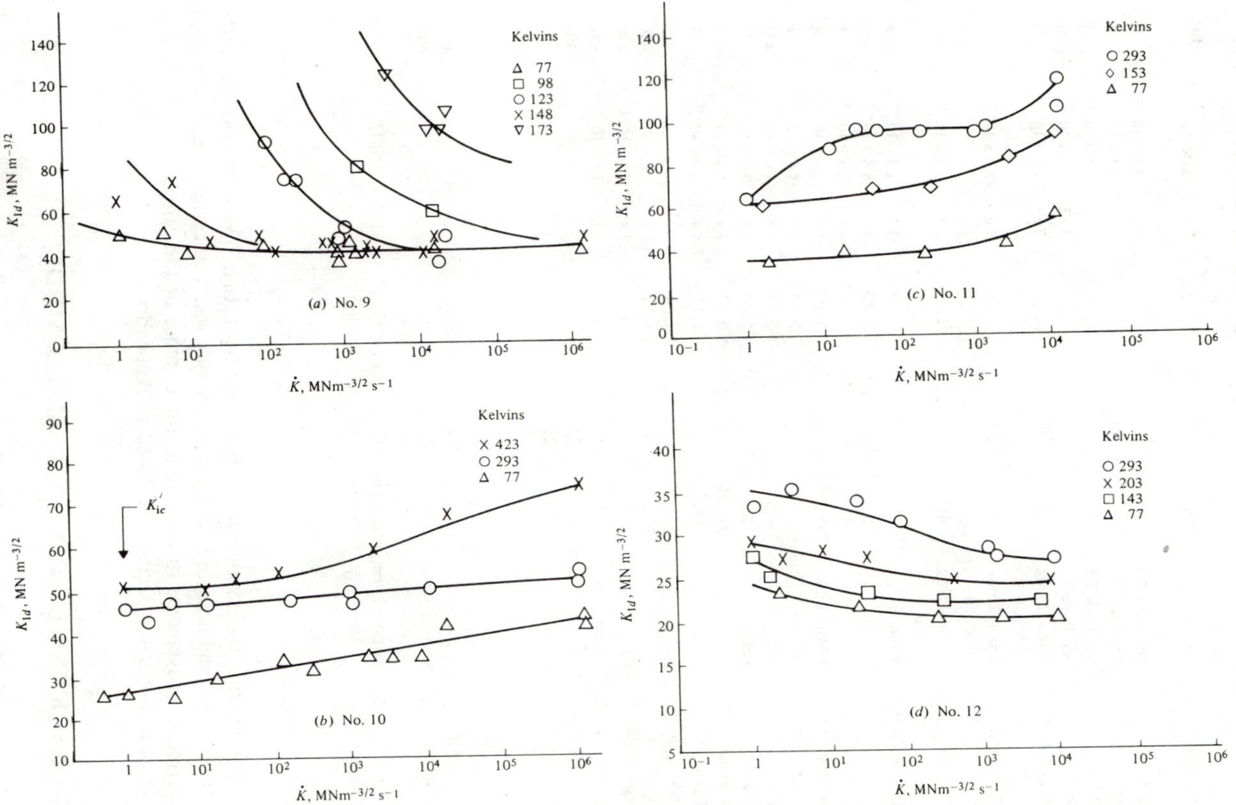

Figure 7.11 Impact fracture toughness as function of temperature and rate of stress intensity factor for metals listed in Table 7.1 [456]. (*Courtesy of A. H. Priest and The Welding Institute.*)

Table 7.3 Crack-growth data for some metals[a]

Material	Weight percent	σ_Y, MN m^{-2}	K_{Ic}, MN m$^{-3/2}$	$\lambda = \dfrac{K_{I,max}}{\Delta K_I}$	n	C_I, MN^{-n} m$^{(3n+2)/2}$	Ref.
2[b,c]	See Table 7.2	....	...	....	2.26	5.1×10^{-11}	38
6[d]	See Table 7.2	....	70	....	6.7	8.5×10^{-18}	38
7	See Table 7.2	620	45	....	3.0	1.5×10^{-11}	38
8[e]	See Table 7.2	....	...	....	3.17	7.2×10^{-12}	38
7075-T6[f]	5.5Zn, 2.5Mg, 1.5Cu, 0.3Cr	620	36	....	3.89	1.1×10^{-11}	38
7075-T651[f]	5.5Zn, 2.5Mg, 1.5Cu, 0.3Cr	525	27	....	4.0	3.5×10^{-12}	38 457
7079-T6[f]	3.8Zn, 0.3Mg, 0.4Cu	450	...	....	2.9	1.0×10^{-10}	457
DTD 687A[f,g]	5.5Zn	495	...	1.2–1.8	3.7	1.3×10^{-10}	458
				2–4.5	4.2	8.8×10^{-11}	458
				5.5–17	4.8	1.6×10^{-10}	458
HS30W[f,g]	1.0Mg, 1.0Si, 0.7Mn	245–280	...	1.3–1.8	3.9	2.4×10^{-11}	458
				2–4.5	4.1	4.3×10^{-11}	458
L71[f,g]	4.5Cu	415	...	1.2–1.8	3.7	3.9×10^{-11}	458
L73[f,g]	4.5Cu	370	...	2–8.3	4.4	3.8×10^{-11}	458
Aluminum[g,h]	....	95–125	...	1.2–7.7	2.9	1.3×10^{-10}	458
Titanium[g,h]	....	440	...	1.1–17	4.4	6.8×10^{-12}	458
18/8 Austenite steel[h]		195–255		1.5–1.8	3.1	3.2×10^{-12}	458
ZWI[i]	1.45Zn, 0.6Zr	165	...	1	3.35	1.2×10^{-9}	459
AM503[i]	1.6Mn	107	...	2	3.35	3.5×10^{-9}	459
2014-T651[f]	....	....	23	......	....	....	460
2024-T851[f]	....	....	23	......	....	....	460
2219-T851[f]	....	....	33	......	....	....	460
2618-T651[f]	....	....	32	......	....	....	460
7001-T75[f]	....	....	25	......	....	....	460
7079-T651[f]	....	....	29	......	....	....	460
7178-T651[f]	....	....	24	......	....	....	460

[a]Unless otherwise noted, temperature is 20°C. [b]$\Delta K_I < 30$. [c]-50°C. [d]0°C. [e]0, 100°C. [f]Aluminum alloy. [g]$\sigma_Y \equiv \sigma_{0.1}$. [h]Pure. [i]Magnesium alloy.

To illustrate the importance of heat treatment, Fig. 7.14 shows simultaneous variations in toughness and strength which may result in certain steels. For each material, improvements in one are at the expense of the other.

Some conversion factors are listed in Appendix G.

7.5 CRACK INITIATION IN A MIXED MODE

We return to the case of combined action and specifically to the asymmetric initial cracking (Sec. 4.3.1) taking place when K_{II} referred to the precrack plane is

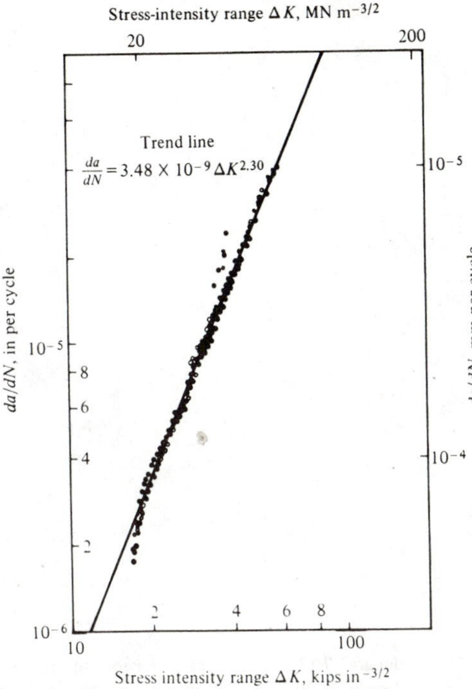

Figure 7.12 A typical record of crack rate as a function of stress intensity range [461].

nonzero. We shall retain the assumption of small-scale yielding as in LFM so that the stress intensity factors govern the process. Analytical and experimental skill have made it possible to address the present problem, where two main considerations are involved:

1. In the Griffith material, which is purely elastic and has the same resistance (2γ) toward cracking in any mode or direction, branching out of the precrack plane must be expected if $K_{II} \neq 0$. A combination of I and II is indicated in Fig. 7.15. The critical crack direction $\theta = -\theta_c$ is the one which maximizes the crack driving force for a given action on the body. The critical action must be the one which makes this crack driving force equal to the cracking resistance.
2. In a real elastoplastic material energy is dissipated beside the ideal fracture surface, and so the total resistance (Sec. 4.3.1) may not be independent of the combination of operative modes. With this, the solution by consideration 1 might turn out to be of limited value, even if yielding were sufficiently localized for overall elasticity to apply.

Consideration 1 has (essentially after Ref. 465) been open to analysis. Experiments performed with real materials have fortunately provided a fair verification of the results thus obtained. The indication in Fig. 7.15 is typical; it

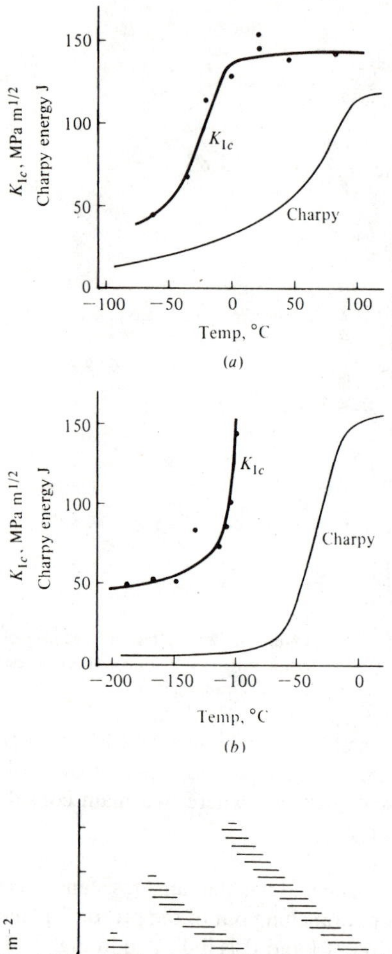

Figure 7.13 Transitional behavior of fracture toughness and Charpy energy for two medium-strength steels; for identification consult Marandet and Sanz [462].

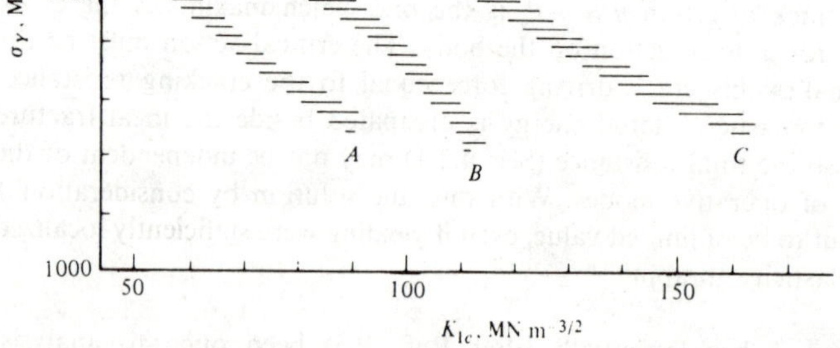

Figure 7.14 Bands of variation in strength and toughness for three steel qualitites; A = precipitation-hardening stainless steel, B = quenched and tempered steel, C = maraging steel; after Pellini et al. [463].

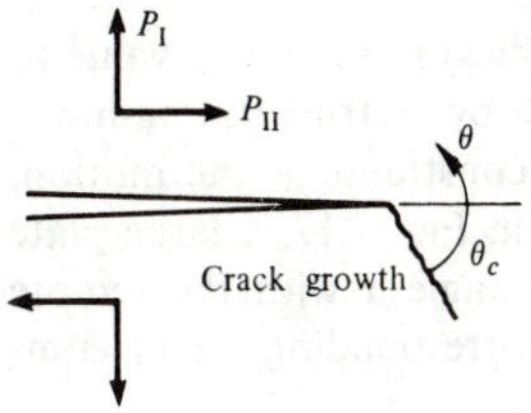

Figure 7.15 Kinked crack initiation in the presence of mode II action.

seems that under all circumstances the crack will seek a direction where opening is favored, so that mode I will be dominant with respect to the *new* plane of the crack. (Indeed, it may be argued to the effect that an unconstrained continued cracking will proceed totally within a current mode I pattern [473].) This may also dismiss some of the criticism in consideration 2.

It was previously noted that (4.12) is a relation which assumes initial motion in the precrack plane, locally tangent at the original crack tip. This is not realistic when $K_{\text{II}} \neq 0$, as indicated above, but it may serve as a starting point for improving the critical condition. Limiting the attention for a mode I–mode II combination, we might look for the substitute of (4.12) in the retained quadratic form

$$a_1\left(\frac{K_{\text{I}}}{K_{\text{I}c}}\right)^2 + a_2\left(\frac{K_{\text{I}}}{K_{\text{I}c}}\right) + a_3\left(\frac{K_{\text{II}}}{K_{\text{I}c}}\right)^2 + a_4\left(\frac{K_{\text{II}}}{K_{\text{I}c}}\right) = a_1 + a_2$$

where a_1 to a_4 are constant. In Fig. 7.16 the remarkable success of taking $a_1 = a_4 = 0$ and $a_3 = 3a_2/2$ [466], implying

$$\frac{K_{\text{I}}}{K_{\text{I}c}} + \left(\frac{K_{\text{II}}}{K_{\text{II}c}}\right)^2 = 1 \qquad K_{\text{II}c} = \sqrt{\tfrac{2}{3}}K_{\text{I}c} \tag{7.4}$$

is illustrated by theory and experiment. Within graphical accuracy (7.4) coincides completely with the theoretical curve. The dotted results will be commented on

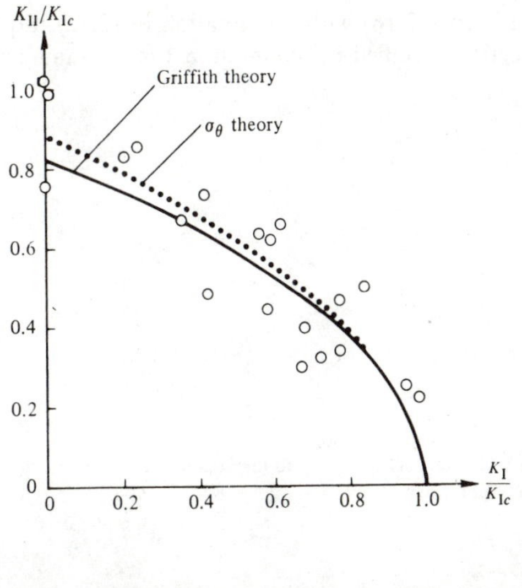

Figure 7.16 Corresponding precrack stress intensity factors at initiation of crack motion; experimental results from Refs. 468 and 469 (*circles*) and theoretical estimates from Refs. 466, 467, and 472 (*solid and dotted lines*).

later. Note that K_{IIc} (as it refers to the precrack intensities) is a *nominal* value to be used in relation (7.4). This would be contrasted by a true sliding-mode toughness if such cracking could be caused by some constraint to the motion.

A specific case of loading and geometry is shown in Fig. 7.17; a large plate has a through-crack, the normal to which makes an angle α with the remote tension axis. By simple transformation we find the corresponding stress components oriented along the crack

$$\sigma_{x\infty} = \sigma_\infty \sin^2 \alpha \qquad \sigma_{y\infty} = \sigma_\infty \cos^2 \alpha \qquad \tau_{xy\infty} = \sigma_\infty \sin \alpha \cos \alpha$$

the last pair contributing to singularities at the crack tip

$$K_I = \sigma_{y\infty}\sqrt{\pi a} = \sigma_\infty \cos^2 \alpha \sqrt{\pi a} \qquad K_{II} = \tau_{xy\infty}\sqrt{\pi a} = \sigma_\infty \sin \alpha \cos \alpha \sqrt{\pi a} \tag{7.5}$$

Inserting (7.5) in (7.4) gives the critical stress

$$\sigma_\infty = \sigma_{\infty c} = \frac{K_{Ic}}{\sqrt{\pi a}} \frac{\sqrt{1 + 6 \tan^2 \alpha} - 1}{3 \sin^2 \alpha} \tag{7.6}$$

The result is shown by the solid line in Fig. 7.18 and compared with experimental data (not carried over from Fig. 7.16). The fracture load is seen to be remarkably insensitive to moderate rotations α. Figure 7.19 shows theoretical and experimental results for the kink angle θ_c, still applying to Fig. 7.17. In contrast to the load, θ_c is highly dependent on α for small rotations—more so, indeed, than would be predicted by a rule of thumb saying that the crack should develop normal to the remote tension direction (see the dashed line in Fig. 7.19).

An earlier investigation by Erdogan and Sih [467] started by assuming (1) that the crack will grow along that direction where the strength $\sqrt{r}\sigma_\theta$ of the singular stress σ_θ is a maximum and (2) that crack growth will be initiated when this strength has the same value as the critical one in pure mode I, $(\sqrt{r}\sigma_\theta)_{\theta=0} = K_{Ic}/\sqrt{2\pi}$.

We refer now to solutions (2.15) and (2.16) with the notation of (2.32). Eq. (A.10b) also indicates that the direction specified by assumption 1 is the same as

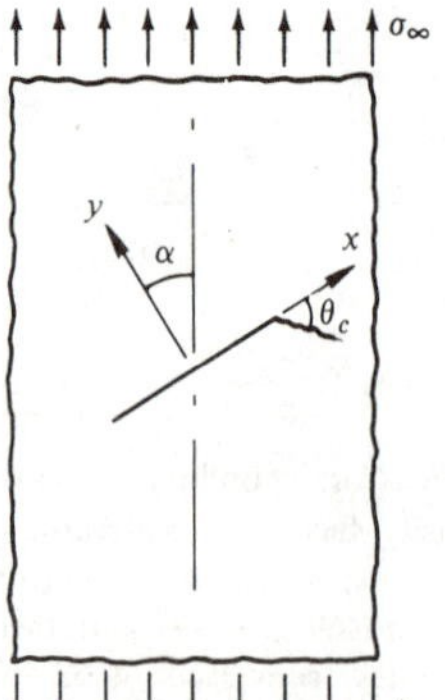

Figure 7.17 Consider geometry to investigate the effects of varying the crack orientation.

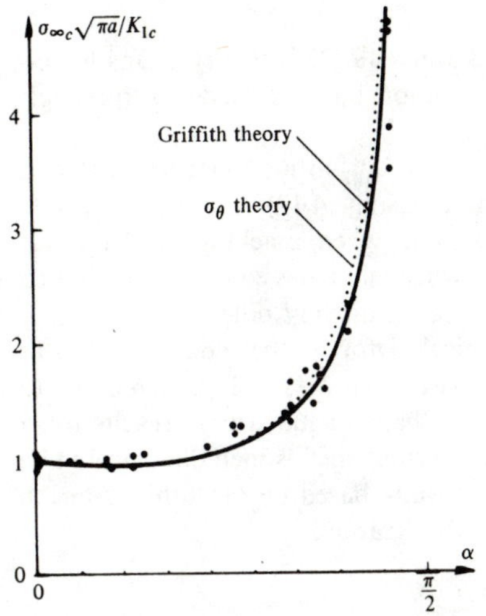

Figure 7.18 Critical remote stress for the situation in Fig. 7.17; experimental results from Ref. 466 (*solid circles*) and theoretical estimates from Refs. 466, 467, and 472 (*solid and dotted lines*).

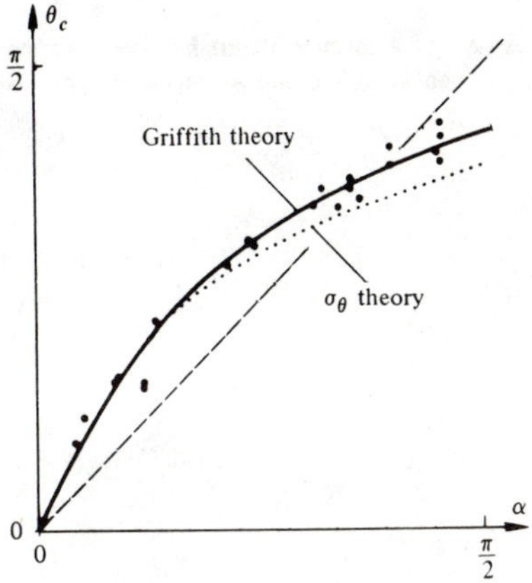

Figure 7.19 Kink angle for the situation in Fig. 7.17; experimental results from Ref. 466 (*solid circles*) and theoretical estimates from Refs. 467 and 472 (*solid and dotted lines*).

that where $\sqrt{r}\tau_{r\theta}$ equals zero. This gives

$$K_{\rm I}\left(\sin\frac{\theta}{2}+\sin\frac{3\theta}{2}\right)+K_{\rm II}\left(\cos\frac{\theta}{2}+3\cos\frac{3\theta}{2}\right)=0$$

$$K_{\rm I}\left(3\cos\frac{\theta}{2}+\cos\frac{3\theta}{2}\right)-K_{\rm II}\left(3\sin\frac{\theta}{2}+3\sin\frac{3\theta}{2}\right)=4K_{\rm Ic} \tag{7.7}$$

which obviously satisfy $\theta=\theta_c=0$ and $K_{\rm I}=K_{\rm Ic}$ if $K_{\rm II}=0$ and which also lead to

$$-\theta=\theta_c=\arccos\tfrac{1}{3}=70.6^\circ \qquad K_{\rm II}=K_{\rm IIc}=\sqrt{\tfrac{3}{4}}K_{\rm Ic}$$

if $K_{\rm I}=0$. Eliminating θ in (7.7) gives the general relationship between $K_{\rm I}$ and $K_{\rm II}$, shown as the dotted curve in Fig. 7.16. For the specific problem defined by (7.5) we further arrive at the results in Figs. 7.18 and 7.19. This theory seems to rely heavily on the details of a near-tip elastic solution but has yielded surprisingly good results.

Experimental data on the fracture toughness $K_{III,c}$ in mode III are scarce; it may be somewhat larger than K_{Ic} for mode I. Using (4.12) with $K_{II} = 0$ would therefore be a reasonably conservative estimate when cracking in the mode I–mode III combination is involved. Further, when all modes contribute but mode I is dominant, (4.12) remains a useful estimate, as already noted.

Despite the rather successful theoretical efforts, the content of this presentation has an essentially experimental basis. The data in Fig. 7.16 apply to steel and aluminum. In Figs. 7.18 and 7.19 we have included only results for a polyurethane, but they are representative for various metals including steel [470, 471, 465]. Further references to a theoretical study based on Griffith's criterion or other assumptions can be found in the Bibliography.

PROBLEMS

7.1 An experiment to determine the fracture toughness K_{Ic} of a given material has been carried out according to the ASTM specification. The test specimen is of the compact-tension type, as

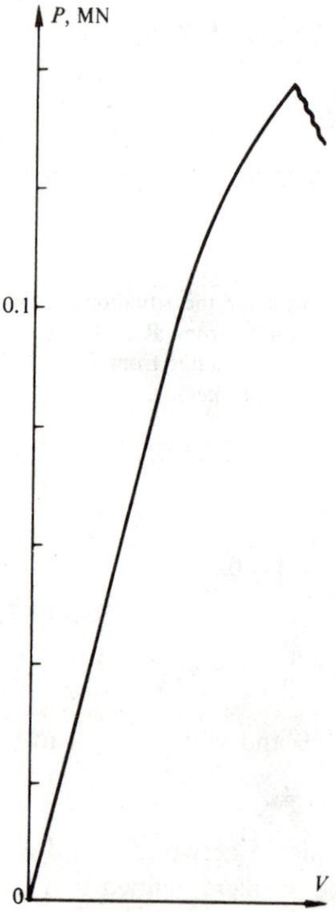

Problem 7.1

in Eq. (C.10), with $a = B = W/2 = 40$ mm. With the load versus crack-mouth opening as shown and with a 0.2 percent offset yield strength $\sigma_Y = 700$ MPa, determine K_{Ic} provided the test is valid.

Answer: The candidate value for K_{Ic} turns out to be $K_Q \approx 95$ MPa m$^{1/2}$, but the test is invalid both by dimensional requirements and by load behavior.

7.2 Fracture toughnesses K_{Ic} and K_{IIc} of a given material are to be determined by experiment. They will be referred to the first of Eqs. (7.4) without anticipating the second. The experiment is performed by testing to incipient fracture two internally cracked plates which are identical except for the crack orientation (see the figure). The observed critical loads σ_∞ are 120 MPa and 130 MPa, respectively, with $a = 20$ mm. Also, $B = 10$ mm and $\sigma_Y = 500$ MPa. From this information derive K_{Ic} and K_{IIc}.

Answer: $K_{Ic} = 30.1$ MPa m$^{1/2}$, $K_{IIc} = 24.1$ MPa m$^{1/2}$.

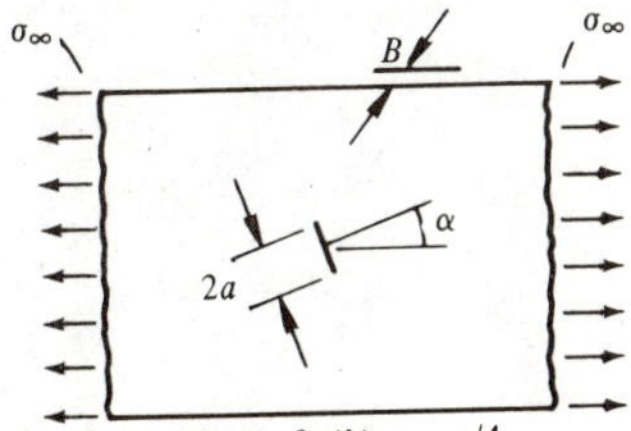

(a) $\alpha = 0$, (b) $\alpha = \pi/4$

Problem 7.2

CHAPTER

EIGHT

THE MICROMECHANICS OF FRACTURE

In this chapter the phenomenon of fracture is considered on a scale where the typical length varies from atomic spacing up to grain size. Simple theories for cleavage strength and void growth will be reviewed; otherwise the presentation will be purely descriptive, referring to the topography of the fractured faces. The last topic must be the outcome of systematic observation, i.e., *fractographic* studies of the fragments. Most important of microfractographic equipment is the electron microscope.

Fractography is an essential discipline in the study of fracture origins and material properties related to fracture. The reader will find important aspects of this topic expertly covered in Broek [7], which has influenced parts of the present exposition. The discussion in the Bibliography lists references on nonmetallic, e.g., ceramic, fracture. The emphasis in the following will be on metals: environmental effects (Sec. 6.4) are omitted.

8.1 CLEAVAGE FRACTURE

We consider here that kind of fracture which dominantly occurs by cleavage of the individual grains. The type is generally associated with a low expenditure of energy and correspondingly small global deformation, such that the fracture is brittle in any meaning of the term (compare Chap. 1). This behavior is possible in body-centered-cubic crystals, e.g., iron or low-carbon steel, and also in some hexagonal-close-packed structures like magnesium. Low temperatures greatly promote its occurrence. First, a simple schematic model of the fracture type is presented.

8.1.1 Ideal Cleavage Strength

The first investigations devoted to fracture strength considered ideal cleavage, in the sense that neighboring planes of the atomic lattice were assumed to be separated. This led to ultimate stresses which far exceed those measured under

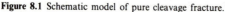

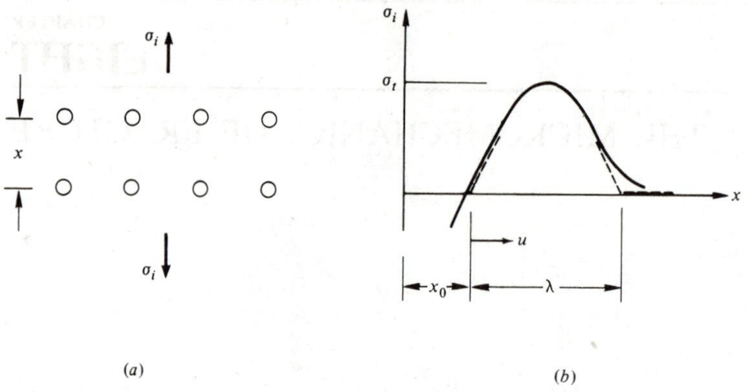

Figure 8.1 Schematic model of pure cleavage fracture.

ordinary conditions (even by orders of magnitude). We review here the approach, which is illustrative and also relevant under special conditions, and finally look into the causes of the discrepancies generally observed.

Figure 8.1 indicates two neighboring planes of atoms a distance x apart when the local stress (force per unit area, positive in separation) is σ_i. A typical relation between σ_i and x is that fully drawn in Fig. 8.1*b*, where x_0 is the spacing in the unloaded lattice and σ_t the theoretical cleavage strength, i.e., the threshold value of the local action necessary for the atomic planes to be separated indefinitely.

A fair approximation of the theoretical curve $\sigma_i(x)$ is the dashed line; a sine curve with the half wavelength λ such that

$$\sigma_i = \begin{cases} \sigma_t \sin\left(\frac{\pi}{\lambda} u\right) & \text{for } 0 \le u \le \lambda \\ 0 & \text{for } \quad u > \lambda \end{cases} \qquad \text{where } u = x - x_0 \tag{8.1}$$

For small values of the displacement u the sine function can be approximated by its argument

$$\sigma_i \approx \sigma_t \frac{\pi}{\lambda} u \tag{8.2}$$

while simultaneously Hooke's law can be assumed to hold

$$\sigma_i \approx E \frac{u}{x_0} \tag{8.3}$$

By (8.2) and (8.3) the cleavage strength can therefore be expressed as

$$\sigma_t = \frac{\lambda}{\pi} \frac{E}{x_0} \tag{8.4}$$

The work expended in a total separation of the lattice planes will be

$$\int_0^\infty \sigma_i \, du \equiv 2\gamma$$

the *surface energy* γ representing the potential energy which can be attributed to one of the free surfaces generated. With the assumed approximation the integral equals the area under the half sine wave, thus

$$\frac{2}{\pi}\sigma_t\lambda = 2\gamma \tag{8.5}$$

whereupon the elimination of λ from (8.4) and (8.5) yields

$$\sigma_t = \sqrt{\frac{E\gamma}{x_0}} \tag{8.6}$$

For many materials the surface energy has an order of magnitude of $Ex_0/100$, which gives

$$\sigma_t \approx \frac{E}{10} \tag{8.7}$$

The value σ_t according to (8.7) *may* be realized as a measurable fracture stress σ_c under very special circumstances, i.e., when the material is tested as extremely thin fibers or whiskers, so that the cross sections are nearly homogeneously strained. In general, however, the fracture stress may be smaller by one to three decades.

It is through the ideas of Griffith that the presence of defects is now recognized as causing the large discrepancies between theoretical and measurable strength. Within such inhomogeneous cross sections stress concentrations will appear, effectively making the average stress $\sigma = \sigma_c$ at the start of fracture far less than the critical peak values $\sigma_i = \sigma_t$.

[On the other hand, from the original Griffith energy balance $\mathcal{G} = 2\gamma$ we can deduce critical stresses which are far too small (compare Fig. 4.2). In order to arrive at realistic predictions, one should also consider those faults or irregularities of the lattice which contribute to defocusing the separation by dissipating mechanisms near the crack front. These effects may therefore imply a total strengthening, insofar as they outweigh that possible reduction of the ultimate cleavage strength which might follow with the local impurities.]

8.1.2 Description of Real Cleavage Fractures

A cleavage fracture proceeds along characteristic planes of the lattice structure so that its orientation changes at the grain boundaries in a polycrystalline material (Fig. 8.2). The local fracture planes are highly reflective, contributing to a shiny

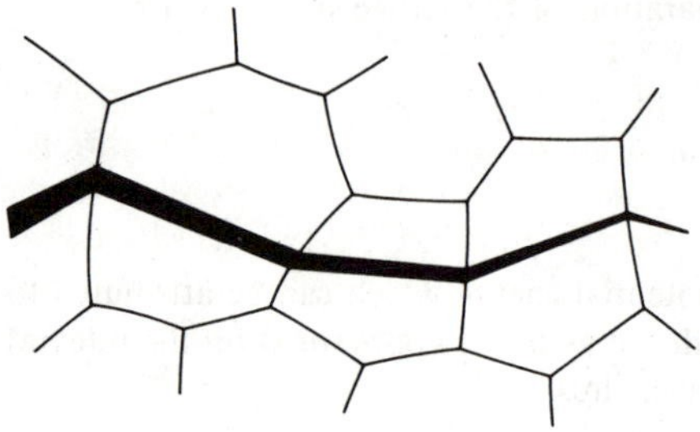

Figure 8.2 Reorientation of the cleavage surface at grain boundaries.

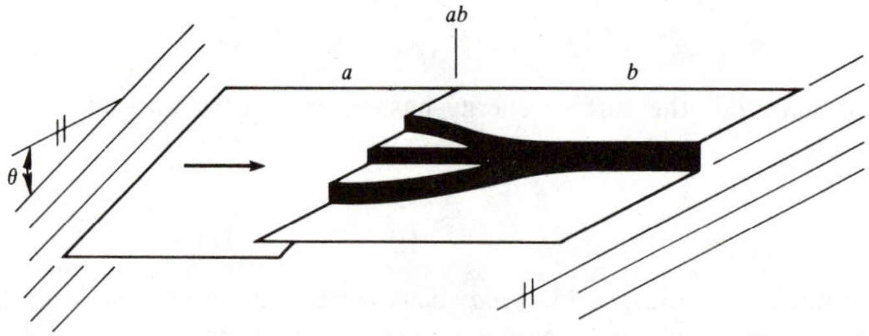

Figure 8.3 The twist element in reorientation, causing steps in the fracture surface. (*Based on material in e.g., Refs. 7 and 493.*)

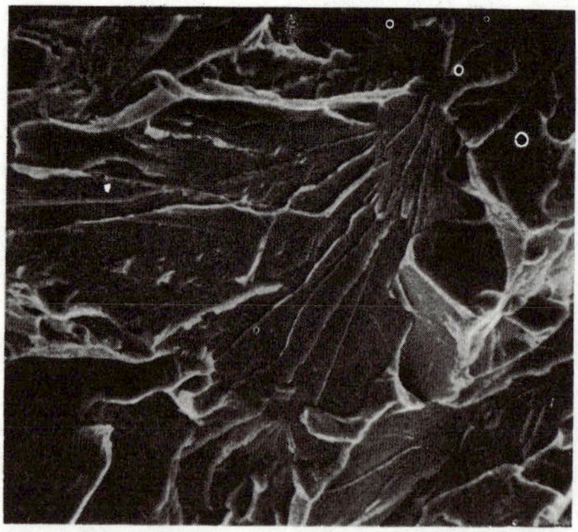

Figure 8.4 Cleavage fracture in mild steel, proceeding upward to the right. (*Courtesy of E. Folmo and N. Ryum.*)

faceted total appearance of the fragment. The reorientation at the grain boundaries has effects other than those apparent in this two-dimensional picture, by rotation of the normals even out of the plane of observation. A typical situation is shown in Fig. 8.3; a crack has advanced in the direction of the arrow through grain *a*, then passed the grain boundary *ab*, and found accommodation to the rotated (through the angle θ) cleavage planes in grain *b*. Cleavage steps have been produced by the passage, and a merging of the steps may follow the continued motion. By consequence, a characteristic *river pattern* is observed, illustrated in the electronmicrograph of Fig. 8.4. The direction of crack motion is like that of a river with its converging tributaries, i.e., up toward the right in the picture. Cleavage steps may also result when the crack has moved along two parallel planes and these have been joined by a normal crack through shearing or secondary cleavage. Such steps are usually parallel to the direction of cracking. By its nature cleavage fracture must be described as transcrystalline, i.e., formed by a separation *within* the individual grains.

8.2 INTERCRYSTALLINE FRACTURE

In the context of metals under monotonic loading, intercrystalline fractures can often be regarded as anomalies. They may be due to a weakened phase of bonding between the grains, as caused by precipitation of oxides, carbides, and sulfides, for example. Excessive grain sizes contribute the same way. The crack may then take the grain boundaries as a path of least resistance, and the fragments will be characterized by unbroken grains, like the "alpine" topography of Fig. 8.5 (where the grains are exceptionally large). Even more than in cleavage fracture grain

Figure 8.5 Intergranular fracture in aluminum alloy partly interrupted by fibrous zones. (*Courtesy of E. Folmo and N. Ryum.*)

boundary fracture may be brittle or low in energy. Together the two forms are often termed granular, in contrast to the fibrous fractures considered below.

Special external conditions (aggressive environment, high temperature) promote intercrystalline fractures because the most sensitive components are often located at the grain boundaries, where they are also most accessible to the atmosphere.

8.3 DUCTILE FRACTURES

Cleavage fractures are normally realized at low temperatures and are typically recognized by small values of fracture toughness and Charpy energy (Figs. 7.9 and 7.13). With heating (past the limit of the lower-energy shelf if such is defined) there will follow a gradual transition to fracture types which are ductile in the microscopic sense. The explanation can be found in the larger mobility of the dislocations, which makes the material plastically compliant to strong local deformation. Slip along characteristic shear directions becomes an active mechanism before separation. This may even lead to a large-scale visible deformation, so that the final fracture can be described as ductile in the macroscopic sense as well.

8.3.1 Description of Fibrous Fractures

Normally manufactured crystalline materials always contain second-phase particles in the matrix forming the basic lattice. The inclusions may occur within the grains or at the grain boundaries, may be small or large (typical lengths range from 10^{-8} to 10^{-5} m), and may have been added intentionally to cultivate specific material properties, such as increased yield strength. Particles with no such positive function (exogeneous inclusions, such as slags) are often relatively large and brittle.

With a monotonically increasing load these large inclusions will fracture or debond at an early stage and from then on contribute to increasing the loading on the adjacent material. This concentration comes in addition to effects from macro or micro notches which might initially exist. If the temperature is sufficiently high, such an action may be accommodated through increased local yielding. Progressively, however, the stage will be reached where the bonding between matrix and the remaining bulk of (smaller) particles is lost, producing pores or voids in large quantities. They may grow rapidly during a continued loading, until they merge or coalesce, internally or with an existing macro crack. In this way a fracture surface is formed with a typical dull or fibrous appearance, caused by a large number of grooves or dimples corresponding to the voids, with torn ridges between, where the plastic deformation has been intense. At the bottom of the dimples the nucleating particles can often be seen (Fig. 8.6).

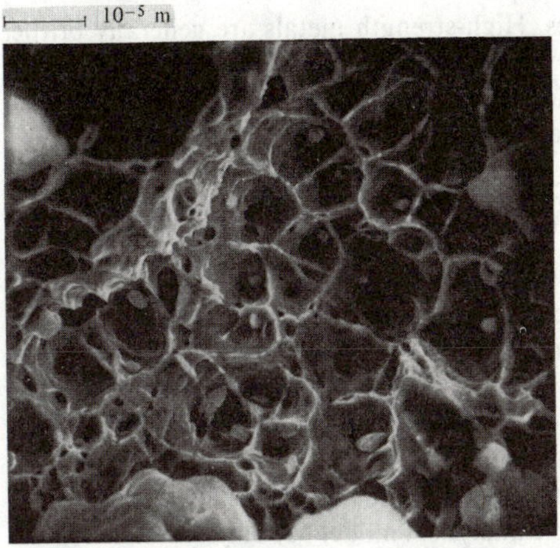

Figure 8.6 Ductile fracture in copper wire, clearly displaying dimples and nucleating particles. (*Courtesy of E. Folmo and N. Ryum.*)

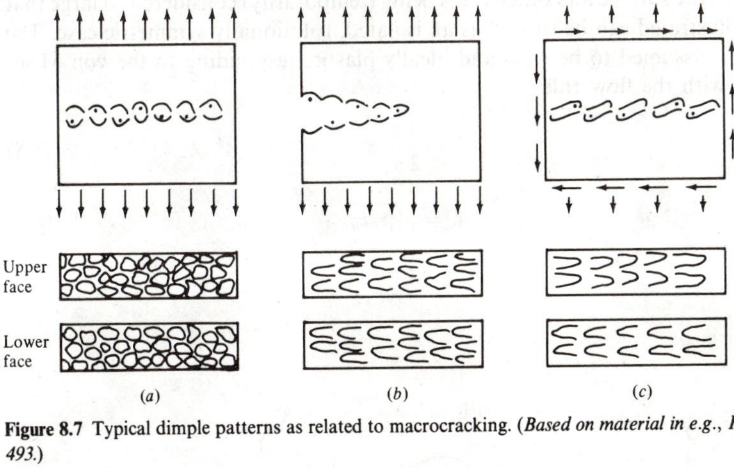

Figure 8.7 Typical dimple patterns as related to macrocracking. (*Based on material in e.g., Refs. 7 and 493.*)

The dimple topography may reveal much about the macroscopic fracture event. With a rather evenly distributed separating action, as in the neck of Fig. 2.29*a*, the dimples will be roughly circular on an average (Fig. 8.7*a*). When the separation progresses one way, as with a crack front in mode I, these markings will be stretched out to something looking like parabolas (Fig. 8.7*b*). Void growth may also contribute to a shear fracture (Fig. 8.7*c*), in which case the markings are roughly the same as above but oppositely oriented in two faces.

Fibrous fractures may consume far more energy than granular ones, as reflected by the upper and lower levels in Fig. 7.13. The transition is typical of low- and medium- strength, body-centered-cubic metals (including low-carbon steels) and has a parallel in ceramics. High-strength metals are generally brittle, i.e., low in fracture energy,† while on the other hand certain metals (low- or medium-strength, face-centered-cubic or hexagonal-close-packed structures) may be tough even at quite low temperatures.

8.3.2 Theoretical Predictions of Void Growth

As stated above, a strong growth of voids is an essential mechanism in fibrous fractures. Therefore, knowing which factors determine such growth and the strength of their effects is of great interest. We shall consider in detail a greatly simplified plane model suggested by McClintock [496] and then refer briefly to the results of more elaborate theories. The discussion will be principally aimed at pointing out qualitatively correct results.

As a specific example to illustrate the problem, Fig. 8.8*a* shows voids in the neck of a tension bar. To have a simple analytical model McClintock assumed the geometry to be cylindrical, independent of the coordinate z, as indicated by the contour shown for one void. The same pore is seen from above in Fig. 8.8*b*, the distance to the surface and other voids being (temporarily) considered so large that the one illustrated can be treated as an isolated, rotationally symmetric case. The material is assumed to be rigid and ideally plastic, responding to the von Mises criterion with the flow rule

$$\dot{\varepsilon}_{ij} = \frac{3}{2}\frac{\dot{\varepsilon}_e}{\sigma_Y} s_{ij} \tag{8.8}$$

where

$$\dot{\varepsilon}_e = \sqrt{\tfrac{2}{3}\dot{\varepsilon}_{ij}\dot{\varepsilon}_{ij}} \tag{8.9}$$

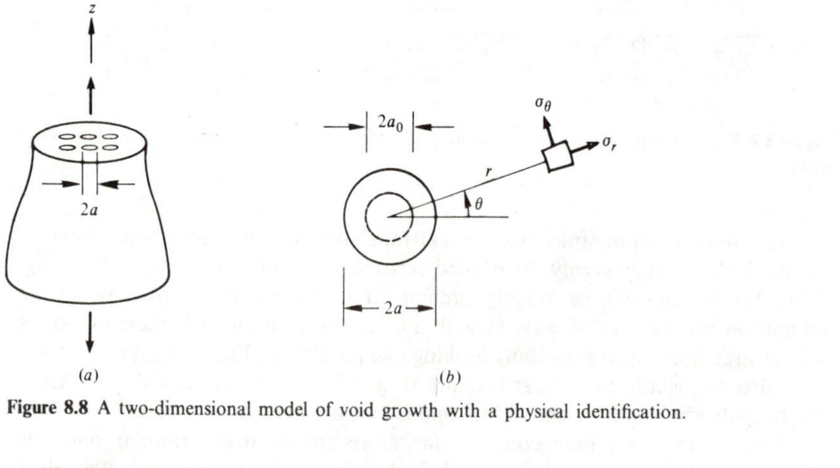

Figure 8.8 A two-dimensional model of void growth with a physical identification.

†For instance, such brittleness is exhibited by certain aluminum alloys even if the topography of fracture is fibrous.

in accordance with (A.42) and (A.50) to (A.52). Equilibrium and the kinematic constraint can be expressed by

$$\frac{d\sigma_r}{dr} = \frac{1}{r}(\sigma_\theta - \sigma_r) \tag{8.10}$$

and

$$\dot{\varepsilon}_r = \frac{d}{dr}(r\dot{\varepsilon}_\theta) \tag{8.11}$$

respectively, as follows from (A.10a) and (A.23) (with $\varepsilon \rightarrow \dot{\varepsilon}$) for the rotationally symmetric case $\partial/\partial\theta = 0$. Boundary conditions are

$$\sigma_r = 0 \qquad \text{for } r = a \tag{8.12}$$

$$\dot{\varepsilon}_\theta = \frac{2\pi\dot{a}}{2\pi a} = \frac{\dot{a}}{a} \equiv \dot{\varepsilon}_a \qquad \text{for } r = a \tag{8.13}$$

where $\dot{\varepsilon}_a$ as defined is the radial extension rate of the pore being sought. The axial strain rate $\dot{\varepsilon}_z$ is considered given, independent of r and θ; further we may know the stresses in the r, θ plane far from the pore

$$\sigma_r \rightarrow \sigma_\theta \rightarrow \bar{\sigma}_\infty \qquad \text{for } r \rightarrow \infty \tag{8.14}$$

From (8.8) we derive

$$s_\theta - s_r = \sigma_\theta - \sigma_r = \frac{2}{3}\frac{\sigma_Y}{\sqrt{\frac{2}{3}(\dot{\varepsilon}_r^2 + \dot{\varepsilon}_\theta^2 + \dot{\varepsilon}_z^2)}}(\dot{\varepsilon}_\theta - \dot{\varepsilon}_r) \tag{8.15}$$

for the introduction in (8.10), and the incompressibility condition

$$\dot{\varepsilon}_\theta = -\dot{\varepsilon}_r - \dot{\varepsilon}_z \tag{8.16}$$

Combining (8.16) with (8.11) gives

$$\dot{\varepsilon}_r = -\frac{d}{dr}(r\dot{\varepsilon}_r) - \dot{\varepsilon}_z$$

yielding upon integration

$$\dot{\varepsilon}_r = \frac{C}{r^2} - \tfrac{1}{2}\dot{\varepsilon}_z \qquad \dot{\varepsilon}_\theta = -\frac{C}{r^2} - \tfrac{1}{2}\dot{\varepsilon}_z \tag{8.17}$$

C is an integration constant to be determined by (8.13)

$$-\frac{C}{a^2} - \tfrac{1}{2}\dot{\varepsilon}_z = \dot{\varepsilon}_a \qquad C = -\frac{a^2}{2}\dot{\varepsilon}_z - a^2\dot{\varepsilon}_a \tag{8.18}$$

When we insert (8.18) in (8.17), (8.17) in (8.15), and (8.15) in (8.10), we find the differential equation

$$\frac{d\sigma_r}{dr} = -\frac{\tau_Y}{\sqrt{1+u^2}}\frac{du}{dr} \tag{8.19}$$

with the introduced variable

$$u \equiv \frac{2a^2}{\sqrt{3}r^2} \frac{\dot{\varepsilon}_a + \frac{1}{2}\dot{\varepsilon}_z}{|\dot{\varepsilon}_z|}$$

and the shear yield stress $\tau_Y = \sigma_Y/\sqrt{3}$. Integration then provides

$$\sigma_r = -\tau_Y \sinh^{-1} u + \bar{\sigma}_\infty \tag{8.20}$$

where (8.14) has been invoked to determine the constant. Finally, (8.20) is inserted into the boundary condition (8.12) to give the wanted rate of extension

$$\dot{\varepsilon}_a = \frac{\sqrt{3}}{2} |\dot{\varepsilon}_z| \sinh \frac{\bar{\sigma}_\infty}{\tau_Y} - \tfrac{1}{2}\dot{\varepsilon}_z \tag{8.21}$$

Provided $\bar{\sigma}_\infty$ is kept constant in time, (8.21) can be further integrated into the total relative extension of the pore

$$\varepsilon_a = \int_0^t \dot{\varepsilon}_a \, dt = \int_{a_0}^{a} \frac{da}{a} = \ln \frac{a}{a_0} = \frac{\sqrt{3}}{2} |\varepsilon_z| \sinh \frac{\bar{\sigma}_\infty}{\tau_Y} - \tfrac{1}{2}\varepsilon_z \tag{8.22}$$

where a_0 is the pore radius before growth.

By entering as the argument of the hyperbolic sine, the remote mean stress $\sigma_r = \sigma_\theta = \bar{\sigma}_\infty$ in the plane has a very strong effect on the void growth. For example, if $\varepsilon_z = 1$, there follows

$\bar{\sigma}_\infty/\tau_Y$	1	2	3	4
a/a_0	1.7	14	~ 3500	$\sim 10^{10}$

The higher values of $\bar{\sigma}_\infty$ can be explained only if σ_z is simultaneously large and positive (because of the yield condition), thus demonstrating the possible detrimental effect of a state of spherical tension. Such conditions may to some extent be realized inside a narrow neck and even more strongly in front of a crack in mode I plane strain (compare Sec. 2.8.1).

Rice and Tracey [497] extended the above analysis by deriving the following result, valid for an isolated *spherical* void,

$$\dot{\varepsilon}_{ia} \approx 2\dot{\varepsilon}_{i\infty} + 0.56\dot{\varepsilon}_{e\infty} \sinh \frac{\sqrt{3}}{2} \frac{\bar{\sigma}_\infty}{\tau_Y} \tag{8.23}$$

where $\bar{\sigma}_\infty$ = remote triaxial mean stress
$\dot{\varepsilon}_{e\infty}$ = remote effective strain rate
$\dot{\varepsilon}_{i\infty}$ = remote strain rate in direction of remote principal stress $\sigma_{i\infty}$
$\dot{\varepsilon}_{ia}$ = void extension rate $\dot{a}_i/a$ in direction of remote principal stress $\sigma_{i\infty}$

The similarity to (8.21) is striking, but (8.23) elucidates deviatoric and isotropic effects. It attributes the void dilatation $\dot{\varepsilon}_{1a} + \dot{\varepsilon}_{2a} + \dot{\varepsilon}_{3a}$ to the remote mean stress even

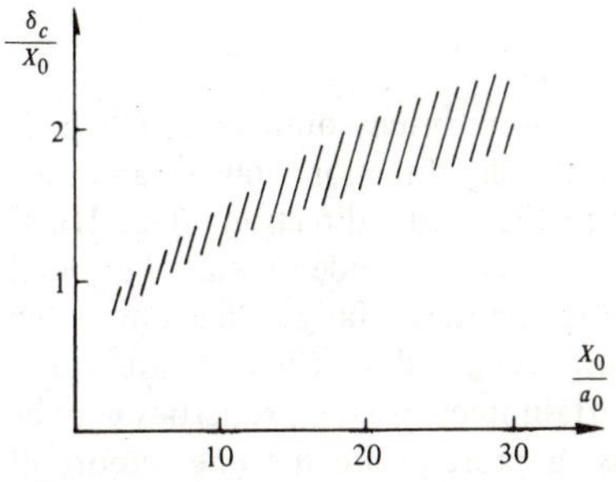

Figure 8.9 Band of predicted critical COD in relation to particle spacing X_0 and radius a_0. (*Based on results in Ref. 206.*)

in the general case (since $\dot{\varepsilon}_{1\infty} + \dot{\varepsilon}_{2\infty} + \dot{\varepsilon}_{3\infty} = 0$), and it predicts that a void located in a deviatoric stress field, such as simple shear, will be distorted roughly twice as much as the remote volume element. Such distortion is confirmed qualitatively in typical shearing fractures, like the slanted final separation indicated in Figs. 2.29*c* and 4.5, and the total separation in pure mode III (or II).†

Quantitative analyses have been undertaken to correlate macroscopic fracture parameters and properties of the void (or inclusion) structure. In a first approximation one might then assume that Eqs. (8.21) to (8.23) govern the void growth until coalescence is imminent, so that the interference of voids is essentially neglected. Rice and Johnson [206] located spherical voids in front of a blunted crack in mode I (like Fig. 4.6) and thus derived a relation between the crack opening δ_c at initiation, the typical void distance X_0, and the typical void radius a_0. The results are sketched in Fig. 8.9. An alternative model was considered by Thomason [503, 504], involving plastic-limit analysis of a regular plane grid of (quadratic) holes in front of the crack. Even if all magnitudes have not been confirmed by experiment, the analyses still served to focus attention on the essential parameters, like the mean stress $\bar{\sigma}_\infty$ and the void distance X_0 above. It has been demonstrated, both analytically and experimentally, that fractures of the fibrous, mostly ductile type, as viewed locally, may well be realized even if the event on a global scale may appear to be brittle (involving small deformation). The reason will be that the stress-concentration effect has focused sufficient energy at the point of separation.

Returning to the neck of Fig. 8.8 and to rather homogeneously strained parts in general, we note that here the analysis of void growth would point to the volume fraction (or a_0/X_0) of the voids as critically related to the macro strain at fracture. Even this observation seems to have been confirmed by experiment.

†However, recent findings by Budiansky et al. [536] indicate that the first term on the right in (8.23) is not generally valid. Indeed, whereas this always predicts void elongation in the direction of a remote extension, a flattening (into the shape of an oblate spheroid) may sometimes be the consequence instead. Therefore it will be safer to apply (8.23) only to the dilatational effect, omitting the first term on the right-hand side. Reference 536 is concerned mainly with void deformation in a creeping material. Since earlier predictions by the present author [499] were strongly confirmed by this analysis, it is tempting to recommend the results in Ref. 499 for void dilatation in a rigid-plastic strain-hardening matrix as well.

8.4 FATIGUE FRACTURE

The typical phases of a normal fatigue fracture were briefly outlined in Chap. 6; initiation is followed by some crack growth in the slip direction (stage I) and then by a further propagation essentially normal to the tensile direction (stage II). If sharp notches existed originally, as is often the case in welded areas, the initial phases may drop out altogether so that stage II dominates fatigue life. This is the type of crack growth for which the relations reviewed in Chap. 6 are valid.

Fatigue fracture is exceptional, in that certain microplastic properties may be decisive for the growth of the crack. Thus, a prerequisite for one important mechanism is that the local planes of slip shift (by cross-slip) to accommodate a continued propagation. In ionic crystals, typical of many ceramics, such compliance is often inhibited and cyclic fatigue will be no serious problem. On the other hand, fatigue is a serious problem in metals, where about 90 percent of all fractures can be traced back to this principal cause.

Despite the importance of the phenomenon, no complete information seems to exist on the detailed micromechanisms involved. On that scale of observation, initiation may be the event which is best understood. It is always located at surfaces or phase transitions (mainly to nonmetallic inclusions), where a degree of freedom for the accumulation of slip exists. The development is illustrated in Fig. 8.10; typical extrusions and intrusions are being formed by repeated slip, as for example seen in monocrystals. Various explanations have been suggested for the inhibition of relative motion on certain planes so that adjacent parallel planes become active, e.g., strain hardening and oxidation of the generated free surface [7, 512, 513] and interference with the first motion by slip on other planes [514]. In any case, it is reasonable that some intrusions may serve as crack starters, both by concentrating stresses at the roots and by admitting aggressives to take effect. We should then expect one of the primary planes of slip to be the first growing crack, i.e., stage I as mentioned. However, to identify a sharp transition between a phase of initiation and a stage I of further growth is obviously almost impossible.

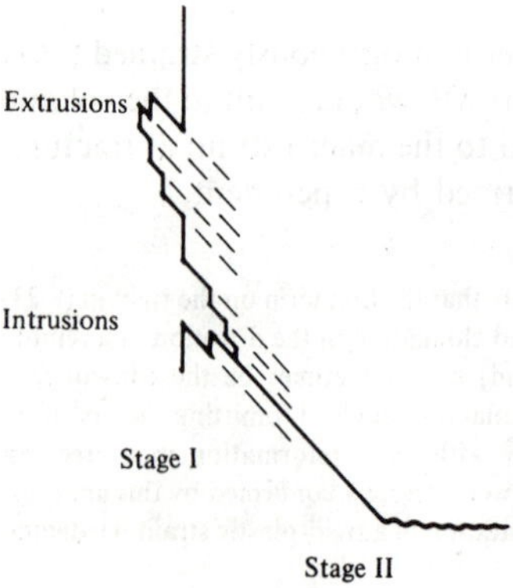

Figure 8.10 Formation of extrusions and intrusions at a free surface; one intrusion plane has developed into a stage I crack, which proceeds into stage II normal to the tensile direction.

During the further progress (stage II) all previously mentioned fracture types may be local forms. Cleavage may occur, and more frequently as the temperature is lowered; intercrystalline fractures are possible, particularly in the lower range of stress within a corrosive environment, and void growth is a mechanism which may be active in the high-strain high-temperature range. Most typical of fatigue in ductile metals, however, is the transcrystalline type of cracking which is easily identified through the *striations* formed, a ripple (or annual-ring) pattern in the fracture surface, which indicates the changing position of the crack front with each new cycle of loading. Aluminum alloys (but not pure aluminum) are metals which may strongly exhibit striations (Fig. 8.11). In steels cleavage mechanisms are often dominant.

A crude idea of striation formation can be obtained as follows. During the opening phase the material in the arch forming the crack front (Fig. 4.6) will be plastically stretched; this arch is therefore too "large" for the adjacent material during the next closure and will consequently be forced to penetrate into the material so that advance of the crack front will be permanent. More detailed models, for example [7, 518–520] can be summarized by Fig. 8.12. The opening steps 1 to 3 are essentially the same as in Fig. 2.23, only the order of displacements being different. By slip on first one and then the other plane a crack opening is formed, as sketched in step 3. The real blunting, step 4, is the result of hardening as well as slip on parallel planes. Following some reversed slip the crack will close (more or less, depending on the amount of unloading), and the permanent displacement and distortion in step 5 are a reasonable consequence of the total past history. During repeated loadings and unloadings the old ripples will remain while new ones are added, so that all of them are permanently left behind in the wake of the moving crack front.

Since the final fracture succeeding a cyclic development is determined by the same set of variables as monotonic loading, that part of the fracture surface may

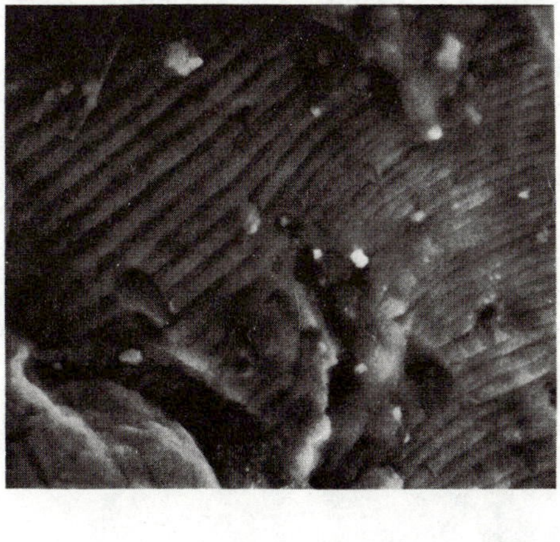

Figure 8.11 Fatigue fracture in aluminum alloy, with typical striation pattern. (*Courtesy of E. Folmo and N. Ryum.*)

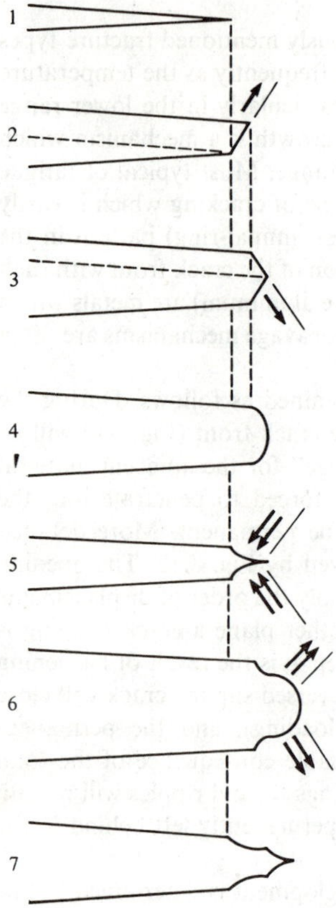

Figure 8.12 A model for the formation of striations; after Broek [7].

have a granular or fibrous (bright or dull) appearance but will be easy to distinguish from a striation pattern in any case. From the striations one can also discover where the crack was initiated, i.e., the "center" of the ripples. A typical fracture face may therefore be like that shown in Fig. 8.13.

By its nature the fatigue fracture is a discrete process, the crack jumping from one atomic position to another and from one load cycle to another. The lowest

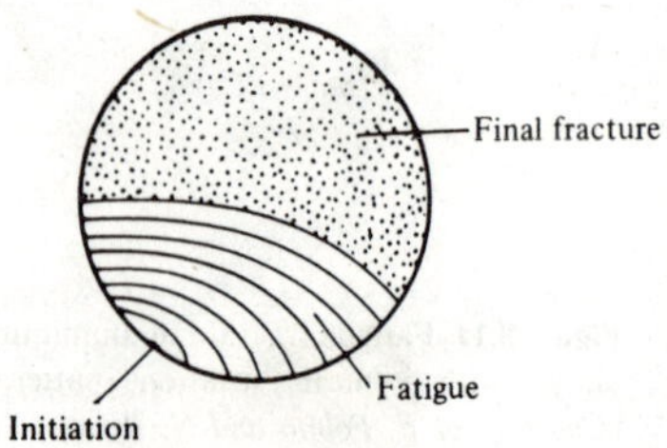

Figure 8.13 The typical events of crack initiation, fatigue fracture, and final fracture as identified in a fragment.

rate possible should therefore correspond to one atomic spacing per cycle, and the threshold value ΔK_t in Chap. 6 should in some way be related to this minimum. When even smaller values are being recorded, they have to be explained as average values along the crack front. The front will then be at rest locally while other parts are moving.

8.5 ADDITIONAL REMARKS

Fracture types have been considered separately above though in practice they may be mixed. In particular, this will be the case in the interval between the upper and lower shelves in energy (Fig. 7.13), such that grains and dimples appear side by side. For example, fibrous regions are evident in Fig. 8.5.

Of particular interest in relation to fracture is the so-called *stretched zone*. Microscopically it can be identified in the surface forming the arc in front of the blunted crack (Fig. 4.6 and steps 4 and 6 of Fig. 8.12). Here the plastification is very intense, and the slip steps can be observed directly under sufficient magnification. The topography is a miniature striation pattern but more irregular (wavy). In this zone fracture may be initiated by monotonic loading. The extent of the zone, normal to the crack front, is comparable to the COD, i.e., to $K_{\mathrm{Ic}}^2/E\sigma_Y$ following fracture; measuring it may give an experimental indication of δ_c.

We have discussed the formation of macro cracks in an originally "faultless" body in the context of fatigue, and in other cases of crack growth we have assumed that sharp macro notches existed from the beginning. Even under noncyclic action the first cracking must obviously be a phenomenon on the micro scale, e.g., as initiated at grain boundaries or inclusions. However, the chances of such initiation are clearly higher where stresses have already been concentrated, i.e., at the sharp end of the macro notch. Normally, this raising of stresses is necessary for cracking upon monotonic loading to become a problem of practical concern. On the other hand, time has an effect in fatigue, so that the initiation period may compensate for the lack of stress raisers.

PROBLEMS

8.1 A rectangular regular grid of initially small spherical voids is shown part (*a*) on the figure. The voids are situated within an average stress field having

$$\sigma_{1\infty} = \sigma_{2\infty} = 2\sigma_Y \qquad \sigma_{3\infty} = 3\sigma_Y$$

as might represent the state in a deep neck of a tension-bar pulled in the 3-direction. The corresponding field of (plastic) strain rates is

$$\dot{\varepsilon}_{1\infty} = \dot{\varepsilon}_{2\infty} = -\tfrac{1}{2}\dot{\varepsilon}_{3\infty} < 0$$

A rough estimation of the axial strain $\varepsilon_{3\infty}$ for cracking to start by void coalescence may be attempted. Assume then that the voids remain spherical by using (8.23) with neglect of the first right-hand term. Pay no heed to the increasing interaction effect and (in that context) the conservation of mass. Verify on this basis the critical strain $\varepsilon_{3\infty} = 0.24$. The critical stage is suggested in part (*b*) of the figure.

Comment: By the present dilatational dominance and the results Ref. 536, the assumption of a continued spherical form before interaction is indeed tenable, but to pursue this evaluation into the stage of contact obviously may lead to no more than a very crude assessment.

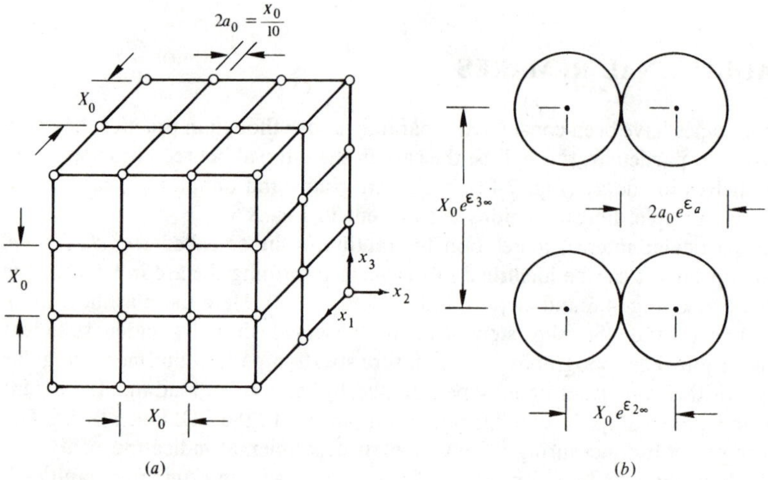

Problem 8.1

8.2 Experiments indicate that the macro strains for ductile fracture in homogeneously tensioned parts may be directly related to the effective volume fraction of voids. Convince yourself that this behavior is in accord with the model of Prob. 8.1.

CHAPTER

NINE

ELEMENTS OF APPLIED FRACTURE MECHANICS

9.1 GENERAL DISCUSSION

Fracture-mechanical evaluations and assessments are available tools to help in planning structures for safe design and operation. We have a set of formulas and critical quantities (stress intensity factors, COD, crack growth rate, etc.) to quantify the risk of cracking, and are expected to provide the corresponding set of pertinent material parameters. Searching the literature and/or performing separate experiments will be necessary efforts to arrive at those resistance data of the material. What to insert in the formulas to represent a local action is determined by the total behavior of the structure and may be even harder to establish than material properties. Furthermore, a severe difficulty is involved in detecting cracks and their geometrical properties, given the product or the structure. When we also realize that even within its own idealized framework fracture mechanics has unsolved problems, e.g., concerning large-scale yielding effects, we become aware that these tools must be handled with caution and a sober regard for the total interplay of economy, safety, and uncertainty. The use of *safety factors* related to the separate elements of the evaluation is obviously motivated by the final aspects.

Among specific benefits we can point to the following achievements of fracture mechanics, related to material selection, design, and operation:

1. The establishment of acceptance norms, e.g., as expressed by a minimum of K_{Ic} or a maximum of flaw size, for materials or products to fit various purposes
2. The development of evaluation procedures which may admit the assessment of a critical load, a service life, or a critical flaw size for an active structural member
3. The suggestion of qualitative guidelines for what is good or bad, safe or unsafe design
4. The development of inspection routines, with a view to method as well as to where and when the inspection should be exercised.

These perspectives, opened up through the gradual development of fracture mechanics, all have an impact on safety and economy. Previously, for instance, it was not clear that strength may be bought at the expense of toughness (Fig. 7.14), with grave implications for defect tolerance and control requirement. For example, if one doubles the service stress σ_∞ in Example 4.2 through doubling the yield stress, and if this would mean a reduction of K_{Ic} by a factor of 2, the reduction of permissible crack size would be to *one-sixteenth* its first value (LFM being invoked for simplicity). In itself, this simple illustration has a bearing on all four items above.

Cracks of critical size are not what would normally be found in the virgin material but are the consequence of working and joining operations as well as service conditions. Critical locations may be the vicinities of holes and notches and, not least, welds. In the last case, the defects may be due to slag or lack of penetration (Fig. 9.1); simultaneously local conditions may favor the growth of cracks. Stresses are concentrated where small slag intrusions are most often formed, i.e., at the toe of a weld. Residual stresses of considerable magnitude may have resulted from phase transitions and cooling. In addition, the material adjacent to the weld may have been unfavorably affected, so that brittle regions occur.

The operational condition for cracking is dominantly fatigue, often combined with corrosive effects. Crack nuclei at the locations mentioned above may then develop into critical size, so that final fracture follows. The tougher the material the larger the critical size and the higher the probability of detecting a crack before fracture. However, a fatigue crack usually grows much faster in its later life, so that one normally does not gain much in total *time* to failure by increasing the toughness. (Drastic exceptions to this rule are mentioned in Sec. 9.4.) If a cyclic loading is expected, it would be more effective to aim at improving the growth-rate parameters, like C_1 and n in (6.10).

Since cracks must be faced in all operation, it is important that they be detected and inspected. Smaller repairs or replacements can then be undertaken to prevent larger damage, or a safe service life without such interference can be estimated.

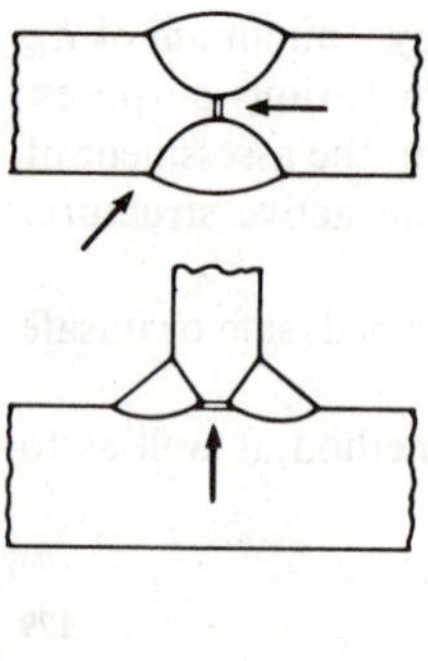

Figure 9.1 Typical sites for cracking in weldments.

Current inspection routines include the following:

1. *Direct inspection by sight.*
2. *Use of penetrant fluid.* After application the fluid is washed off the surface and replaced by a developing suspension. If there is a crack, the developer will extract the penetrant, which produces coloring.
3. *Use of magnetic particles.* Fluid containing iron particles covers the surface and is subjected to a strong magnetic field. The field lines will be distorted by the presence of a crack, as revealed by the pattern of particles. The base material must be magnetic.
4. *Ultrasonic recordings.* Excited high-frequency waves are reflected by the crack. The time from emission to incidence is interpreted as distance from the observer.
5. *Eddy-current measurements.* Eddy currents are induced in the metal by a coil; the corresponding current by induction in the coil will be influenced by the presence of a crack.
6. *X-ray photography.* The crack absorbs less radiation than the surrounding material and is therefore recognized as a line on the film. This is normally considered a reliable method of inspection, often applied to important weldments, but may still prove incapable of providing detailed information.

Besides these sources of crack detection, mechanical signals emitted at the crack tip due to yielding and separation can be recorded by mobile or permanent equipment. This *acoustic-emission* technique has been considered promising for some time and may hopefully meet expectations. So far a serious problem has been separation of emitting effects. One attractive feature is the possibility of continuous supervision of critical components.

An indirect safety routine may also go through *proof testing*, which implies that the structure, say a pressure vessel, is being loaded to a higher level than the service condition. If this load can be carried, one knows that the service condition is not immediately dangerous. But, more information can be extracted from the test, with regard to a possible crack growth under cyclic loading.† The fact that the test load can be carried implies that all cracks are smaller than a critical one which can be determined by fracture-mechanical methods, e.g., through $K_I = K_c$, where K_I depends on test load and crack size. Conservatively, one can then assume that a crack almost as large as the critical one does exist and then evaluate how long it will take for that crack to grow to critical size under the actual loading conditions. Before the end of this predicted life a new test can be undertaken, and so on. The method is suitable for pressure vessels when a liquid (not a gas) is the pressure fluid during the test.

Strength evaluations can be carried out with reference to known, real cracks in the field, but also on the basis of *nominal* defects of estimated size and shape at reasonable locations. This can be seen as the threshold into a purely

†A beneficial effect under all loading conditions may be the removal of high residual stresses.

probabilistic analysis, proceeding from assumed distributions (frequency functions) of (1) initial size, density, orientation, location, etc., of the cracks, (2) amplitude, frequency, mean level, etc., in the load spectra, (3) crack growth properties, and (4) toughness properties of the material. Such a point of departure must be extensively based on inspection, measurement, and some experience. With a confidence which depends heavily on the quality and quantity of the input data one predicts a probability of failure (within a given time if fatigue contributes). Present efforts are directed at implementing detailed fracture-mechanical concepts in statistical analysis, whereas earlier investigations, e.g., by Weibull [546–548], were gross evaluations leaving parameters to be assessed empirically. The success of the probabilistic approach has been confined to brittle fractures (prominently in ceramics, until recently), where the first event of crack initiation can be interpreted as total failure. For some further information see the discussion for Chap. 9 in Bibliography.

It has been mentioned that cracks often appear in regions where stresses would be concentrated for other geometrical reasons. This makes evaluation of their effect even more difficult. Formulas of stress intensity factors often contain a "remote" stress σ_∞, equal to the stress which would appear uniformly if the crack did not exist. Such uniformity is lacking at the above locations, however, and one either has to estimate what should replace σ_∞ or carry out separate total analyses. Some examples will be discussed below. Reference will be made to numerical analyses of crack growth during cyclic loading and to the concept of leak before break. Further, some views on the practice of nonlinear fracture mechanics will be related to crack initiation and to continued stable growth, and some aspects of the thickness effect will be discussed. Finally, causes and effects of crack instability in certain applications will be assessed.

9.2 CRACKS EMANATING FROM NOTCHES

We start by interpreting Bowie's solution [25] for a crack emanating from a circular hole in a plate normal to the direction of a remote tension σ_0 [compare (C.11) of Appendix C]. If the crack is quite small compared with radius r of the hole, as in Fig. 9.2*a*, it can be viewed as an edge crack in a large plate subject to the remote tensile stress

$$\sigma_\infty = C_t\sigma_0 \tag{9.1}$$

where C_t is the concentration factor for the stress along the hole boundary in the absence of a crack. For uniaxial remote stress, as specified, we have

$$C_t = 3 \tag{9.2a}$$

while

$$C_t = 2 \tag{9.2b}$$

would have to be inserted if an equally large remote stress were applied parallel

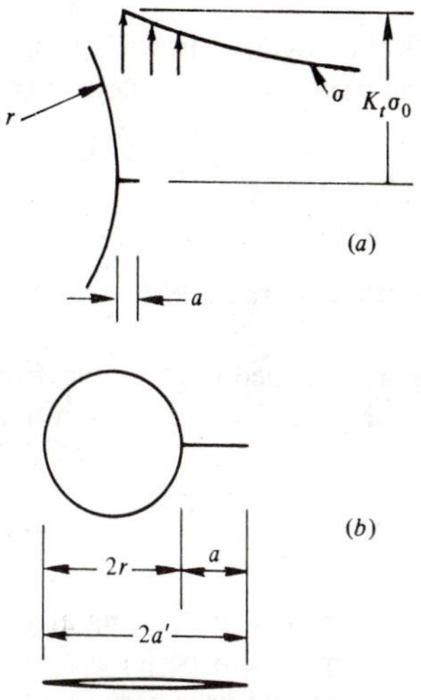

Figure 9.2 Cracking from a circular hole; small or larger crack length.

to the crack. This leads to the stress intensity factor

$$K_{\mathrm{I}} = 1.12\sigma_{\infty}\sqrt{\pi a} = 1.12C_t\sigma_0\sqrt{\pi a} \tag{9.3}$$

according to (C.5) and $a/W \rightarrow 0$. The result agrees with the first row in the table of $F(a/r)$ values related to (C.11). On the other hand, if the crack length becomes comparable to r or larger, we can expect the state at the crack tip to differ only moderately from that produced if the hole were replaced by a crack along its diameter (Fig. 9.2*b*). Internally in a large plate this would lead to the stress intensity factor

$$K_t = \sigma_{\infty}\sqrt{\pi a'} = \sigma_0\sqrt{\pi a'} \tag{9.4}$$

with (C.3) and $a/W = 0$. Since σ_{∞} is the same as σ_0 here, the effective crack has a total length of

$$2a' = 2r + a \tag{9.5a}$$

For *two* symmetric cracks the effective length should be

$$2a' = 2r + 2a \tag{9.5b}$$

If, furthermore, the plate width is not much larger than $2a'$, and if the effective crack is centrally located, the stress intensity factor will correspond to the general case (C.3); thus

$$K_{\mathrm{I}} = \sigma_0\sqrt{\pi a'}f\left(\frac{a'}{W}\right) \tag{9.6}$$

The function f is then the ratio in (C.3) between K_{I} and $\sigma_{\infty}\sqrt{\pi a}$, where a' replaces a.

The validity of (9.4) and (9.5) can easily be checked against Bowie's results [25]. Figure 9.3 represents the last set of values in the uniaxial-loading case in the form

$$K_{\mathrm{I}} = \sigma_0 \sqrt{\pi a'}\, g\left(\frac{a}{r}\right) \tag{9.7}$$

In this loading case the correction factor $g(a/r)$ to (9.4) is remarkably close to unity, even for very small cracks.

A related problem consists of a crack entering a V-shaped edge notch (Fig. 9.4). For a remote tensile stress σ_0 parallel to the edge a plausible form of solution [corresponding to (9.7)] is

$$K_{\mathrm{I}} = 1.12\sigma_0 \sqrt{\pi(a+b)}\; h\left(\frac{a}{b}\right) \tag{9.8}$$

where the function $h(a/b)$ is probably close to unity for smaller opening angles γ or smaller ratios b/a. The correctness of the last estimate can be judged by a comparison with the exact solution [560] (Fig. 9.5). We should note here that the limit $a/b \to 0$ of cracking from a sharp notch theoretically poses a particular problem of its own. An expression like (9.3) would then include an infinite concentration factor C_t, and forms like (9.3) and (9.8) would also break down because the stress singularity of a V notch is not strictly of the type $r^{-1/2}$. The conflict is a minor one for small opening angles, however, such as $\gamma < 45°$ [561],

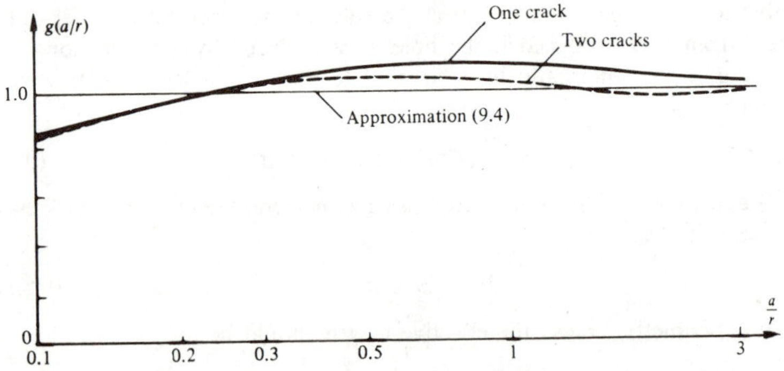

Figure 9.3 Results for the Fig. 9.2 situation, illustrating the accuracy of replacing the hole by a slit over its diameter; uniaxial loading normal to the crack. (*Based on results in Ref. 25.*)

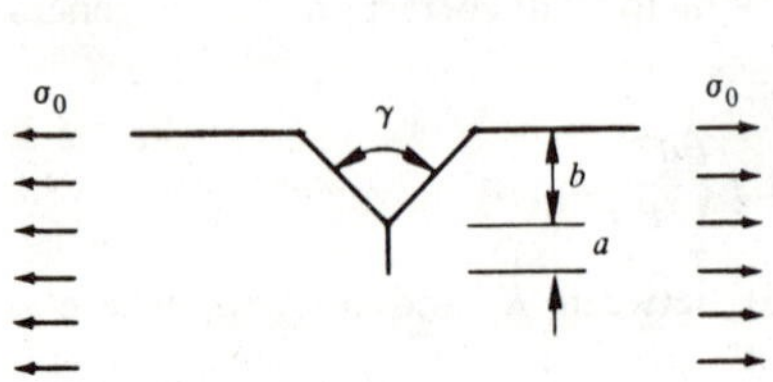

Figure 9.4 Cracking from a V-shaped notch.

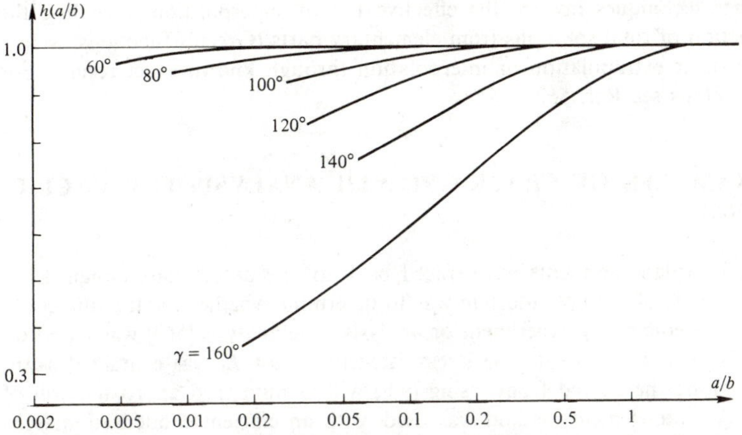

Figure 9.5 Results for the Fig. 9.4 situation, illustrating the accuracy of replacing the notch by a crack over its depth. (*Based on results in Ref. 560.*)

and we can therefore apply (9.8) with $h \approx 1$ arbitrarily close to $a/b = 0$ in those geometries.

If the plate width W is limited, expressions like (C.4) to (C.6) or (C.9) should be used instead, $a + b$ replacing a.

When direct handbook references are not available and no simple estimation is justified, a numerical analysis using the finite-element method may give the stress intensity factor. As a practical means we suggest the energy basis

$$\beta \frac{K_I^2}{E} = -\left(\frac{\partial \Phi}{\partial A}\right)_u = +\left(\frac{\partial \Phi}{\partial A}\right)_P \tag{9.9}$$

according to (3.18), (3.21), and (3.30) if mode I is considered. The strain energy Φ is determined for neighboring crack sizes, whereby either the loading-point displacements or the loads are kept constant and the slopes of the $\Phi(A)$ curve are determined by a suitable numerical method.

For a small crack in the interior of a plate a possible alternative may be to use (B.37)

$$K_I = \frac{1}{\sqrt{\pi a}} \int_{-a}^{a} g_2(x) \sqrt{\frac{a + x}{a - x}}\, dx \tag{9.10}$$

where $g_2(x)$ is the stress which would appear normal to the plane of the crack in its closed configuration; it can therefore be determined analytically or numerically for the intact body. Modes II or III may be similarly included. These cases represent just that type of problem where the efficiency of (9.9) may be rather limited.

Other techniques involve the effective use of superposition, including the construction of total solutions from elementary parts (Green's functions, weight functions), or extrapolation or interpolation through known exact results. For further details see Ref. 562.

9.3 EXAMPLES OF CRACK-GROWTH ANALYSIS FOR CYCLIC LOADING

Repeated airplane accidents were traced back to one unfortunate design detail (Fig. 9.6). Part of the post mortem was to determine whether the fracture could have been predicted by experiment or analysis. The analysis [563] was based on Forman's version of (6.13). The stress intensity factor K_I was evaluated as an average along the curved front, using (9.9) with a numerical approximation of $\partial\Phi/\partial A$. A finite-element method was used, with an efficient condensation technique to minimize repetition of computing steps. The crack front was approximated by a circle (Fig. 9.7).

Results of experiment and analysis are shown in Fig. 9.7. The measurements started by observing an existing crack (0), whereas the calculations were based on

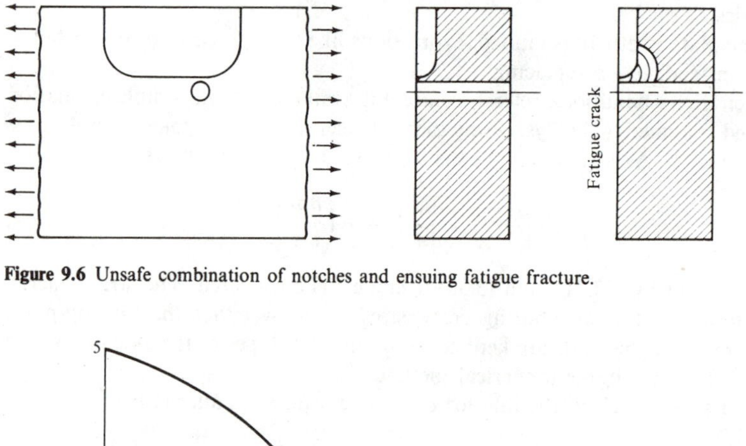

Figure 9.6 Unsafe combination of notches and ensuing fatigue fracture.

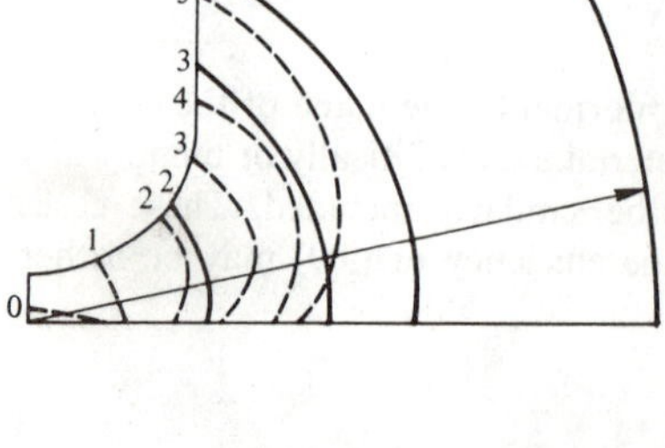

Figure 9.7 Experiments and analysis for the situation in Fig. 9.6; solid curve = estimated crack front, dashed curve = measured crack front, curve 0 = initial crack, curve 1 = 239 h, curve 2 = 557 h, curve 3 = 1073 h, curve 4 = 1550 h, and curve 5 = 2067 h; Ref. 563.

the developed state 1 of initial cracking. As seen, the latter results were conservative (predicting too rapid growth), but the difference in times for the nearly equal crack sizes, 5 as measured and 4 as calculated, is still acceptable. Incidentally, it is reasonable for the present analysis to predict too large a crack since the real load spectrum was more of the type shown in Fig. 6.3 and the evaluation did not include the retarding effect by residual stresses upon reducing the load amplitude.

The example confirms that fracture mechanics can lead to fair predictions. It also illustrates that certain designs are destined for failure with almost mathematical consistency.

Critical defects can often be idealized into flat elliptical cracks internally or semielliptical cracks at the surface of a plate (compare cases 1 and 2 in Table C.1). Some average value of the stress intensity factor K_I may be associated with changes in the minor and the major axis, respectively, when the evaluation of Φ in (9.9) is based approximately on the elliptic crack growing into another ellipse. Reference 564 analyzed the growth of an internal crack when its shape and location and the parameters describing a cyclic normal load were varied. One of

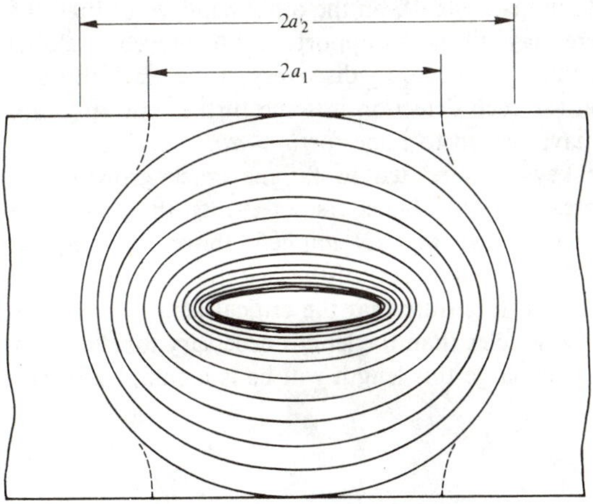

Figure 9.8 A numerical estimation of the cyclic growth of an embedded elliptical crack [564].

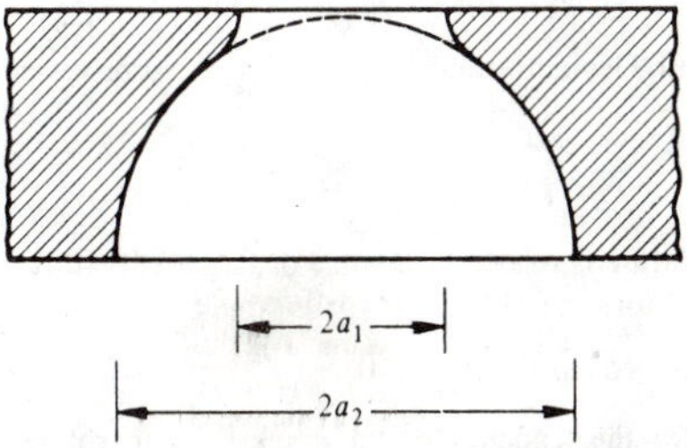

Figure 9.9 Crack penetrating a wall.

the results (Fig. 9.8) pertains to the growth of a central crack due to a central normal load. When the crack front is sufficiently close to the surface, it will actually penetrate to or through a configuration like that indicated by dashed lines. The same applies to cracks starting from a surface (Fig. 9.9).

9.4 LEAK BEFORE BREAK

The plate in Figs. 9.8 or 9.9 can be identified as part of the wall of a pressure vessel. From a crack-growth evaluation like the preceding one the event of penetration can be conjectured as in the figures.

Figure 9.10 shows an axial crack through the wall of a cylinder having a length $2a$. For the given maximum pressure value p, with corresponding circumferential stresses $\sigma_\phi = pR/B$, one can determine a critical crack length $2a_c$.† If $a > a_c$, unstable crack growth along the wall is the reasonable consequence of effectively keeping the pressure at its value p.

Therefore, if $a_1 > a_c$ is being produced (Figs. 9.8 and 9.9), we expect cracking to continue unstably along the wall after penetration. This will be a matter of grave concern if the contained fluid is a gas. If, on the other hand, a_2 of Figs. 9.8 and 9.9 is smaller than a_c, there may still be an opportunity to prevent a global failure. Reasonably, this opportunity is offered by discovery of the crack through leakage. Perhaps the condition for such detection is some further cracking into the size $a_3 > a_2$, and we then have reason to hope for $a_c > a_3$.

"Leak before break" are keywords related to fatigue crack growth in a pressure vessel. Since they express the desirable course of events, they provide a guideline for design and material selection. As implied above, the in-plate toughness K_c (going to K_{Ic} for larger thicknesses) of the material should be high enough and the stresses low enough to ensure that the critical length of a crack through the thickness will not be smaller than the length necessary for discovery of the crack through leakage. How large this length will be is a complex matter

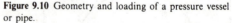

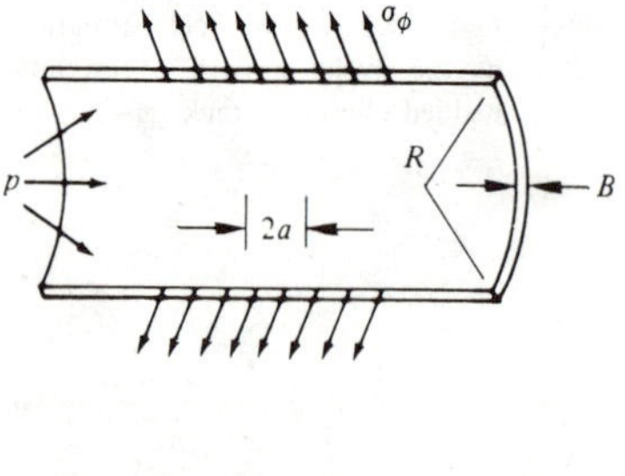

Figure 9.10 Geometry and loading of a pressure vessel or pipe.

†$a_c = (K_c/\sigma_\phi)^2/\pi$ if yielding is localized and the curvature is relatively small. For larger curvatures the condition $K_I = K_c$ should be combined with Folias' formula [73, eq. (44)]

$$K_I = \sigma_\phi \sqrt{(1 + 1.6a^2/RB)\pi a}$$

(or modifications) which contains a correction term for the tendency of the crack sides to rotate.

of empiricism involving many variables, but efforts toward establishing useful criteria are being made.

A simplified model of the above evaluation appears if the precrack, say as developed internally, along a weld, has a large longitudinal extent compared with the wall thickness. Then in a first stage of cracking the front will move rather uniformly in the radial direction, and a time to leakage can be estimated by equating the developed crack-tip intensity factor K_{I} (for example, case 5 in Table C.2) with a transverse material toughness $K_{\mathrm{I}ct}$. With the penetration a longitudinal through-crack has been generated, its length similar to that of the precrack. For this crack the stress intensity factor K_{I} is next evaluated and compared with the in-plate toughness K_c.

In any case, it turns out that an effective anisotropy $K_c/K_{\mathrm{I}ct} > 1$ is usually necessary for the developed through-crack to be stable. If the wall is suitably thin, this condition for leak before break may well be satisfied even with basic isotropy of the material (compare Fig. 4.5*a*).

9.5 CRACK INITIATION UNDER LARGE-SCALE YIELDING

Our earlier experiences with crack-initiation assessments within the frame of nonlinear fracture mechanics were related to Figs. 4.7, 4.10, and 4.12. In particular, Fig. 4.7 was arrived at by combining the COD condition

$$\delta = \delta_c = \frac{K_c^2}{E\sigma_Y} \tag{9.11}$$

with the expression for the tip opening of a through-crack normal to the remote tension σ_∞ in a large plate

$$\delta = \frac{8\sigma_Y a}{\pi E} \ln\left[\left(\cos\frac{\pi\sigma_\infty}{2\sigma_Y}\right)^{-1}\right] \tag{9.12}$$

The analytical result was the initiation condition (4.25), or

$$\ln\left[\left(\cos\frac{\pi\sigma_\infty}{2\sigma_Y}\right)^{-1}\right] = \frac{\pi^2}{8}\left(\frac{\sigma_{\mathrm{lin}}}{\sigma_Y}\right)^2 \tag{9.13}$$

introducing the critical stress which would follow through formal use of quasi-linear fracture mechanics, i.e.,

$$\sigma_{\mathrm{lin}} = \frac{K_c}{\sqrt{\pi a}} \tag{9.14}$$

The plate being thin, Tresca type, and ideally plastic was a prerequisite for the form (9.12). In this case the simple relation

$$J = \sigma_Y \delta$$

also holds. Thus, as previously pointed out, the J integral might equally well have been the basis of Fig. 4.7 under these special conditions.

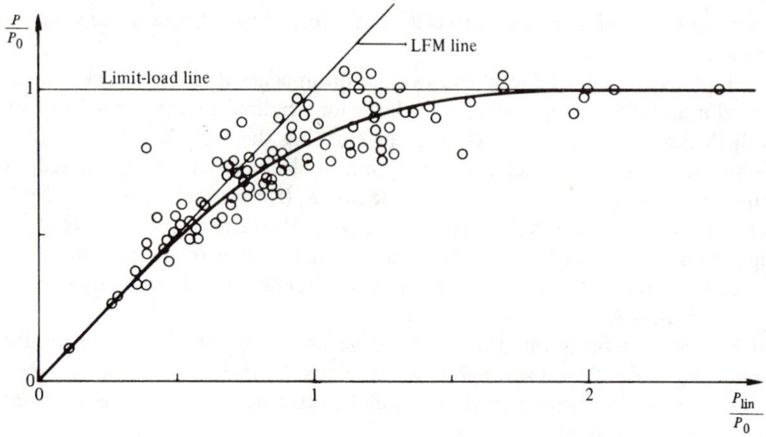

Figure 9.11 Failure by combined yielding and cracking: experimental results (circles) and a suggested master curve for average estimation. (*Based on material in Ref. 568.*)

More generally, a failure diagram looking like Figs. 4.12 and 9.11 can be envisaged. Here the ordinate is the ratio of the critical load (or load factor) P and P_0, P_0 being the carrying capacity assuming infinite fracture toughness (no further cracking) of the body. The abscissa is the ratio between the load P_{lin}, which might be deduced by formal use of linear fracture mechanics, and P_0. Figure 9.11 depicts the reasonable conclusion that for brittle materials ($P_{\text{lin}} \ll P_0$) the result $P = P_{\text{lin}}$ should apply, whereas for tougher materials $P = P_0$ sets the limit. Between these extremes the interpolated curve will result from some failure mode by combined yielding and cracking. The shape of the curve depends on the particular structure in question, but its margins may not be too widely separated in practice.

The very special property of the curve in Fig. 9.11 lies in its exact duplication of Fig. 4.7 with P, P_{lin}, and P_0 replacing σ_∞, σ_{lin}, and σ_Y, respectively. A British research group [567–569] suggested that this particular relation

$$\ln\left[\left(\cos\frac{\pi P}{2P_0}\right)^{-1}\right] = \frac{\pi^2}{8}\left(\frac{P_{\text{lin}}}{P_0}\right)^2 \tag{9.15}$$

should be considered as some average or master curve applying in an approximate way to a variety of situations. No reason can be adduced for this choice, apart from the fact that the asymptotes are correct, beyond simplicity and the historical backing of (9.13). However, a set of experimental points can be taken to support the suggestion. As included in Fig. 9.11, they represent such different cases as compact-tension or cylindrical test specimens, flywheels, and pressure vessels. We record the suggestion without further comment, except for recapitulating a simple use of this or similar curves, i.e., an evaluation of P_0 and P_{lin} for the given material and structure with its (nominal or actually identified) crack, followed by a direct reading of the failure load P in the ordinate value.

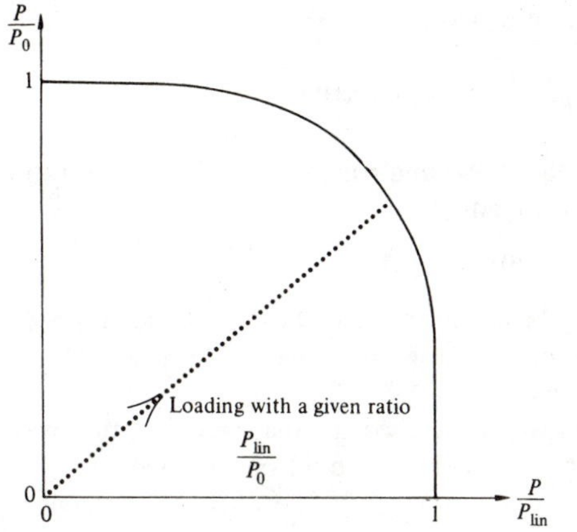

Figure 9.12 Failure–assessment diagram, a recast version of Fig. 9.11.

Another way to present the same information is indicated by the failure-assessment diagram of Fig. 9.12. With a known ratio of P_{lin} and P_0, monotonic loading must proceed along a straight trajectory as indicated by the dotted line. Intersection with the solid master curve indicates failure, whereas P values corresponding to points between that curve and the coordinate axes are loads that can be carried. The ratio of the ray lengths from the origin to the curve and from the origin to the actual loading point can be interpreted as the factor of safety.

Finally we note the suggestion in literature that P_0 be estimated by formal limit-load analysis even in cases of moderate strain hardening. This is to be achieved by interpreting σ_Y as an average of the nominal yield stress ($\sigma_{0.2}$) and ultimate stress.

Example 9.1 For a cylindrical pressure vessel that pressure p is sought under which an axial through-crack of length $2a$ becomes critical (Fig. 9.10). We consider the case $a = B = 50$ mm, $R = 500$ mm, $K_c = 120$ MPa m$^{1/2}$, $\sigma_Y = 300$ MPa, assuming almost ideally plastic behavior. Obviously, conditions (4.13) are not satisfied, and nonlinear fracture mechanics must be applied.

With the pressure p in the role of the typical load P, the limit load p_0 will be determined first. The Tresca condition, for instance, provides

$$p_0 = \frac{B}{R}\sigma_\phi = \frac{B}{R}\sigma_Y = 30 \text{ MPa}$$

assuming that the crack is too small to influence the limit load. On the other hand, linear fracture mechanics would mislead us into finding a critical load p_{lin} through

$$\sigma_\phi\sqrt{(1 + 0.16)\pi(0.05)} = K_c$$

according to the footnote in Sec. 9.4. Thus

$$p_{\text{lin}} = \frac{B}{R}\sigma_\phi = \frac{B}{R} 234 = 28.1 \text{ MPa}$$

This means that $p_{\text{lin}}/p_0 = 28.1/30 = 0.94$ and (from Fig. 9.11) the ratio $p/p_0 = 0.8$. Thus the critical pressure equals

$$p = 0.8(30) = 24 \text{ MPa}$$

by the present assessment. It may be added that the Dugdale model has been applied to pressure vessels for some time, for which reason this evaluation is in accord with established concepts.

Given p and p_0, we might as well have used the diagram the other way to obtain p_{lin}. On this basis a critical crack size could be specified.

9.6 STABLE CRACK GROWTH

We have already been concerned more than once with cracks growing stably under monotonic action. In Sec. 4.3.2 the conditions of linear fracture mechanics were accepted, and the resistance or R curve was postulated as a unique function of crack-tip translation. In Sec. 4.4 the modification to smaller thicknesses which would influence R was admitted. In Chap. 3 the energy balance during crack growth was discussed; in particular, a numerical simulation of crack growth by nodal release was suggested in Sec. 3.3.4. In Sec. 4.5 the apparent constancy of the crack-opening angles after an initial transient was briefly commented on. Finally, the last part of Sec. 4.6 was devoted to the J integral as a parameter which might be uniquely related to some limited crack-tip translation.

We have also noted uncertainties involved in a standard fracture test, e.g. case 3 of Fig. 7.4, due to possible stable crack growth. What might be aimed at, ideally speaking, is a simple datum of the material which relates to the initiation of cracking. As it is, the occurrence of some (up to 2 percent) crack growth implies that even effects pertaining to the geometry may have been incorporated. On the other hand, those ideal data for initiation might be far too conservative if applied later to the prediction of total failure. Clearly, with a general characterization of the material or a sheet of it in mind, we should try instead to establish the complete resistance function as primarily dependent on crack length traveled.

Testing procedures for the determination of R curves have been specified by the ASTM [488]. Equating this resistance to the crack driving force by LFM gives a valuable tool for the assessment of stable crack growth when yielding is highly localized at the crack tip. However, it remains a difficult and still partially unsolved problem to identify critical parameters and to describe such growth when yielding is extensive. These ductile cases of stable crack growth may also be the most challenging to investigate from a practical point of view.

As previously mentioned, stable crack growth is often experienced in thin plates. One example for a sheet of 2024-T3 alloy is seen in Fig. 9.13. After crack

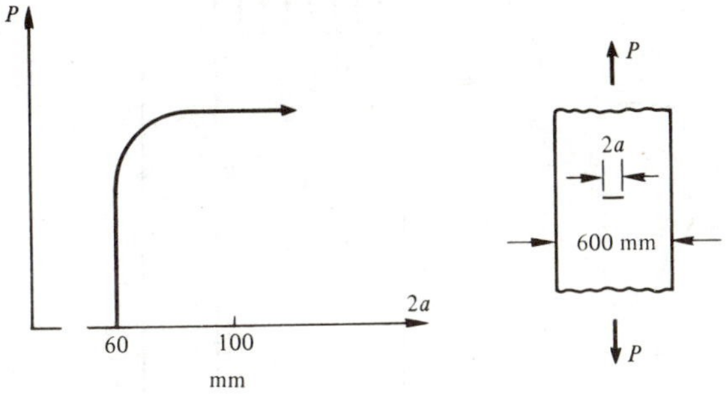

Figure 9.13 Record of stable crack growth in a metal sheet [573].

initiation both the load and the crack length could be increased by about 50 percent before stability was lost. An attempt to trace a similar development numerically is presented in Ref. 166. On the basis of the observed load as a function of crack length, $\sigma_\infty(a)$ of Fig. 9.14, the work was aimed at computing the separation work C according to the suggestion in Sec. 3.3.4, simulating crack growth by successive nodal release. The solution was the fairly constant value shown in the figure. Conversely, the result by assuming such a value of C as a parameter independent of crack length would have been a load curve quite like the one measured (only slightly steeper) in the present case. An obvious criticism of the above analysis must be that the convergence of the separation work with decreasing mesh size has not been explored. A similar set of results in Fig. 9.15 is more revealing in this respect, however. Shown are the values obtained with varying coarseness of the grid; indeed, this is a case when the crack-tip stresses were bounded and C should therefore vanish in the limit. Nevertheless, it is remarkable that in each case the derived C is at a nearly constant value, possibly after a short initial transient. If, conversely, the constants had been used as input data within the respective meshes, nearly the same correct load variation would have been predicted. This fact was duly noted in Ref. 167, with the suggestion that any relevant parameter, e.g., the J integral, could be used to trigger the initiation, after which the separation work should be kept constant at its enforced initial value to govern the later propagation.

Also indicated in Fig. 9.15 is an ordinate level by open symbols. These are values derived by modeling the process zone specifically (Secs. 3.3.4 and 4.2). Here, a well-defined, mesh-independent constant value $\tilde{C}$ has emerged, and the predictive capability seems to be at least as good as above. "At least" refers to the fact that such numerical invariance makes irregular meshes more feasible and even to the possibility that the critical value $\tilde{C}$ may be invariant to further geometries, given the material and the thickness. Of course, a coupling of $\tilde{C}$ to the particular modeling of the process zone still has to be accepted.

A further parameter, which seems to have the "basic" property of being a constant pertaining to the material, is the crack-tip opening angle CTOA (Sec.

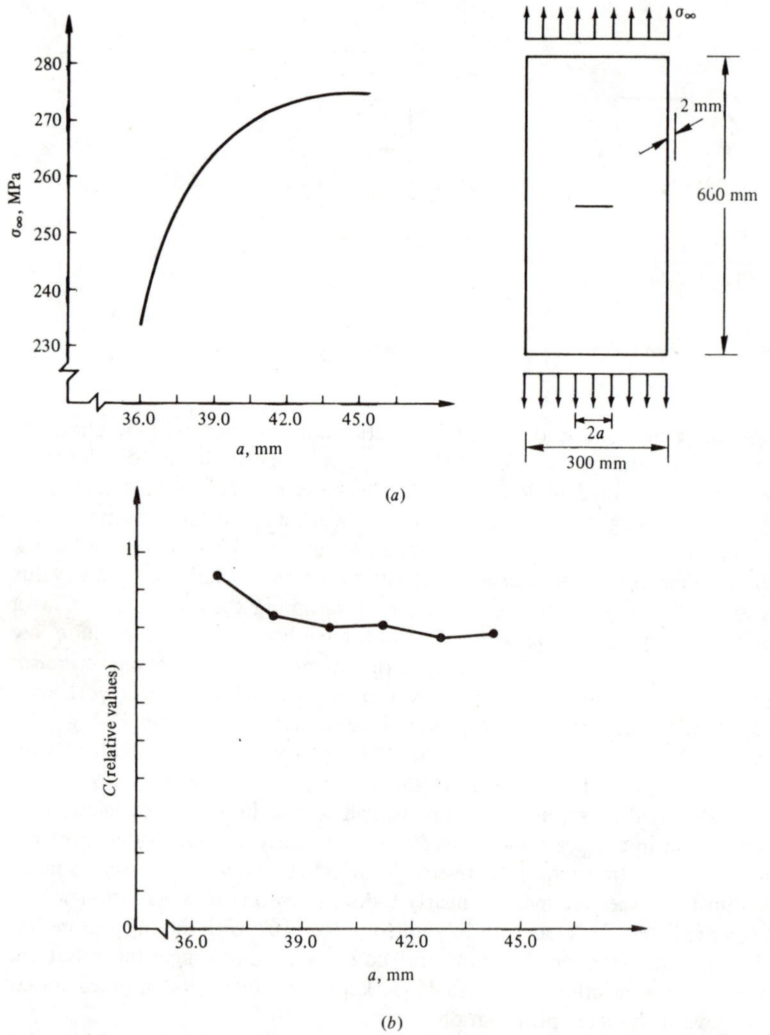

Figure 9.14 (*a*) Observed crack growth [478] and (*b*) numerical assessment of the corresponding trend in the separation work C [166].

4.5). Sufficient thickness and some developed cracking will be conditions necessary for realizing the characteristic value. The proposition conforms rather well to experimental results, as duly interpreted. Figure 9.16 reproduces a typical result. We should also recall that the J integral may be a governing parameter during the initial stages of growth, just the period when CTOA appears to have transient

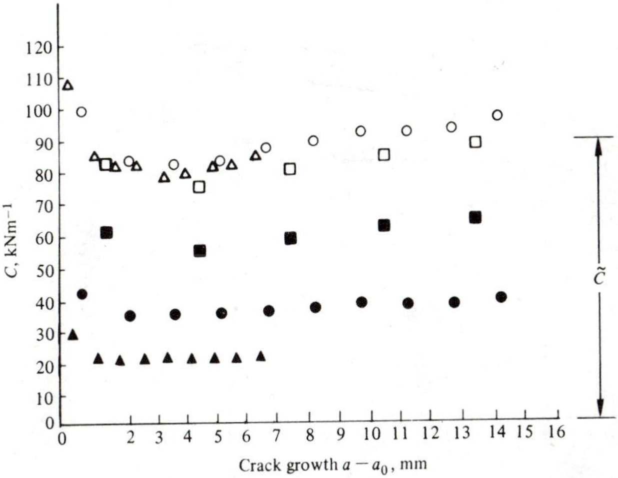

Figure 9.15 Mesh-size dependence of the separation work as derived numerically from a load record; mesh size: squares = 3.00 mm, circles = 1.50 mm, triangles = 0.75 mm [167].

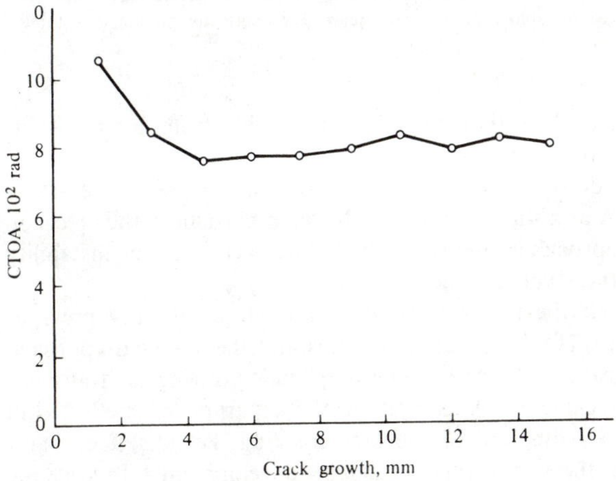

Figure 9.16 Crack tip opening angle CTOA deduced from a record of stable crack growth [220].

properties. This fact has led Kanninen et al. [220] to suggest the following two-parametric characterization of the propagation history. First, J is calculated and compared with the resistance R curve to predict the initial cracking while CTOA is simultaneously evaluated. Then when CTOA has attained a constant value and J (roughly at the same stage, by assumption or by some numerical check on path independence) has lost validity, continued cracking is predicted by

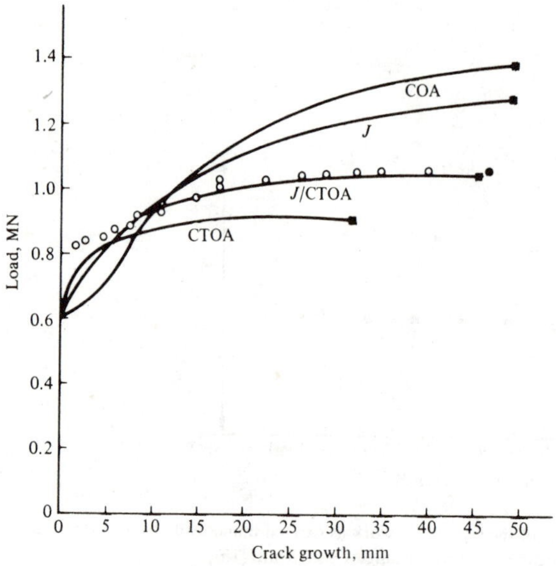

Figure 9.17 Predictions of stable growth by prescribing paths of crack opening angle COA, *J* integral, CTOA, or a combination of the last two [220]. Open circles = experimental measurements, solid circle = experimentally measured instability point, and squares = numerically predicted instability points.

keeping the CTOA numerically at that value. Figure 9.17 illustrates a remarkable success with this type of prediction, comparing it with the respective results if the crack-length dependency derived from *another* type of test specimen were attributed to *J*, COA, or CTOA as a single parameter. Note, in particular, the precision of this two-parametric approach in predicting the failure event, i.e., the instability point which terminates the stable regime.

It is also notable that the last method does not anticipate any previous knowledge of the critical CTOA. Operationally, this and the *C*-based approach above (where the *J* integral might also have been appended during the transient) are therefore in the same category. A calibration of the parameter is effected in each case, say through *J* estimation of the first cracking; keeping it constant thereafter is proposed as the way to predict a later development.† Indeed, the physical content of the parameter chosen may seem rather immaterial in this line of reasoning, with the important qualification that its constancy should reflect a stationary, autonomous state which has been attained quite locally with respect to the moving crack tip. This last condition is in accord with fundamental views on the development of stable fracture, e.g., [121], and may be the key to describing large stable cracking situations. Of course, focusing on one particular local

†These are examples of the calibration and comparison mentioned in Secs 4.2 and 3.3.4.

parameter is still of relevance: if the constant value to be attained is a known material property (possibly assigned to thickness and a computational model), this will certainly aid and direct the estimation.

The discussion of stable fracture has not been concluded yet, and the reader is advised to be critically open to further suggestions.

9.7 THICKNESS AS A DESIGN PARAMETER

Within economic bounds the design of a structural part should anticipate the risk of failure through mechanisms like buckling, plastic collapse, and fracture. Even though these modes may be jointly involved in a general situation, a single one is often dominant in a specific performance. In any case, a rational design philosophy could be to address the modes separately and to assess their interaction by some final synthesis. Aspects of such an evaluation were illustrated in Sec. 9.5, relating to the combination of plastic collapse and fracture.

For simplicity we discuss here platelike specimens under essentially unidirectional loading. The effect of varying the thickness is the topic of present concern. First, the trivial observation is made that in a compression member the reduction of thickness has an adverse effect by reducing the buckling load. In a tension member the matter is more complex, the tendency being rather to the contrary [574]. If defects are small and/or the toughness high, overall yielding may set the limit to the carrying capacity, in which case the failure stress must be fairly independent of the thickness. On the other hand, when activated, the fracture mode may be thickness-sensitive to a high degree. We have already seen (Fig. 4.5) that the effective fracture toughness K_c for through-cracks may increase considerably if the thickness is reduced below the plateau regime. This is tied to a relaxation of the local stress triaxiality and a transition toward more dissipative yield mechanisms when the crack penetrates and moves *parallel* to the plate surface. A similar increase in carrying capacity with decreasing thickness should reasonably be expected when the yielding is more extensive than implied above.

The last observation has been verified by experiment. Indeed, the risk of failure even by cracks growing *normal* to the surface (from external flaws) may increase with thickness, precrack size and remote stress being given. We illustrate this for a case significant in practice, when stress concentrations in weldments contribute (Fig. 9.18). Cracking is often observed to occur at the toe of a fillet weld or a butt weld with some overfill. The stress intensity factor at the current crack tip can be stated in the form

$$K_{\mathrm{I}} = M_k K_{\mathrm{I}}(\theta = 0) = \sigma_\infty a^{1/2} f\left(\theta, \frac{\rho}{B}, \frac{a}{B}\right) \tag{9.16}$$

where M_k is a multiplier which incorporates the stress-concentration effect of a nonzero toe (maximum) angle θ. Typically M_k behaves as in Fig. 9.19, the toe radius ρ affecting only the values for smaller crack lengths. Obviously, with a given

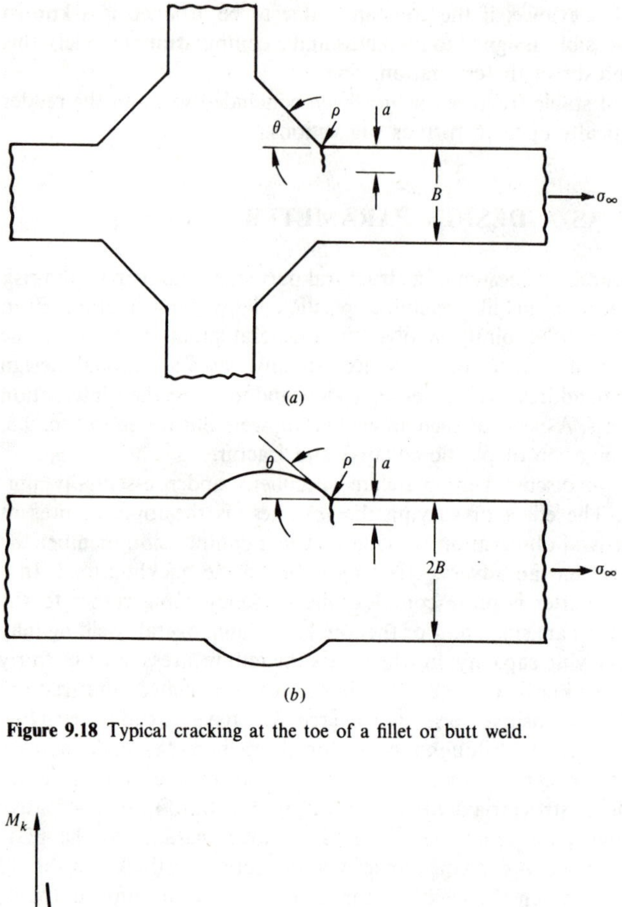

Figure 9.18 Typical cracking at the toe of a fillet or butt weld.

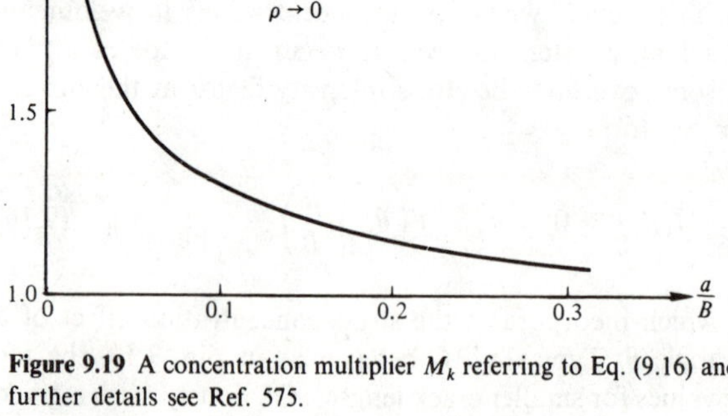

Figure 9.19 A concentration multiplier M_k referring to Eq. (9.16) and the geometry in Fig. 9.18*b*; for further details see Ref. 575.

precrack size the crack tip in the thinner plate is situated relatively farther from the entrant surface, which means that the crack in the thicker plate normally experiences the highest stress intensity and is the one most likely to grow. This is recognized as a significant effect even when fracture by fatigue is considered. Numerically and experimentally it has been demonstrated that the thicker plate may then have the shorter life. Figure 9.20 illustrates the effect, summarizing experimental results relating to the geometry in Fig. 9.18*a*. N is the number of cycles to failure in repeated tension ($\lambda = 1$) when the stress range $\Delta\sigma_\infty$ is kept at a constant value throughout and the toe parameters and the cracks are first as produced by the welding operation. The shown scatter band partly reflects the variation of the latter parameters. Besides, numerical simulations based on the accurately measured initial geometries were able to confirm the experimental trend to a satisfactory degree.

Above we reviewed some arguments (on the grounds of deterministic fracture mechanics) for keeping the thickness down. As an alternative to increasing another dimension, one could achieve this aim by splitting a compact member into parallel parts. The designer may find this observation worthy of attention for the following qualitative reasons as well: (1) the risk that a contained defect will become critical is higher in general in a compact member [546]; (2) the risk that a serious defect will not be discovered must normally increase with thickness; and (3) the level of built-up residual stresses is generally higher in compact parts.

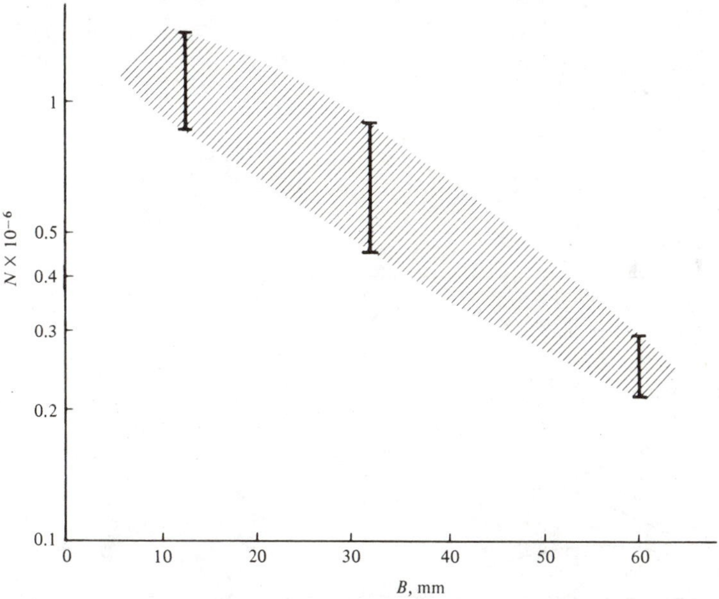

Figure 9.20 Band of experimental results for the geometry in Fig. 9.18*a* with endurance N dependent on the thickness B. Repeated tension with the stress range $\Delta\sigma_\infty = 150$ MPa in a C–Mn steel (St 42) [578].

9.8 CRACK INSTABILITY IN THERMAL OR RESIDUAL-STRESS FIELDS

Section 9.7 dealt with cases where cracks might traverse thick sections faster than thin ones. This section deals with another paradox, related to brittle components that have undergone severe thermal action. In a similar frame of reference effects of residual stresses in weldments will be considered next.

First we review the energy-balance equation in crack dynamics

$$\mathcal{G} \equiv -\frac{dT}{dA} - \frac{d\Phi}{dA} = R \tag{9.17}$$

where attention is limited to cases where no external work is performed; $W = 0$ in (5.1). If the operative resistance R and the thickness B are known constants, the kinetic energy T of a through-crack can be formally derived by integration of (9.17), to give

$$T(a) = \int_{a_0}^{a} \left(-\frac{d\Phi}{da} - BR \right) da = \Phi(a_0) - \Phi(a) - BR \times (a - a_0) \tag{9.18}$$

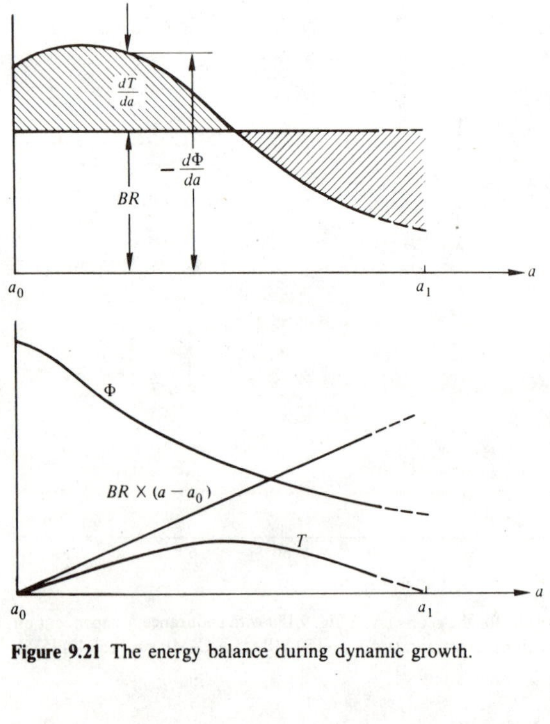

Figure 9.21 The energy balance during dynamic growth.

in terms of current crack length $a = A/B$. The relation is illustrated in Fig. 9.21, a_0 being the crack length before motion and $T(a > a_0) = 0$ defining a crack length a_1 such that the two hatched areas are equal.

This energy consideration is not sufficient for predicting a dynamic crack jump. A complete dynamic analysis would have to provide the strain energy Φ, as well as T, so that crack arrest could be placed at the event when R exceeds the left-hand side in (9.17). However, two simplifying assumptions have been suggested whereby (9.18) itself can be used to assess the size of the jump. First, the crack length at arrest a_a is taken to be defined by the fully expended kinetic energy, $T(a_a) = 0$ or $a_a = a_1$ above. Second, Φ is replaced by its static value Φ^* for the same boundary conditions, defining $-d\Phi^*/dA \equiv \mathcal{G}^*$. Figure 7.8 illustrated the non-validity of the first assumption, in general, and the figure to Prob. 5.3 clearly shows that the second may be far from correct. (Indeed, in that case $\Phi^* = \Phi_0 a_0/a$, which in no way resembles the bilinear true function Φ.)

One would expect the semistatic assessment above to apply better if impulsive rapid motions were excluded. It is notable, nevertheless, that in Probs. 5.1 to 5.3 the same strategy using the "wrong" function Φ^* will provide the *correct* jump $\Delta a = a_a - a_0$. The operative resistance R would have to be known a priori, however, and the success of the approach remains unexplored in general, even with that restriction.

9.8.1 Cracking by Thermal Action

Materials subjected to a severe thermal environment, e.g., in refractory linings and blast furnaces, Bessemer converters, and cyclicly operated kilns and ovens, operate under such conditions that generally thermal stress cracking cannot be avoided. Those considered for use, specifically ceramics, should be selected to have a reasonably high and time-independent strength, minimizing the recession of the furnace wall.

Relying on a semistatic assessment as above, Hasselman [579] conducted a study related to such design. The effect of cooling the surface was roughly simulated by considering a plane specimen normally clamped to an upper and a lower edge before cooling (Fig. 9.22). This is a kinematic constraint not unlike that exerted from the supporting walls in a real situation. A regular pattern of cracks was introduced, and a temperature drop equal to $\Delta\Theta$ was assumed. The analysis from this point proceeded effectively as follows.

After the temperature drop the average vertical stress equals

$$\sigma_\infty = \alpha E_e \Delta\Theta \tag{9.19}$$

where α is the linear expansion coefficient and E_e the effective Young's modulus of the cracked structure (e.g., Ref. 225)

$$E_e = \frac{E}{1 + 2\pi a^2/hl} \tag{9.20}$$

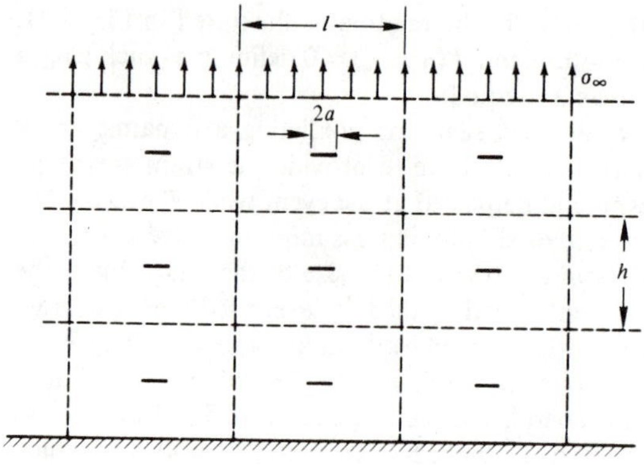

Figure 9.22 Plane rectangular model of a multicracked specimen; vertical motion is prevented at the lower and upper edge (latter not shown). After Ref. 579.

This form assumes the crack length a to be small compared with the cell dimensions h and l, with provision for making the average stress mainly uniaxial. Under the same conditions, the static stress intensity factor and crack driving force are $K_I^* = \sigma_\infty \sqrt{\pi a}$ and

$$\mathscr{G}^* = \frac{K_I^{*2}}{E} = \frac{E\alpha^2(\Delta\Theta)^2\pi a}{(1 + 2\pi a^2/hl)^2} \tag{9.21}$$

With an original crack length equal to a_0 and a known constant resistance R also governing the initial motion (no barrier), the temperature drop to start cracking is

$$\Delta\Theta = \sqrt{\frac{R}{E\alpha^2\pi a_0}}\left(1 + \frac{2\pi a_0^2}{hl}\right) \tag{9.22}$$

from the condition $\mathscr{G}^* = R$. In particular, a maximum of $\mathscr{G}^*$ or a minimum of $\Delta\Theta$ as a is varied appears at the crack size

$$a_m = \sqrt{\frac{hl}{6\pi}} \tag{9.23}$$

The result (9.22) including (9.23) is shown in Fig. 9.23*b* as the solid curve.

When a_0 is smaller than a_m, the first increase of $\mathscr{G}^*$ with crack length implies that the initial cracking is unstable. A sudden crack jump from a_0 to the arrest length a_a will ensue, the size of the latter being tentatively given by the condition

$$0 = \int_{a_0}^{a_a} \mathscr{G}^* \, da - R \times (a_a - a_0) \tag{9.24}$$

Recall that $B\mathscr{G}^* = -d\Phi^*/da$; then this equation comes from (9.18) and the subsequent assumptions. Straightforward integration using (9.21) and (9.23) yields

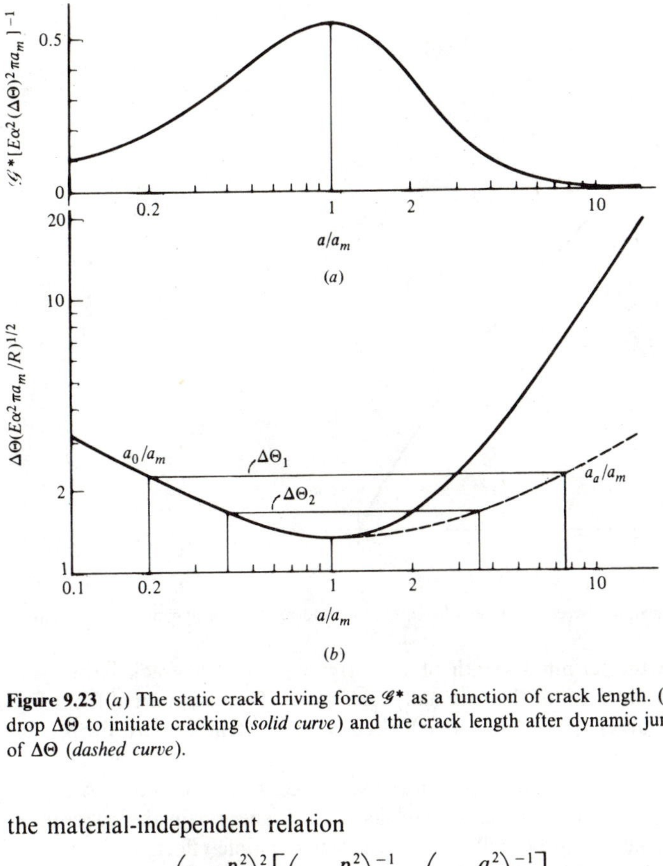

Figure 9.23 (*a*) The static crack driving force $\mathscr{G}^*$ as a function of crack length. (*b*) The temperature drop $\Delta\Theta$ to initiate cracking (*solid curve*) and the crack length after dynamic jump at various levels of $\Delta\Theta$ (*dashed curve*).

the material-independent relation

$$3\left(1+\frac{p^2}{3}\right)^2\left[\left(1+\frac{p^2}{3}\right)^{-1}-\left(1+\frac{q^2}{3}\right)^{-1}\right]=2p(q-p) \qquad (9.25)$$

where $$p=\frac{a_0}{a_m} \qquad q=\frac{a_a}{a_m}$$

The solution of (9.25) is shown in Fig. 9.24 or as the dashed line in Fig. 9.23.

It is clear that the assumption of a relatively small crack is easily violated in the a_a configuration. This fact, on top of the error involved in the dynamic estimation, implies that numerical detail should be viewed with caution. However, the qualitative discussion of the results is still relevant and interesting.

During continued operation a specimen having smaller cracks is preferable thanks to its higher mechanical strength under otherwise identical conditions. To achieve this alternative when a severe thermal environment is concerned, however, it may be unwise to aim at producing a material with *initially* small cracks. Consider two alternative crack lengths in the regime $a_0 < a_m$, and assume that the prospective temperature drop exceeds the level $\Delta\Theta_1$ necessary to make the smallest

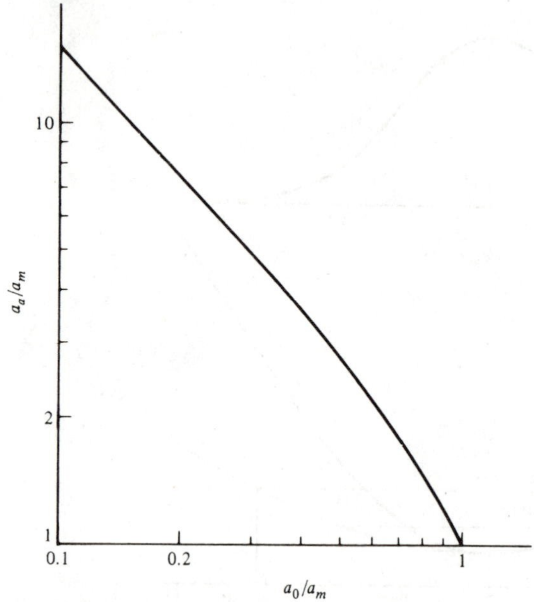

Figure 9.24 Relationship between crack lengths before, a_0, and after, a_a, a dynamic crack jump.

one grow. Then the terminal length of the originally smaller crack is likely to exceed the terminal length of the originally larger one. This is apparent from Fig. 9.23, where the larger crack has moved already at the temperature drop $\Delta\Theta_2$. Therefore, the originally weaker material has turned out to be the permanently stronger, insofar as the assumed operational sequence is representative. A simple explanation of this paradox is that, once released, the higher strain energy required for the smaller crack to grow will have the larger dynamic effect.

There is experimental and practical evidence in support of this conclusion. Typical curves of tensile strength as a function of temperature difference, through quenching into water from various temperatures, are shown in Fig. 9.25. Specimen A has small precracks relative to a_m, and a marked drop from high initial strength at the critical abscissa ($\Delta\Theta \approx 300$ K). Specimen B has somewhat larger precracks, while for C (the materials being chemically different) the precrack size has exceeded a_m. In the last case crack growth proceeds stably along the rising part of the solid curve in Fig. 9.23, with no discontinuities reflecting crack jump in the strength recording. Hasselman's recommendation [579] is that materials exhibiting small (if any) discontinuities, like C or cases when a_0 is exceeded only slightly by a_m, should generally be preferred in severe thermal environments.

9.8.2 Cracking Initiated by Residual Stress

Locations near welds are known to be sites for brittle cracking. This is not due only to material embrittlement by the welding process in such zones, where cracks

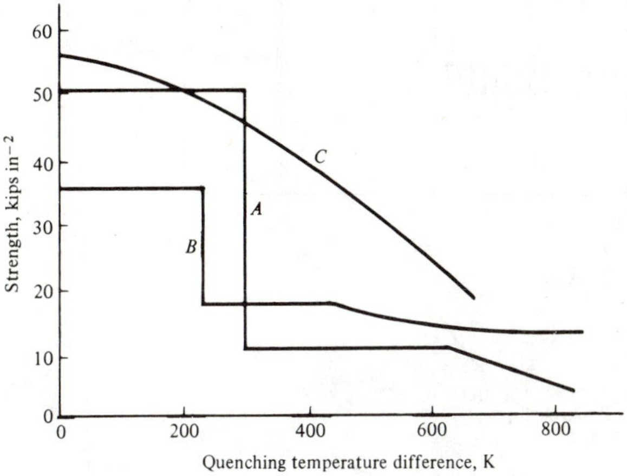

Figure 9.25 Typical dependence of tensile strength on temperature difference during previous quenching. After Ref. 579.

are often present. Residual stresses caused by previous gradients in temperature and shrinkage often contribute more critically. We shall briefly review a typical pattern of such cracking, much of the discussion proceeding along lines suggested in Ref. 539.

A dominant residual stress field close to a butt weld is often of the type shown in Fig. 9.26; i.e., a tensile stress of considerable magnitude (approaching the yield stress) parallel to the weld in its near neighborhood which changes into compressive stresses away from the weld. Assuming that a small crack has developed centrally and normal to the weld, we can try to predict its course, first by evaluating the corresponding static crack driving force $\mathscr{G}^*$ as a function of the crack length. An estimate of this kind is shown by curve I, as effected by the residual stresses alone. Similarly, curve II is the $\mathscr{G}^*$ function which might derive solely from an external loading parallel to the weld, and curve III indicates the combined action of residual stress and external load.

If residual stresses were ignored, a critical crack length a_{02} would have been predicted, given the initial resistance $\mathscr{G}_c$. This estimate might be severely in error on the nonconservative side. By the residual stresses alone the much shorter crack a_{01} is critical. In the nonloaded configuration, a crack of this length would jump dynamically to a length like a_{a1} if only slightly disturbed (assuming $R \approx \mathscr{G}_c$ and adopting the foregoing dynamic estimation). This may be a supercritical crack with respect to later external loading, however, which means that a total unstable fracture could thus originate from residual-stress damage. The simultaneous effect of residual stress and loading is even more drastic; with the $\mathscr{G}^*$ function along curve III the a_{03} crack has already become critical, and a total fracture may result if larger cracks are present.

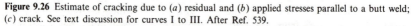

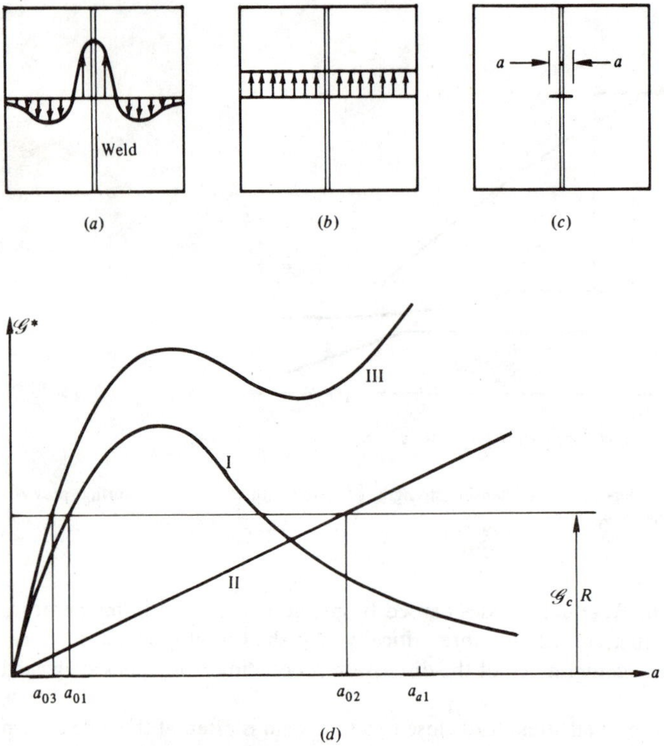

Figure 9.26 Estimate of cracking due to (*a*) residual and (*b*) applied stresses parallel to a butt weld; (*c*) crack. See text discussion for curves I to III. After Ref. 539.

Obviously, for this kind of total fracture to occur it is necessary that a sufficiently brittle material contain a critically situated and oriented crack, with residual stresses to trigger the event and the external loading to drive it through. Such unfortunate combinations are unlikely to occur frequently, but in view of the serious consequences thermal stress relieving is commonly applied to important structures. Since it is quite expensive, however, and may even be impracticable in larger parts, the alternative of recording cracks and residual stresses and estimating their consequences by fracture mechanics should not be overlooked.

PROBLEMS

9.1 A plate under the action of an in-plane bending moment has a central hole through the thickness. A crack extending toward the tension side from the hole and through the thickness is discovered, the ratio of crack length and hole radius being $a/r = 2$. Assuming brittle behavior and a known fracture toughness K_{Ic}, estimate the largest bending moment M that can be carried without further cracking. *Hint:* First replace the hole by a crack extension over its diameter, as in Fig. 9.2*b*. Keeping this equivalent crack closed will require a longitudinal stress

$$\sigma_y = \frac{12M(x+r)}{BH^3} \equiv g_2(x)$$

observing beam theory and part (*b*) of the figure. By (9.10), this determines the stress intensity factor at the tip $x = a = 2r$, which should not exceed K_{Ic}.

Answer:
$$M = \frac{BH^3}{12a\sqrt{\pi a}} K_{Ic}$$

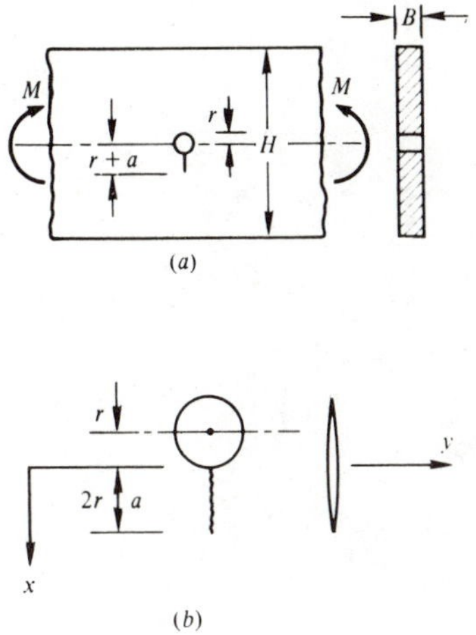

Problem 9.1

9.2 Referring to Prob. 6.3, assume that the crack just before penetration can be approximated by a rectangle of length $L = 100$ mm along the wall. With maximum pressure still at $p_0 = 10$ MPa, assess the further prospects of cracking if the toughness toward growth *along* the wall equals $K_c = 60$ MPa m$^{1/2}$.

Answer: $K_{I,\max} = 80$ MPa m$^{1/2}$ is found at $x = \pm L/2$, exceeding K_c. The crack will therefore run along the wall immediately upon penetration. In the contrary case, say $K_c = 100$ MPa m$^{1/2}$, leakage might lead to detection of the crack before it could grow to critical size by fatigue.

APPENDIX

A

ELEMENTS OF SOLID MECHANICS

The following short review covers the necessary background in the mechanics of solids. General definitions and relations applying to continua are considered first, and the relevant elastic and plastic properties follow.

A.1 THE GEOMETRY OF STRESS AND STRAIN

Through any imaginary surface dividing a body a force is transmitted between the adjacent parts. The stress vector, the local area intensity of that force, can be represented by its scalar components normal and tangential to the surface. The cartesian stress components at a point are contained in the matrix

$$[\sigma_{ij}] \equiv \begin{bmatrix} \sigma_{11} & \sigma_{12} & \sigma_{13} \\ \sigma_{21} & \sigma_{22} & \sigma_{23} \\ \sigma_{31} & \sigma_{32} & \sigma_{33} \end{bmatrix} \equiv \begin{bmatrix} \sigma_x & \tau_{xy} & \tau_{xz} \\ \tau_{yx} & \sigma_y & \tau_{yz} \\ \tau_{zx} & \tau_{zy} & \sigma_z \end{bmatrix} \qquad i, j = 1, 2, 3 \tag{A.1}$$

where the ith row represents the components on a plane normal to the x_i axis, parallel to x_1, x_2, and x_3, respectively. The form with Latin indices refers to the alternative coordinate notation which uses (x, y, z) instead of (x_1, x_2, x_3), implying identities like $\sigma_{11} \equiv \sigma_x$ and $\sigma_{12} \equiv \tau_{xy}$. The diagonal terms are normal stresses, i.e., the components normal to the respective planes, and the remaining terms are shear stresses, tangential to the surfaces. A symmetry property usually found is expressed by $\tau_{xy} = \tau_{yx}$, etc., or $\sigma_{ij} = \sigma_{ji}$. The relationship between two sets of stress components at a point differing by a rotation of the coordinate axes is that typical of a tensor of the second order. By varying this orientation to arrive at the extreme normal stresses we find that they are two of the three *principal stresses* at the point; these act through orthogonal planes which are free of shear stress, their values σ_1, σ_2, and σ_3 being the roots of the cubic equation

$$\sigma^3 - I_1\sigma^2 - I_2\sigma - I_3 = 0 \tag{A.2}$$

with coefficients in terms of the (A.1) components

$$
\begin{aligned}
I_1 &= \sigma_{ii} \equiv \sigma_{11} + \sigma_{22} + \sigma_{33} \\
I_2 &= \tfrac{1}{2}(\sigma_{ij}\sigma_{ij} - I_1^2) \equiv \tfrac{1}{2}(\sigma_{11}^2 + \sigma_{22}^2 + \sigma_{33}^2 + 2\sigma_{12}^2 + 2\sigma_{23}^2 + 2\sigma_{31}^2 - I_1^2) \qquad \text{(A.3)} \\
I_3 &= \det[\sigma_{ij}]
\end{aligned}
$$

The summation rule adopted here is to the effect that the repetition of an index denotes summation through the whole interval of that index. Because the principal stresses are physical quantities at the point considered, the coefficients I_1 to I_3 must be independent of the orientation of the axes to which the stress components have been referred. For this reason I_1 to I_3 are termed the *stress invariants.*

Figure A.1 illustrates the plane which is at an angle of 45° with the directions of the extreme normal stresses [$\sigma_{max} \equiv \max(\sigma_1, \sigma_2, \sigma_3)$, $\sigma_{min} \equiv \min(\sigma_1, \sigma_2, \sigma_3)$] and is parallel to the direction of the intermediate principal stress σ_{int}. This plane can be shown to carry the maximum value τ_{max} of all shear stresses through the point while also having a normal stress equal to $\tilde{\sigma}$. The magnitudes are

$$
\tau_{max} = \tfrac{1}{2}(\sigma_{max} - \sigma_{min}) \qquad \tilde{\sigma} = \tfrac{1}{2}(\sigma_{max} + \sigma_{min}) \qquad \text{(A.4)}
$$

as can be inferred from *Mohr's stress circles.* Each circle is the locus of corresponding normal- and shear-stress values in planes parallel to one principal direction, in such a way that a given angle of rotation in the Mohr plane (Fig. A.2) maps one-half of that angle in the physical plane (Fig. A.1). Stress coordinates in the region bounded by the largest and the two smallest circles represent planes which are not parallel to any principal direction.

By subtracting their mean value $I_1/3$ from the normal stresses we define the stress deviation

$$
s_{ij} = \sigma_{ij} - \frac{\delta_{ij} I_1}{3} \qquad \text{where } \delta_{ij} = \begin{cases} 1 & \text{for } i = j \\ 0 & \text{for } i \neq j \end{cases} \qquad \text{(A.5)}
$$

To this deviatoric state of stress correspond the invariants J_1 to J_3, just like I_1 to I_3 above. Of special concern is the second of that set

$$
J_2 = \tfrac{1}{2}(s_{ij}s_{ij} - J_1^2) = \tfrac{1}{2}s_{ij}s_{ij} \qquad \text{(A.6)}
$$

since $J_1 = s_{ii} = \sigma_{ii} - I_1 = 0$.

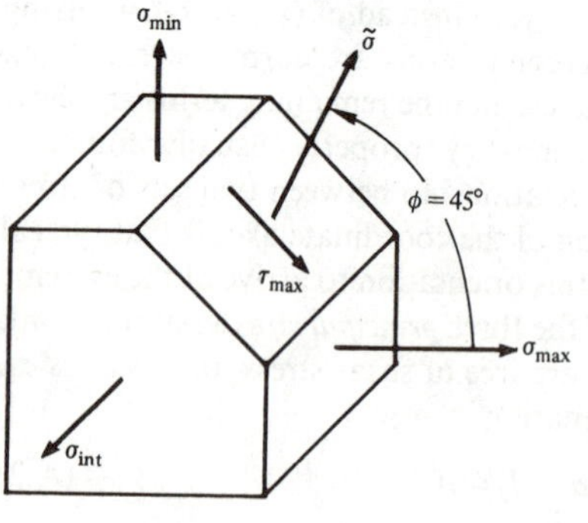

Figure A.1

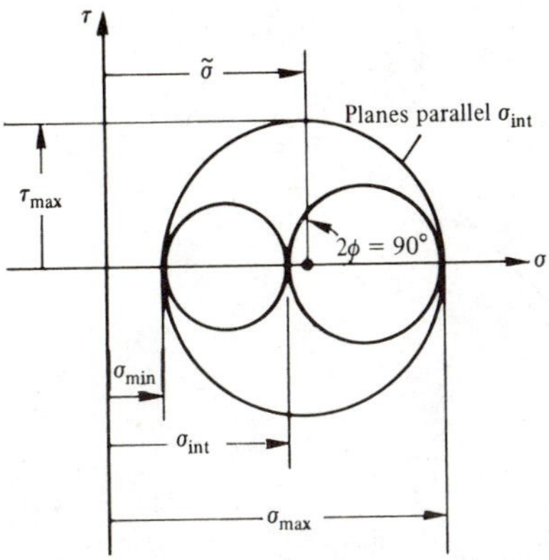

Figure A.2

A volume element can be visualized as being bounded by infinitesimally adjacent coordinate surfaces, such that the sides are subject to the stresses σ_{ij} in a cartesian frame. The gradients of σ_{ij} then lead to a resulting force on the element, with components equal to $(\partial\sigma_{ji}/\partial x_j)\,dV$, where dV is the volume ($i = 1, 2, 3$). If the local density is ρ, if the components of body force (in volume) are b_i, and if those of acceleration are $\ddot{u}_i$, the *equations of motion* of the element become

$$\left(\frac{\partial \sigma_{ji}}{\partial x_j} + b_i - \rho\ddot{u}_i\right) dV = 0$$

or

$$\frac{\partial \sigma_{ji}}{\partial x_j} + b_i = \rho\ddot{u}_i \qquad i = 1, 2, 3 \tag{A.7}$$

These reduce to the *equilibrium equations* if the right-hand sides are zero. Equation (A.7) has been referred to cartesian coordinates but can be formally transformed to other systems. We find the equilibrium equations in a cylindrical reference in the following by considering more directly a volume element naturally defined in that system. Even body forces will be considered to be negligible below.

A *plane state* of stress or strain, e.g., with x_1, x_2 as independent variables, is characterized by $\partial/\partial x_3 = 0$ and $\sigma_{13} = \sigma_{23} = 0$. Then when $b_i = \ddot{u}_i = 0$, Eq. (A.7) gives the nontrivial equilibrium equations

$$\frac{\partial \sigma_{11}}{\partial x_1} + \frac{\partial \sigma_{21}}{\partial x_2} = 0 \qquad \frac{\partial \sigma_{12}}{\partial x_1} + \frac{\partial \sigma_{22}}{\partial x_2} = 0 \tag{A.8}$$

In polar coordinates the normal stresses σ_r and σ_θ and the shear stress $\tau_{r\theta} = \tau_{\theta r}$ act in the r, θ plane (Fig. A.3). Using the operators

$$L_r = 1 + dr\frac{\partial}{\partial r} \qquad \text{and} \qquad L_\theta = 1 + d\theta\frac{\partial}{\partial \theta} \tag{A.9}$$

we can then prescribe radial and circumferential equilibrium, respectively,

$$L_r(\sigma_r r)\,d\theta - \sigma_r r\,d\theta + L_\theta(\tau_{\theta r})dr - \tau_{\theta r}\,dr - \sigma_\theta\,dr\,d\theta = 0$$

$$L_\theta(\sigma_\theta)\,dr - \sigma_\theta\,dr + L_r(\tau_{r\theta} r)\,dr - \tau_{r\theta} r\,d\theta + \tau_{\theta r}\,dr\,d\theta = 0$$

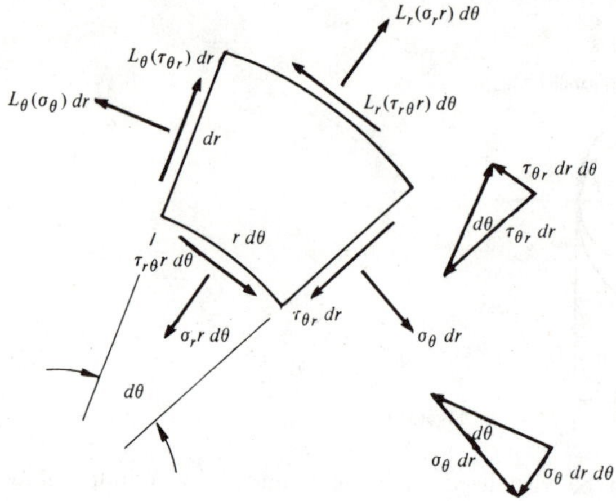

Figure A.3

giving

$$\frac{\partial}{\partial r}(\sigma_r r) + \frac{\partial \tau_{\theta r}}{\partial \theta} - \sigma_\theta = 0 \qquad \text{(A.10}a\text{)}$$

$$\frac{\partial \sigma_\theta}{\partial \theta} + \frac{\partial}{\partial r}(\tau_{r\theta} r) + \tau_{\theta r} = 0 \qquad \text{(A.10}b\text{)}$$

An *antiplane* state is characterized by $\sigma_{11} = \sigma_{22} = \sigma_{33} = \sigma_{12} = 0$. Only $\sigma_{13} = \sigma_{31}$ and $\sigma_{23} = \sigma_{32}$ remain; they will be independent of x_3 according to the first two of Eqs. (A.7) under equilibrium and are governed by the third, $(i = 3)$ through

$$\frac{\partial \sigma_{13}}{\partial x_1} + \frac{\partial \sigma_{23}}{\partial x_2} = 0 \qquad \text{(A.11)}$$

In polar coordinates the shear stresses $\tau_{rz} = \tau_{zr}$ and $\tau_{\theta z} = \tau_{z\theta}$ (Fig. A.4) are nonzero. Equilibrium is expressed by

$$L_r(\tau_{rz} r)\,d\theta - \tau_{rz} r\,d\theta + L_\theta(\tau_{\theta z})\,dr - \tau_{\theta z}\,dr = 0$$

giving

$$\frac{\partial}{\partial r}(\tau_{rz} r) + \frac{\partial \tau_{\theta z}}{\partial \theta} = 0 \qquad \text{(A.12)}$$

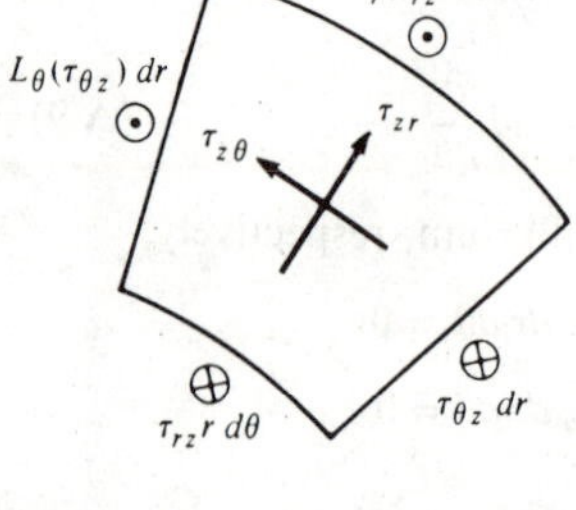

Figure A.4

Equations (A.7) to (A.12) govern the variation of stresses inside the body. Boundary values are at the surface, where the stresses should correspond to given or derived tractions p_i ($i = 1, 2, 3$). Let ABC in Fig. A.5 be an element of the surface, bounding (together with the faces $x_i = \text{const}$) a small volume element of the body; $\mathbf{n} = \{n_i\}$ is the unit vector on ABC normally out of the body. Equilibrium of the volume element under the action of p_i and the internal stresses can then be expressed by the static boundary condition

$$p_i = n_j \sigma_{ji} \qquad i = 1, 2, 3 \tag{A.13}$$

Aiming at normal applications (small elastic and large plastic deformation), we imply that the stresses above are "true" or Cauchy-defined values; i.e., the intensities within the deformed body as referred to axes fixed in space. With small deformation and rotation, as considered in a linear or first-order theory, these values differ insignificantly from nominal stresses having their reference in the undeformed configuration.

Further, let $\dot{u}_i(x_j)$ represent the field of particle velocities within the body at a chosen instant t. Through the local gradients of this field the current deformation close to x_i and t will be defined. In a cartesian frame this is expressed by the rates of deformation

$$\dot{\varepsilon}_{ij} = \frac{1}{2}\left(\frac{\partial \dot{u}_i}{\partial x_j} + \frac{\partial \dot{u}_j}{\partial x_i}\right) \qquad i, j = 1, 2, 3 \tag{A.14}$$

which also constitute a second-order tensor, just as σ_{ij} did. The matrix form will be

$$[\dot{\varepsilon}_{ij}] = \begin{bmatrix} \dot{\varepsilon}_{11} & \dot{\varepsilon}_{12} & \dot{\varepsilon}_{13} \\ \dot{\varepsilon}_{21} & \dot{\varepsilon}_{22} & \dot{\varepsilon}_{23} \\ \dot{\varepsilon}_{31} & \dot{\varepsilon}_{32} & \dot{\varepsilon}_{33} \end{bmatrix} \equiv \begin{bmatrix} \dot{\varepsilon}_x & \frac{1}{2}\dot{\gamma}_{xy} & \frac{1}{2}\dot{\gamma}_{xz} \\ \frac{1}{2}\dot{\gamma}_{yx} & \dot{\varepsilon}_y & \frac{1}{2}\dot{\gamma}_{yz} \\ \frac{1}{2}\dot{\gamma}_{zx} & \frac{1}{2}\dot{\gamma}_{zy} & \dot{\varepsilon}_z \end{bmatrix} \tag{A.15}$$

in alternative notation. The diagonal terms $\dot{\varepsilon}_{11} \equiv \dot{\varepsilon}_x$, etc., represent stretching of line elements along the axes; the remaining terms, the rates of shear $2\dot{\varepsilon}_{12} \equiv \dot{\gamma}_{xy}$, etc.,

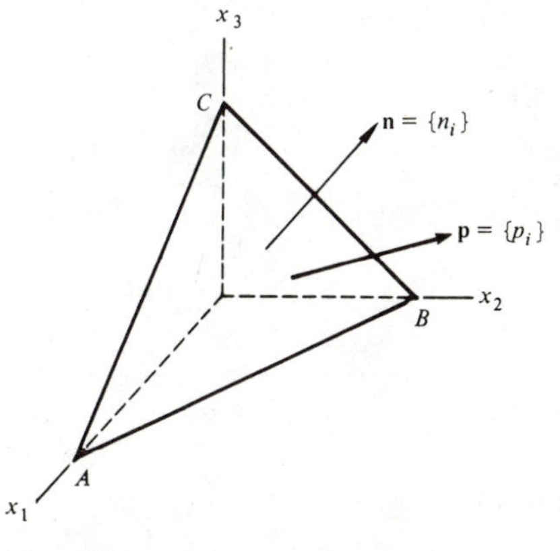

Figure A.5

denote relative angular velocities of lines along two axes (Fig. A.6). By the tensorial property, Mohr's circles of deformation can also be defined (Fig. A.7). They indicate the maximum rate of shear (Fig. A.8)

$$\dot{\gamma}_{max} = \dot{\varepsilon}_{max} - \dot{\varepsilon}_{min} \tag{A.16}$$

A material element having faces normal to x_1, x_2, x_3 at time t will in the next interval dt change its volume from $dx_1\, dx_2\, dx_3$ into

$$dx_1\,(1 + d\varepsilon_{11})\, dx_2\,(1 + d\varepsilon_{22})\, dx_3\,(1 + d\varepsilon_{33}) = dx_1\, dx_2\, dx_3\,(1 + d\varepsilon_{ii})$$

when differentials of a higher order are ignored. Thus, the specific volume increment during dt is given by the invariant

$$d\varepsilon_{ii} \equiv \dot{\varepsilon}_{ii}\, dt \equiv (\dot{\varepsilon}_1 + \dot{\varepsilon}_2 + \dot{\varepsilon}_3)\, dt \tag{A.17}$$

the last form referring to axes along the principal directions of $[\dot{\varepsilon}_{ij}]$.

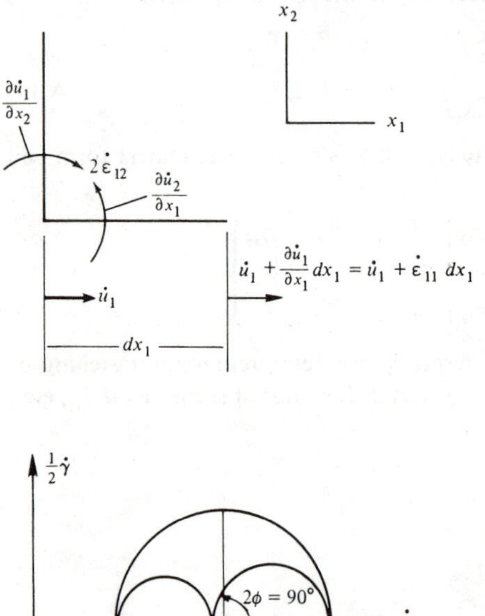

Figure A.6

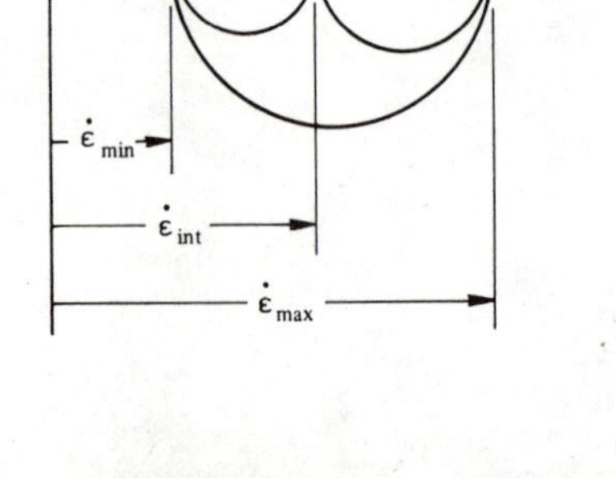

Figure A.7

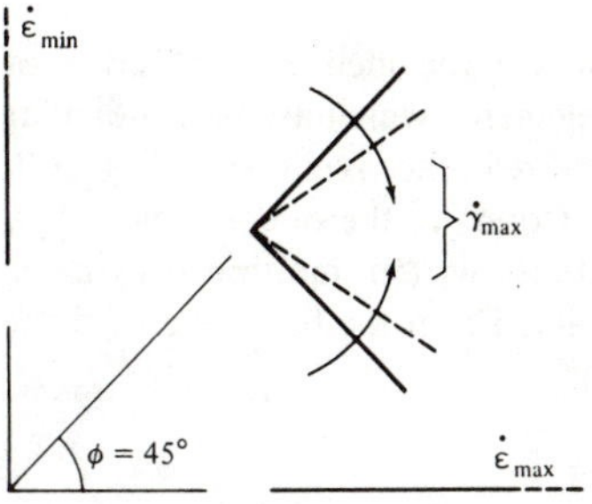

Figure A.8

When an individual element is followed, (A.14) can be formally integrated into the "total" strains

$$\varepsilon_{ij} = \int_0^{\varepsilon} d\varepsilon_{ij} \equiv \int_0^{t} \dot{\varepsilon}_{ij}\, dt \tag{A.18}$$

Then if the displacements are small within the time interval considered, we can replace the current coordinate $x_i(t)$ of each particle by its initial value $\bar{x}_i \equiv x_i(0)$ and thus express the total deformation [observing (A.14)]

$$\varepsilon_{ij} = \frac{1}{2}\left(\frac{\partial}{\partial \bar{x}_j}\int_0^t \dot{u}_i\, dt + \frac{\partial}{\partial \bar{x}_i}\int_0^t \dot{u}_j\, dt\right) = \frac{1}{2}\left(\frac{\partial u_i}{\partial \bar{x}_j} + \frac{\partial u_j}{\partial \bar{x}_i}\right) \tag{A.19}$$

where u_i is the displacement of the particle. In the theory of elasticity the total strains are essential tools for describing such small deformations. They are also of some interest even for large deformation, through (A.18), if the ratios of the components $\dot{\varepsilon}_{ij}$ (and hence ε_{ij}) are kept constant throughout the deformation. Only a scalar measure will then be varied, and the situation is described as *proportional strain.* Under the above special circumstances the transformation of the total strains ε_{ij} will be the same as for the rates of deformation.

Through (A.19) six small strain components have been expressed in terms of three displacements. This means that the strains must be interrelated in certain ways. For the components in the x_1, x_2 plane (Fig. A.6) we have in particular, with $x_i \equiv \bar{x}_i$,

$$\varepsilon_{11} = \frac{\partial u_1}{\partial x_1} \qquad \varepsilon_{22} = \frac{\partial u_2}{\partial x_2} \qquad 2\varepsilon_{12} = \frac{\partial u_1}{\partial x_2} + \frac{\partial u_2}{\partial x_1} \tag{A.20}$$

which shows that

$$\frac{\partial^2 \varepsilon_{22}}{\partial x_1^2} - 2\frac{\partial^2 \varepsilon_{12}}{\partial x_1\, \partial x_2} + \frac{\partial^2 \varepsilon_{11}}{\partial x_2^2} = 0 \tag{A.21}$$

must be satisfied. This is one of the *compatibility equations*, i.e., the one which will be of further interest. Of course, similar relations apply between the rates of deformation, through (A.14).

As implied above, the current deformation can be identified in terms of stretching and relative rotation of material line elements which are perpendicular at the moment. Such a pair of lines in a cylindrical reference is shown in Fig. A.9, where it is followed from the position at time t (solid) to the one at time $t + dt$ (dashed). The displacements in the interval are those written multiplied by dt, $\dot{u}_r$ being the radial and $\dot{u}_\theta$ the circumferential velocity. From the figure we read the stretchings and rate of shear in the r, θ plane as

$$\dot{\varepsilon}_r = \frac{\partial \dot{u}_r/\partial r\, dr}{dr} = \frac{\partial \dot{u}_r}{\partial r}$$

$$\dot{\varepsilon}_\theta = \frac{(r + \dot{u}_r)\, d\theta - r\, d\theta + \partial \dot{u}_\theta/\partial \theta\, d\theta}{r\, d\theta} = \frac{\dot{u}_r}{r} + \frac{1}{r}\frac{\partial \dot{u}_\theta}{\partial \theta} \tag{A.22a}$$

$$\dot{\gamma}_{r\theta} = \frac{\partial \dot{u}_r/\partial \theta\, d\theta}{r\, d\theta} + \frac{\partial \dot{u}_\theta/\partial r\, dr}{dr} - \frac{\dot{u}_\theta}{r} = \frac{1}{r}\frac{\partial \dot{u}_r}{\partial \theta} + \frac{\partial \dot{u}_\theta}{\partial r} - \frac{\dot{u}_\theta}{r}$$

Together with the rates of deformation in the other surfaces

$$\dot{\gamma}_{rz} = \frac{\partial \dot{u}_r}{\partial z} + \frac{\partial \dot{u}_z}{\partial r} \qquad \dot{\gamma}_{\theta z} = \frac{\partial \dot{u}_\theta}{\partial z} + \frac{1}{r}\frac{\partial \dot{u}_z}{\partial \theta} \qquad \dot{\varepsilon}_z = \frac{\partial \dot{u}_z}{\partial z} \tag{A.22b}$$

this provides the set of components corresponding to those in (A.14).

The same formal expressions apply to total small strains and displacements when the dots are removed and r, θ, z are considered as the coordinates of the undeformed body. Again we have six deformation components expressed by three (rates of) displacement, a corresponding compatibility equation in the r, θ plane

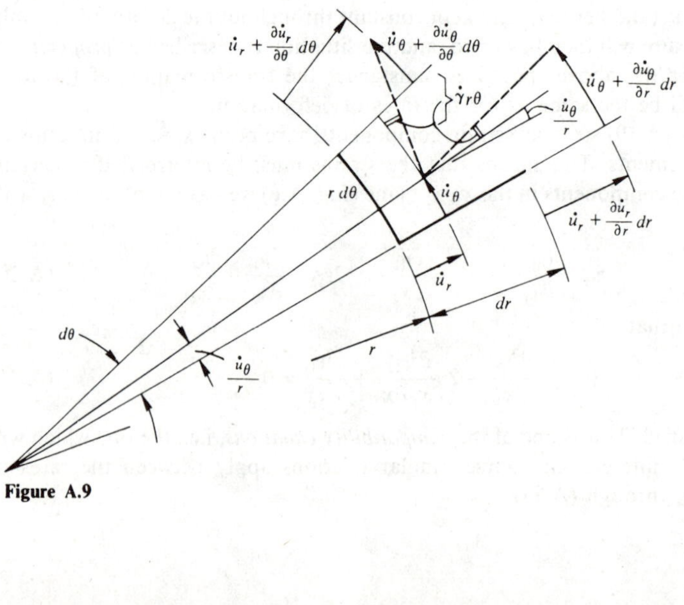

Figure A.9

being

$$\frac{\partial^2 \varepsilon_\theta}{\partial r^2} + \frac{2}{r}\frac{\partial \varepsilon_\theta}{\partial r} - \frac{1}{r}\frac{\partial^2 \gamma_{r\theta}}{\partial r\,\partial \theta} - \frac{1}{r^2}\frac{\partial \gamma_{r\theta}}{\partial \theta} + \frac{1}{r^2}\frac{\partial^2 \varepsilon_r}{\partial \theta^2} - \frac{1}{r}\frac{\partial \varepsilon_r}{\partial r} = 0 \tag{A.23}$$

as can be verified by direct substitution.

On any material element the stresses perform work when the body is deformed. The power per unit of volume will be

$$\dot{w} = \sigma_{ij}\dot{\varepsilon}_{ij} \tag{A.24}$$

which can be meaningfully integrated through a small deformation into

$$w = \int_0^t \dot{w}\,dt = \int_0^t \sigma_{ij}\dot{\varepsilon}_{ij}\,dt \equiv \int_0^\varepsilon \sigma_{ij}\,d\varepsilon_{ij} \tag{A.25}$$

the total work of deformation per unit of material volume. By integrating (A.24) over the whole volume V of the body one can derive the identity

$$\int_V \sigma_{ij}\dot{\varepsilon}_{ij}\,dV = \int_S p_i\dot{u}_i\,dS + \int_V (b_i - \rho\ddot{u}_i)\dot{u}_i\,dV \tag{A.26}$$

This involves formal transformations to the surface S and the use of the equations of motion (A.7), the boundary condition (A.13), and the kinematic constraint (A.14). Equation (A.26) is the *principle of virtual work*, which can be interpreted as expressing the equality of internal and external work.

A.2 ELASTIC DEFORMATION

In a (hyper)elastic body the work of deformation is a single-valued function in the strain space; i.e.,†

$$w = \int_0^\varepsilon \sigma_{ij}\,d\varepsilon_{ij} = \phi(\varepsilon_{ij}) \tag{A.27}$$

where ϕ denotes the *strain energy density*. This implies that $\sigma_{ij}\,d\varepsilon_{ij} = d\phi$ for arbitrary variations of ε_{ij}, so that

$$\sigma_{ij} = \frac{\partial \phi}{\partial \varepsilon_{ij}} \tag{A.28}$$

†Such uniqueness depends on conditions being either adiabatic or isothermal; then ϕ equals the internal energy or the free energy per unit of volume, respectively.

With *linear* elasticity and isotropy (directional indifference) the energy density is given by

$$\phi = G\left[\varepsilon_{ij}\varepsilon_{ij} + \frac{\nu}{1-2\nu}(\varepsilon_{kk})^2\right] \tag{A.29}$$

in terms of two material constants, namely the shear modulus G and Poisson's ratio ν $(0 \le \nu \le 0.5)$, which define Young's modulus E through

$$E = 2(1+\nu)G \tag{A.30}$$

By (A.28) and (A.29) the stress-strain relations

$$\sigma_{ij} = 2G\left(\varepsilon_{ij} + \frac{\nu}{1-2\nu}\delta_{ij}\varepsilon_{kk}\right) \quad \text{or} \quad \varepsilon_{ij} = \frac{1}{2G}\left(\sigma_{ij} - \frac{\nu}{1+\nu}\delta_{ij}\sigma_{kk}\right) \tag{A.31}$$

follow, where the second set is obtained through inversion. Hooke's law (A.31) can be specialized in x_1, x_2 components to states of *plane stress* $\sigma_{33} = 0$ or *plane strain* $\varepsilon_{33} = 0$, respectively,

$$E\varepsilon_{11} = \sigma_{11} - \nu\sigma_{22} \qquad E\varepsilon_{22} = \sigma_{22} - \nu\sigma_{11} \qquad 2G\varepsilon_{12} \equiv G\gamma_{xy} = \tau_{xy} \equiv \sigma_{12} \tag{A.32}$$

$$2G\varepsilon_{11} = (1-\nu)\sigma_{11} - \nu\sigma_{22} \qquad 2G\varepsilon_{22} = (1-\nu)\sigma_{22} - \nu\sigma_{11} \qquad 2G\varepsilon_{12} = \sigma_{12} \tag{A.33}$$

In such plane states the equations of equilibrium (A.8) apply in cartesian coordinates (body forces taken to zero). They are seen to be satisfied when the stress components are expressed by Airy's stress function χ through

$$\sigma_{11} = \frac{\partial^2\chi}{\partial x_2^2} \qquad \sigma_{22} = \frac{\partial^2\chi}{\partial x_1^2} \qquad \sigma_{12} = -\frac{\partial^2\chi}{\partial x_1\,\partial x_2} \tag{A.34}$$

When (A.34) is substituted in (A.32) or (A.33) and this is further substituted in the compatibility equation (A.21), the result is a bipotential equation in χ

$$\nabla^2(\nabla^2\chi) = 0 \tag{A.35}$$

where

$$\nabla^2 \equiv \frac{\partial^2}{\partial x_1^2} + \frac{\partial^2}{\partial x_2^2} \tag{A.36}$$

This equation therefore implies equilibrium and the satisfaction of kinematic constraints and Hooke's law. A total solution can thus be characterized as satisfying (A.35) and the prescribed boundary conditions.

In polar coordinates there corresponds to (A.34)

$$\sigma_r = \frac{1}{r}\frac{\partial\chi}{\partial r} + \frac{1}{r^2}\frac{\partial^2\chi}{\partial\theta^2} \qquad \sigma_\theta = \frac{\partial^2\chi}{\partial r^2} \qquad \tau_{r\theta} = -\frac{\partial}{\partial r}\left(\frac{1}{r}\frac{\partial\chi}{\partial\theta}\right) \tag{A.37}$$

whereby (A.10) will be identically satisfied. Reinterpreting the indices 11, 22, and xy of (A.32) and (A.33) as r, θ, and $r\theta$, respectively, we can further invoke the compatibility equation (A.23) to obtain (A.35) with the polar laplacian

$$\nabla^2 \equiv \frac{\partial^2}{\partial r^2} + \frac{1}{r}\frac{\partial}{\partial r} + \frac{1}{r^2}\frac{\partial^2}{\partial\theta^2} \tag{A.38}$$

This last result also follows from a direct transformation of (A.36).

A complex-variable technique for solving the bipotential equation is discussed in Appendix B.

A.3 PLASTIC AND ELASTOPLASTIC DEFORMATION

Elastic and plastic deformations are both time-independent, in the sense that the stresses do not depend explicitly on the rate of deformation. Beyond this the deviations are significant. During plastic yielding the deformation may be very large; no natural geometric reference exists here (compare the stress-free configuration of an elastic body), and isochoric (or incompressible) flow is often realized to a good approximation. The last set of properties may point to similarities between plastic flow and the flow of a viscous fluid instead. A final characteristic of the plastic deformation is that it proceeds only if some measure of the intensity of the state of stress satisfies a critical condition, i.e., the *yield condition.*

In this section we outline a plasticity theory conforming with these properties, which incorporates the principle of maximum plastic resistance and the assumption of isotropic hardening.

The yield condition will be represented in the form

$$f(\sigma_{ij}) = \sigma_e - \sigma_Y = 0 \tag{A.39}$$

where σ_Y is the yield stress in uniaxial tension and the function f is written in terms of the *effective stress* σ_e, our measure of the stress intensity. This measure should be invariant with respect to the orientation of axes and so normalized that $\sigma_e = \sigma$ if a uniaxial tension σ is prescribed. Equation (A.39) must be satisfied for yielding to occur while all other possible states $f < 0$ are elastic. The power of deformation is assumed to be nonnegative

$$\dot{w} = \sigma_{ij}\dot{\varepsilon}_{ij} \equiv \boldsymbol{\sigma} \cdot \dot{\boldsymbol{\varepsilon}} \geq 0 \tag{A.40}$$

which defines the material to be *dissipative* in the sense that net work has to be exerted *on* the body during any plastic deformation. Alternatively, the dissipation above can be expressed by the nine-component vectors $\boldsymbol{\sigma} \equiv \{\sigma_{ij}\}$ and $\dot{\boldsymbol{\varepsilon}} \equiv \{\dot{\varepsilon}_{ij}\}$. The yield condition can be visualized as in Fig. A.10, in the space where σ_{ij} and ε_{ij} are

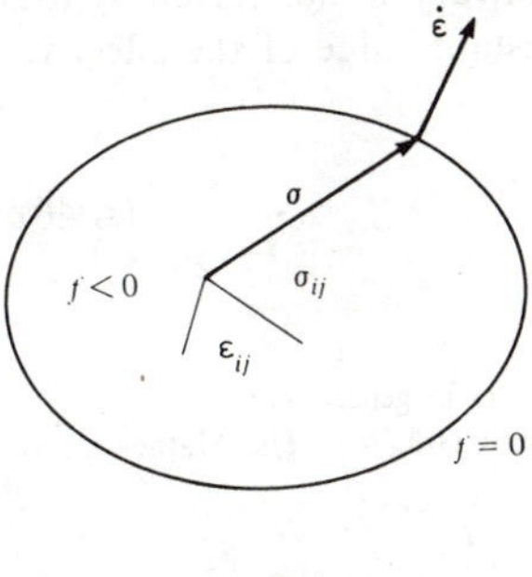

Figure A.10

superposed axes, such that the satisfaction of (A.40) corresponds to contact between the vector $\boldsymbol{\sigma}$ and the yield surface $f = 0$.

The *flow rule* asserts the manner and (possibly) the extent of yielding. With reference to (A.40) it can be formulated in terms of the principle of *maximum plastic resistance*. It is postulated, then, with a physical basis in crystalline aggregates, that to a given rate of deformation $\dot{\varepsilon}_{ij}$ the state of stress corresponds which makes the dissipation $\dot{w}$ as *large as possible*.† In the frame of Fig. A.10 this implies that the current stress point will be such that $\dot{\boldsymbol{\varepsilon}}$ is outward normal to the yield surface (assumed convex). This is reflected by the flow rule, which in vector form is

$$\dot{\boldsymbol{\varepsilon}} = \dot{\varepsilon}_e \operatorname{grad} f = \dot{\varepsilon}_e \operatorname{grad} \sigma_e \tag{A.41}$$

or, in cartesian components,

$$\dot{\varepsilon}_{ij} = \dot{\varepsilon}_e \frac{\partial \sigma_e}{\partial \sigma_{ij}} \tag{A.42}$$

where $\dot{\varepsilon}_e$ is a positive scalar. To attach a physical meaning to this last quantity we insert (A.42) in (A.40), to obtain

$$\dot{w} = \sigma_{ij}\dot{\varepsilon}_e \frac{\partial \sigma_e}{\partial \sigma_{ij}} = \sigma_e \dot{\varepsilon}_e = \sigma_Y \dot{\varepsilon}_e \tag{A.43}$$

since σ_e must be a homogeneous function of the first order in σ_{ij} (Euler's theorem). Thus, since the scalar $\dot{\varepsilon}_e$ is associated with the effective stress in the dissipation function, it can naturally be viewed as an *effective stretching*. By (A.41) and (A.42) it can be expressed as

$$\dot{\varepsilon}_e = \frac{|\dot{\boldsymbol{\varepsilon}}|}{|\operatorname{grad} \sigma_e|} = \left[\frac{\dot{\varepsilon}_{ij}\dot{\varepsilon}_{ij}}{(\partial\sigma_e/\partial\sigma_{kl})(\partial\sigma_e/\partial\sigma_{kl})}\right]^{1/2} \tag{A.44}$$

which must specialize into the stretching $\dot{\varepsilon}$ along the direction of a uniaxial tension, according to (A.43). For material volume to be conserved, (A.17) and (A.42) require

$$\dot{\varepsilon}_{ii} = 0 = \frac{\partial \sigma_e}{\partial \sigma_{ii}} \tag{A.45}$$

such that assumed incompressibility will pose another restraint on the form of σ_e.

Ideal (or perfect) plasticity is characterized by arbitrary strains (given by the kinematic constraints) being achieved under one constant value of the effective stress; i.e.,

$$\dot{\varepsilon}_e \begin{cases} \geq 0 & \text{when } \sigma_e = \sigma_Y = \text{const} \\ = 0 & \sigma_e < \sigma_Y \end{cases} \tag{A.46}$$

†For this proposition in particular and fundamentals of plasticity in general consult: Bishop, J. F. W., and R. Hill.: *Phil. Mag.*, **42**:414 (1951), and Hill, R.: "The Mathematical Theory of Plasticity," Clarendon, Oxford, 1950, respectively.

With *strain-hardening* plasticity the flow rule should generalize the known relation under a uniaxial tension (Fig. A.11)

$$\varepsilon = g(\sigma) \qquad \frac{\dot{\varepsilon}}{\dot{\sigma}} = \frac{d\varepsilon}{d\sigma} = g'(\sigma) \equiv \frac{1}{h(\sigma)}$$
$$\text{if} \quad \begin{matrix} \sigma = \sigma_{Y1} > \sigma_{Y0} \\ \dot{\sigma} > 0 \end{matrix} \quad \text{where} \quad \sigma_{Y1} \equiv \max \sigma \tag{A.47}$$

Since effective stress and stretching generalize σ and $\dot{\varepsilon}$ in (A.47), it is reasonable to assume the relation

$$\varepsilon_e = g(\sigma_e) \qquad \dot{\varepsilon}_e = \frac{\dot{\sigma}_e}{h(\sigma_e)}$$
$$\text{if} \quad \begin{matrix} \sigma_e = \sigma_{Y1} > \sigma_{Y0} \\ \dot{\sigma}_e > 0 \end{matrix} \quad \text{where} \quad \sigma_{Y1} \equiv \max \sigma_e \tag{A.48}$$

With (A.42), (A.46), and (A.48) plastic behavior has been specified, assuming the functions $\sigma_e(\sigma_{ij})$ and $h(\sigma)$ to be known, both for ideal plasticity and *isotropic hardening*. The last notion conveys the fact that the adopted measure σ_e is only a scalar, otherwise incapable of distinguishing between different stress states in the past or present. This corresponds to a uniform expansion of the yield surface $f = 0$ during hardening and does not allow for such behavior as the Bauschinger effect.

If the ratios of the stress components are kept constant during the deformation, so are the coefficients $\partial\sigma_e/\partial\sigma_{ij}$ and also, according to (A.42), the ratios between the rates of deformation $\dot{\varepsilon}_{ij}$. This last equation can then be integrated through a monotonic action to give the total strains

$$\varepsilon_{ij} = \frac{\partial\sigma_e}{\partial\sigma_{ij}} \int_0^t \dot{\varepsilon}_e \, dt = \varepsilon_e \frac{\partial\sigma_e}{\partial\sigma_{ij}} = g(\sigma_e) \frac{\partial\sigma_e}{\partial\sigma_{ij}} \tag{A.49}$$

Such a state of stress and strain is termed *proportional* and can be viewed as a magnification without rotation of the vector $\boldsymbol{\sigma}$ of stresses in Fig. A.10. It has sometimes been argued that (A.49) should be accepted as a constitutive relation,

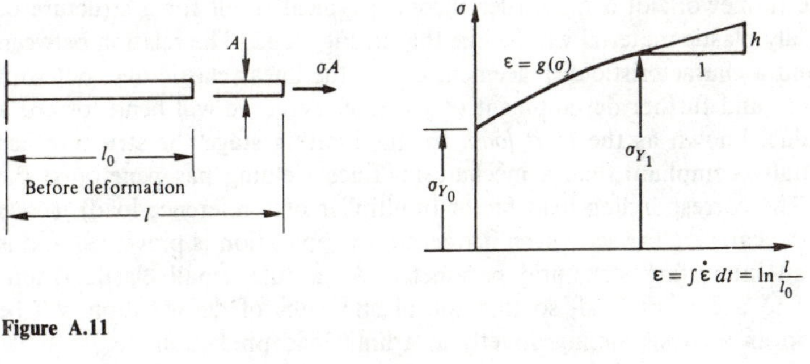

Figure A.11

under the name of *total-strain theory*, even with deviations from proportionality in the stress path.

Assumptions of present concern on the yield function are the von Mises and the Tresca hypotheses. Von Mises assumed

$$\sigma_e = \sqrt{3J_2} = \sqrt{\tfrac{3}{2}s_{ij}s_{ij}} = \sigma_Y \tag{A.50}$$

using (A.6), so that

$$\frac{\partial \sigma_e}{\partial \sigma_{ij}} = \frac{3}{2\sigma_e} s_{ij} \tag{A.51}$$

to be introduced into (A.42), (A.49), and (A.44). The last operation leads to

$$\dot{\varepsilon}_e = \sqrt{\tfrac{2}{3}\dot{\varepsilon}_{ij}\dot{\varepsilon}_{ij}} \tag{A.52}$$

Tresca considered the largest shear stress to be critical, or

$$\sigma_e = 2\tau_{\max} = \sigma_{\max} - \sigma_{\min} = \sigma_Y \tag{A.53}$$

Then if x_1, x_2, and x_3 are taken along the directions of $\sigma_{\max}$, σ_{int}, and $\sigma_{\min}$, respectively, (A.42) and (A.44) lead to

$$\{\dot{\varepsilon}_{11}, \dot{\varepsilon}_{22}, \dot{\varepsilon}_{33}\} = \{\dot{\varepsilon}_1, \dot{\varepsilon}_2, \dot{\varepsilon}_3\} = \dot{\varepsilon}_e\{1, 0, -1\} \tag{A.54}$$

and

$$\dot{\varepsilon}_e = \dot{\varepsilon}_1 = -\dot{\varepsilon}_3 = \sqrt{\tfrac{1}{2}\dot{\varepsilon}_{ij}\dot{\varepsilon}_{ij}} \tag{A.55}$$

Here the principal stresses are assumed to be unequal, since equalities will define *singularities* of the yield surface, where the gradient is indeterminate within certain bounds. To illustrate, in a tension bar with x_1 taken along the axis (so that $\sigma_{\text{int}} = \sigma_{\min} = 0$) all contractions $\dot{\varepsilon}_{22} < 0$ and $\dot{\varepsilon}_{33} < 0$ are permissible which satisfy $\dot{\varepsilon}_{22} + \dot{\varepsilon}_{33} = -\dot{\varepsilon}_{11}$.

With simultaneous elastic and plastic deformation, rates of deformation and here even total strains can be considered directly additive. This is asymptotically correct when the total deformation is small or when only the elastic strains are small enough to be defined with a basis in the current configuration.

A.4 LIMIT ANALYSIS

Within the framework of a first-order theory a typical result for a structure of elastic–ideally plastic material will be like that in Fig. A.12. The relation between the load and a characteristic displacement is first the linear elastic one, but with the initiation and further development of yielding the curve will bend toward a plateau value, known as the *limit load.* In this limiting stage the structure has become totally compliant, like a mechanism, since yielding has penetrated the structure. The corresponding load factor (multiplier of a reference load) μ_0 sets a limit to the carrying capacity even if fracture by separation is prevented and is therefore an interesting structural parameter. As a rule, small elastic strains correspond to a current load, so that the plastic rates of deformation will be dominant in an analysis aiming directly at a limit-load prediction.

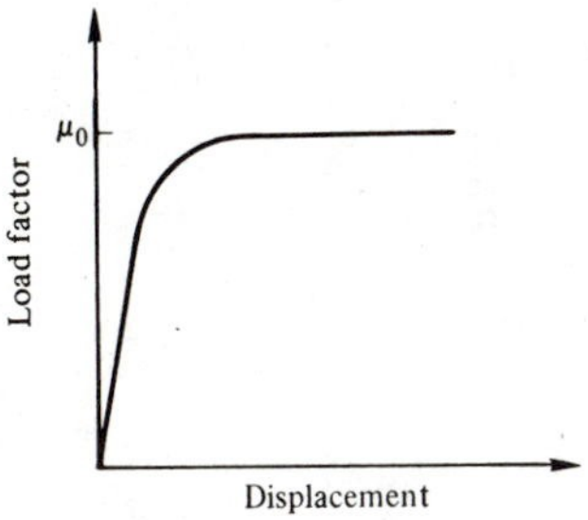

Figure A.12

Because of the general complexity of such limit analyses, approximate methods have been advanced for estimating the load between upper and lower bounds. This implies the inequalities

$$\mu^- \leq \mu_0 \leq \mu^+ \tag{A.56}$$

The defining and derived properties of a lower and an upper bound of the load factor, μ^- and μ^+, respectively, are presented below without proof.

Property 1 μ^- is the load factor corresponding to a stress field which is statically permissible in the sense that equilibrium is satisfied everywhere, and $\sigma_e \leq \sigma_Y$; $\mu^- \leq \mu_0$ always holds.

Property 2 μ^+ is the load factor corresponding to a kinematically permissible velocity field $\dot{u}_i$ with derived rates $\dot{\varepsilon}_{ij}$. Such a field must comply with the kinematic constraints and lead to a positive power of the external loads. Compatible with that field are stresses σ_{ij} in accord with (A.39) and (A.42). μ^+ is defined by

$$\int_V \sigma_{ij}\dot{\varepsilon}_{ij}\, dV = \int_S p_i \dot{u}_i\, dS + \int_V b_i \dot{u}_i\, dV \tag{A.57}$$

formally related to the principle of virtual work. μ^+ is implicit in (A.57) through the assumption of proportional loading, $p_i = \mu^+ p_i'$ and $b_i = \mu^+ b_i'$, where the primed fields are given reference distributions. $\mu^+ \geq \mu_0$ always holds.

In (A.57) the left-hand side is the power of dissipation within the total body, uniquely determined by the velocity field considered.

Since an exact solution is permissible in stress as well as in velocity, the two sets of bounding factors μ^- and μ^+ must have the one value μ_0 in common.

Upper-bound solutions are often constructed on discontinuities in the velocity field. Assume that such a discontinuity (in the limit $\delta \rightarrow 0$) has been formed, as in Fig. A.13, in a plate where stresses and strains are considered independent of the coordinate normal to the plate. $\dot{\mathbf{u}}_{\text{rel}} = \dot{\mathbf{u}}_b - \dot{\mathbf{u}}_a$ denotes the difference in velocities of particles in zones b and a; each zone is considered to be a (relatively) rigid boundary of the deforming material in zone c. The rate of shear and the stretching

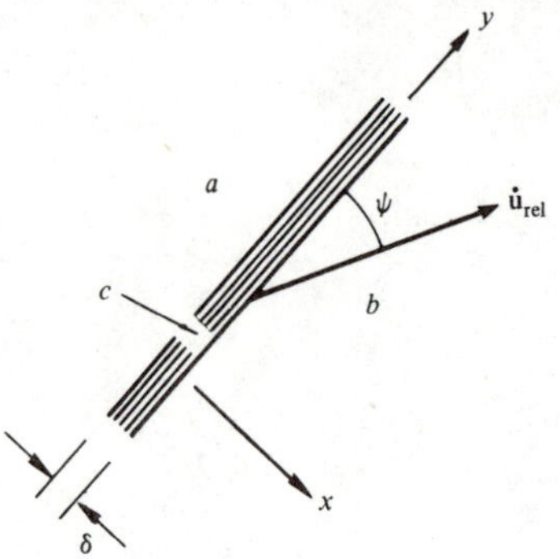

Figure A.13

normal to the yield strip are then given by

$$\dot{\gamma}_{xy} = \frac{\dot{u}_{rel}\cos\psi}{\delta} \qquad \dot{\varepsilon}_x = \frac{\dot{u}_{rel}\sin\psi}{\delta} \tag{A.58}$$

referring to axes x and y normal to and along the strip. $\dot{\varepsilon}_y$ is negligibly small due to the contact with the rigid zones a and b. These rates of deformation correspond to the Mohr's circle of Fig. A.14 and therefore to the principal rates within the strip

$$\dot{\varepsilon}_1 = \frac{\dot{u}_{rel}}{2\delta}(\sin\psi + 1) \qquad \dot{\varepsilon}_2 = \frac{\dot{u}_{rel}}{2\delta}(\sin\psi - 1) \qquad \dot{\varepsilon}_3 = -\dot{\varepsilon}_1 - \dot{\varepsilon}_2 \tag{A.59}$$

The last equality assumes incompressibility, and those preceding it can be solved with respect to the angle ψ of discontinuity

$$\sin\psi = \frac{\dot{\varepsilon}_1 + \dot{\varepsilon}_2}{\dot{\varepsilon}_1 - \dot{\varepsilon}_2} \tag{A.60}$$

The 1-direction will be as in Fig. A.15, making angles $\pi/4 + \psi/2$ with the y

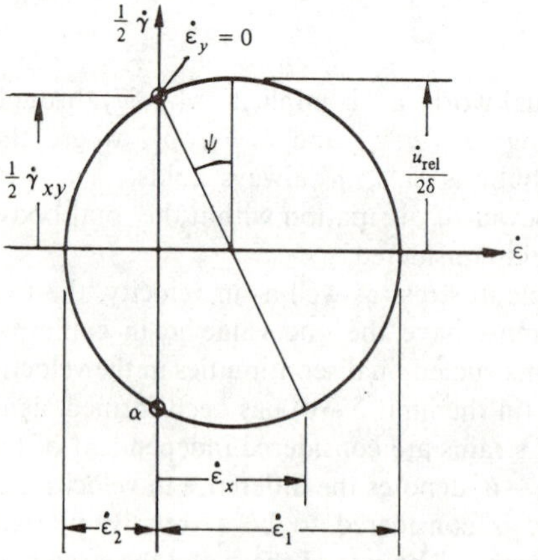

Figure A.14

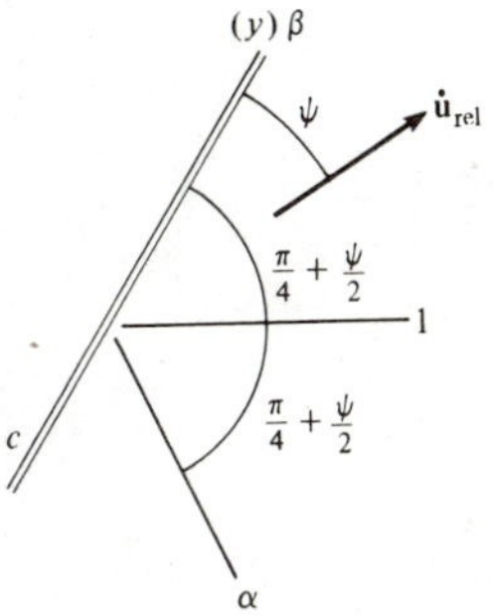

Figure A.15

direction and with the other nonstretching direction α. Clearly, this α direction must also be normal to $\dot{\mathbf{u}}_{rel}$. The dissipation per unit of thickness and length along the strip c becomes, using (A.40),

$$\dot{w}\delta = (\sigma_1\dot{\varepsilon}_1 + \sigma_2\dot{\varepsilon}_2 + \sigma_z\dot{\varepsilon}_z)\delta = -\frac{\dot{u}_{rel}}{2}[\sigma_1 - \sigma_2 + \sin\psi\,(\sigma_1 + \sigma_2 - 2\sigma_z)] \quad \text{(A.61)}$$

when (A.59) is observed. In particular, for a Mises material we further obtain

$$\dot{w}\delta = \sigma_Y\dot{\varepsilon}_e\delta = \sigma_Y\sqrt{\tfrac{2}{3}(\dot{\varepsilon}_1^2 + \dot{\varepsilon}_1^2 + \dot{\varepsilon}_z^2)\delta^2} = \sigma_Y\dot{u}_{rel}\sqrt{\tfrac{1}{3} + \sin^2\psi} \quad \text{(A.62)}$$

in view of (A.43) and (A.52). The same expression, multiplied by $\sqrt{3}/2$, would apply to a Tresca material if the principal stresses σ_1, σ_2, and σ_z were unequal; compare (A.55).

For *plane deformation*, when $\dot{\varepsilon}_z = 0$ and therefore $\dot{\varepsilon}_2 = -\dot{\varepsilon}_1$, it follows from (A.60) that

$$\psi = 0 \quad \text{(A.63)}$$

implying that the velocity discontinuity in that case can only be parallel to the interface. Along this *shear line c*, where $\dot{\gamma}_{xy}$ is intense, the dissipation will be

$$\dot{w}\delta = \tau_Y\dot{u}_{rel} \quad \text{(A.64)}$$

in terms of the shear yield stress, $\tau_Y = \sigma_Y/\sqrt{3}$ or $\tau_Y = \sigma_Y/2$, for the Mises or the Tresca material, respectively.

In general ψ must be varied in an upper-bound analysis, together with the orientation of the yield line c, in order to minimize the load factor μ^+. In exceptional cases one may even arrive at a compatible stress field which is statically permissible, implying that the limit load has been determined exactly. Some examples follow.

Example A.1 A thin sheet of constant thickness t is stretched with a *fixed width b*, such that $\dot{\mathbf{u}}_{rel}$ can only be parallel to the guided edges. This can be consistently modeled by taking

$$\psi = \frac{\pi}{2}$$

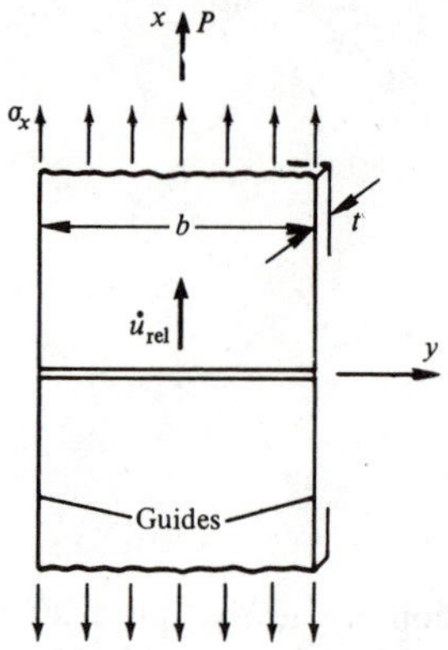

Example A.1

giving a transverse yield strip c as shown, with stretching

$$\dot{\varepsilon}_1 \equiv \dot{\varepsilon}_x = \frac{\dot{u}_{rel}}{\delta} \qquad \text{and} \qquad \dot{\varepsilon}_2 \equiv \dot{\varepsilon}_y = 0$$

The compatible principal stresses are

$$\sigma_1 = \sigma_x = \frac{2}{\sqrt{3}}\sigma_Y \qquad \sigma_y = \frac{1}{\sqrt{3}}\sigma_Y \qquad \sigma_z = 0$$

for the Mises material, by (A.42), (A.51), and (A.50), or

$$\sigma_1 = \sigma_x = \sigma_Y \qquad 0 < \sigma_y < \sigma_Y \qquad \sigma_z = 0$$

under Tresca assumptions (A.54) and (A.55). These stress fields can be continued into the rigid zones and are permissible under the limit load

$$P = P_0 = \sigma_x bt$$

Alternatively, the same load can be derived from (A.57), which here takes the form

$$\dot{w}\delta bt = P\dot{u}_{rel}$$

We conclude that the assumed fields with a localized yielding contribute to a *possible* solution. Experiments have shown that the case may indeed be realized. $\dot{\varepsilon}_z < 0$ signifies that the thickness decreases locally in the yield strip, such that a neck is formed and a fracture will follow with continued motion.

In the following examples, x and y are not used to mean coordinates normal and parallel to the yield line.

Example A.2 A plate is stretched with two *free* lateral sides as shown. First assume that

$$\dot{\varepsilon}_1 \equiv \dot{\varepsilon}_x > 0 \qquad \dot{\varepsilon}_2 \equiv \dot{\varepsilon}_y = \dot{\varepsilon}_z = -\tfrac{1}{2}\dot{\varepsilon}_x$$

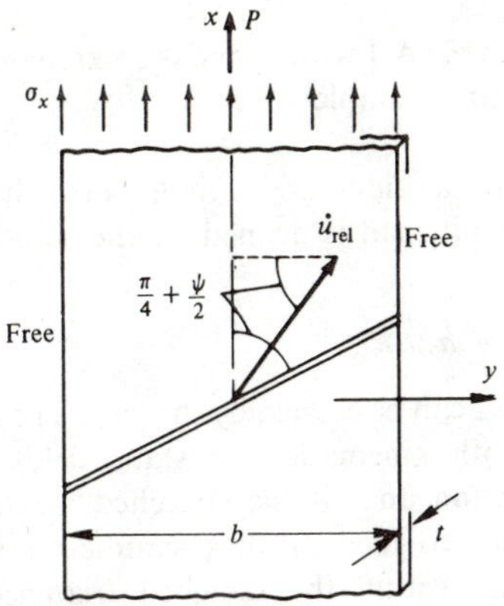

Example A.2

giving, by (A.60),

$$\sin\psi = \tfrac{1}{3} \qquad \text{or} \qquad \psi = 19.5^\circ \qquad \frac{\pi}{4}+\frac{\psi}{2} = 54.7^\circ$$

and the compatible stress field

$$\sigma_1 = \sigma_x = \sigma_Y \qquad \sigma_y = \sigma_z = 0$$

for both Mises and Tresca material. This field is permissible throughout the plate, provided that even the surfaces parallel to the paper are nonloaded (as Example A.1). Thus, we have again arrived at a limit-load value

$$P = P_0 = \sigma_x bt = \sigma_Y bt$$

for both materials. Alternatively, this might follow from (A.57) in its present form

$$\frac{\dot{w}\delta bt}{\sin 54.7^\circ} = P\dot{u}_{rel} \sin 54.7^\circ$$

The configuration is a neck including shear, indicating a skew fracture in the wake of a limit state in thin sheets. Even this result has experimental support. Within the Tresca idealization there is also an infinity of alternative solutions [compare the discussion following (A.55)], all having

$$-\dot{\varepsilon}_x - \dot{\varepsilon}_z = \dot{\varepsilon}_y \leq 0 \qquad 0 \leq \psi \leq \frac{\pi}{2}$$

with the same limit load P_0 as above.

By considering instead

$$\dot{\varepsilon}_1 = \dot{\varepsilon}_x = -\dot{\varepsilon}_y > 0$$

we model the plane deformation discussed through (A.63) and (A.64). Thus, $\psi = 0$ applies here, the shear line c will be at 45° with the line of loading, and the stresses can be carried over from Example A.1 with σ_y and σ_z interchanged. The limit load therefore is the same as in Example A.1.

Example A.3 A plate with two free lateral sides and a hole through the thickness t is stretched as shown. The limit load is defined by the solutions to Example A.2, with the value

$$P_0 = \sigma_x(b_1 + b_2)t$$

This is because the stresses of Example A.2 can be carried by the parallel zones on each side of the hole such that both kinematic and static fields are permissible. A corresponding line of action and (if the stretched length is short) distribution of the load P must be provided. A more symmetric yield pattern is possible, in that the material within the sketched triangles is displaced toward the hole while the zones above and below are in axial separation.

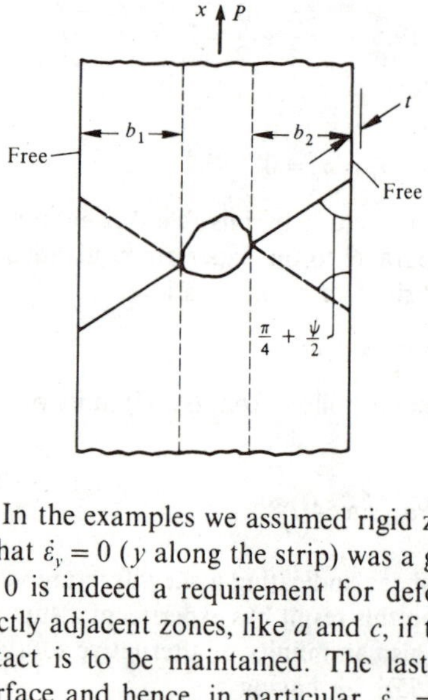

Example A.3

In the examples we assumed rigid zones in the plate outside the yield strip c, so that $\dot{\varepsilon}_y = 0$ (y along the strip) was a given condition. It will now be shown that $\dot{\varepsilon}_y = 0$ is indeed a requirement for deformation discontinuities to exist between directly adjacent zones, like a and c, if the flow rule is to apply in both zones and contact is to be maintained. The last condition implies $\dot{\mathbf{u}}_a = \dot{\mathbf{u}}_c$ along the a-c interface and hence, in particular, $\dot{\varepsilon}_{ya} = \dot{\varepsilon}_{yc}$. The flow rule can be stated as

$$\frac{\dot{\varepsilon}_x}{\partial f/\partial \sigma_x} = \frac{\dot{\gamma}_{xy}}{\partial f/\partial \tau_{xy}} = \frac{\dot{\varepsilon}_y}{\partial f/\partial \sigma_y}$$

[compare (A.42)] in terms of the one yield function f applying to both zones. Here, then, $\partial f/\partial \sigma$ and $\partial f/\partial \tau$ are equal by pairs in the two zones. It follows that even $\dot{\varepsilon}_x$ and $\dot{\gamma}_{xy}$ must be continuous, with the possible exception of the particular case $\dot{\varepsilon}_y = 0 = \partial f/\partial \sigma_y$, as proposed above.

The *characteristics* in an x_1, x_2 plane are curves defined by the property that a given system of partial differential equations in x_1 and x_2 can be reduced to a set of ordinary differential equations if the independent variables are taken as arc lengths along the characteristics rather than x_1 and x_2. This means that if a characteristic is known, the pertinent dependent variable F can be determined along the whole curve if the necessary initial conditions are prescribed at a point on the curve. In the direction normal to the characteristic such a function F and/or its derivatives may be discontinuous because prescribed conditions in adjacent characteristics may be different. This implies, conversely, that a nonstretching trajectory, as discussed above, may have the properties of a characteristic since it can carry discontinuities. Two such possible directions were distinct in Figs. A.14 and A.15, one which we have already called the α direction and the other (along c) to be denoted by β. A convention is that a line along the principal direction 1 (for the largest stress and stretching) shall be intersected next to the α line, before the β line, in a mathematically positive circuit.

We summarize the above results as follows:

1. A discontinuity in rates of deformation may appear at a line of zero stretching, locally in the α or β direction.
2. A discontinuity in the velocity field may appear at a yield strip oriented along one nonstretching direction, β or α, such that the relative velocity is normal to the other direction, α or β. If the deformation corresponds to a widening of the yield strip, as often happens in thin sheets, total fracture will be the immediate consequence unless the widening is bounded by kinematic constraints.

As implied above, the differential equations governing the velocity field will be ordinary when formulated in terms of arc lengths along the nonstretching lines, s_α and s_β. It can be shown that these α and β lines are also the characteristics of the stress field. In particular, under plane-deformation conditions the α and β lines are orthogonal ($\psi = 0$) and generally referred to as the *slip lines*. Then the variation of stresses is simply expressed by the Hencky equations

$$\frac{\partial \tilde{\sigma}}{\partial s_\alpha} - 2\tau_Y \frac{\partial \theta}{\partial s_\alpha} = 0 \qquad \frac{\partial \tilde{\sigma}}{\partial s_\beta} + 2\tau_Y \frac{\partial \theta}{\partial s_\beta} = 0 \qquad \tilde{\sigma} \equiv \frac{\sigma_{11} + \sigma_{22}}{2} \tag{A.65}$$

where θ = local slip-line orientation
$\tilde{\sigma}$ = in-plane mean stress
τ_Y = shear yield stress

See Figs. A.1, A.2, and A.16. If the slip-line field is known, and thus θ, $\tilde{\sigma}$ follows from (A.65) by

$$\begin{aligned} \tilde{\sigma} - 2\tau_Y\theta &= \text{const along } \alpha \text{ line} \\ \tilde{\sigma} + 2\tau_Y\theta &= \text{const along } \beta \text{ line} \end{aligned} \tag{A.66}$$

Equation (A.66) has been applied to plane-strain analyses in Secs. 2.8 and 2.9. For the complete record we also show in Fig. A.17 the characteristics pertaining

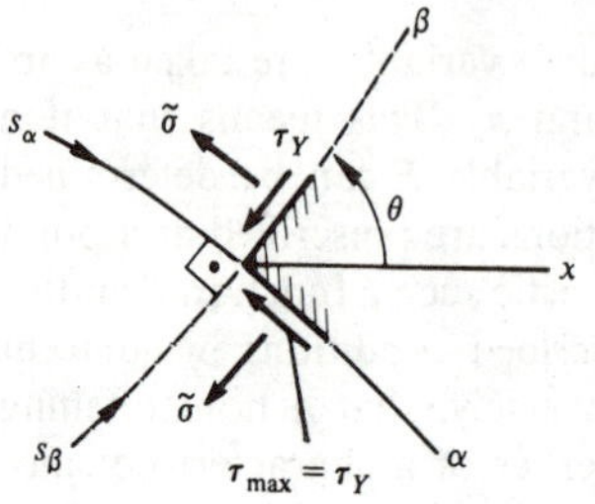

Figure A.16

to crack-tip local yield in a Mises material under plane-stress conditions. Comparable to this graph is Fig. 2.20 covering plane strain, Tresca and Mises types, and the Dugdale yield strip applying to Tresca yielding in plane stress. In addition to being nonstretching trajectories so that principal directions bisect the angles between α and β, the α and β lines in Fig. A.17 have an additional property: the normal stress perpendicular to the line is twice that along the line. This is a simple consequence of von Mises' flow rule.

The field has three typical stress regimes:

$$\begin{aligned}
&\text{In } DOC\text{:} \quad \sigma_x = \tau_Y \cos^3\theta \qquad \sigma_y = \tau_Y \cos\theta\,(2+\sin^2\theta) \qquad \tau_{xy} = -\tau_Y \sin^3\theta \\
&\text{In } COB\text{:} \quad \sigma_x = 0.0057\tau_Y \qquad \sigma_y = 0.5307\tau_Y \qquad \tau_{xy} = -0.9524\tau_Y \\
&\text{In } BOA\text{:} \quad \sigma_x = -\sigma_Y = -\sqrt{3}\tau_Y \qquad \sigma_y = \tau_{xy} = 0
\end{aligned} \tag{A.67}$$

They are connected with full continuity along the ray $\theta = \phi_1 = 79.7°$ and a permissible discontinuity (in the parallel normal stress) along the ray $\theta = \phi_1 + \phi_2 = 151.4°$. In contrast to the Dugdale model, this is a diffuse yield pattern. The above values (though not the figure in all detail) originate from a study by Hutchinson [113], where earlier results by Hill [123] were utilized.

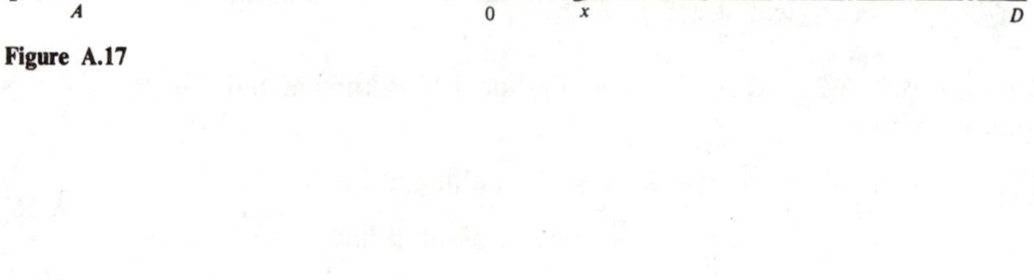

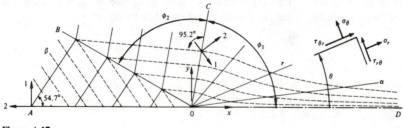

Figure A.17

APPENDIX

B

ANALYTICAL SOLUTIONS FOR AN INTERNAL CRACK

Here we show how closed solutions can be found for stresses and displacements in the vicinity of an isolated internal crack when the basis is linear elasticity. Complex-variable techniques will be presented first, in their more general context.

B.1 PLANE STATES: A GENERAL FORMULATION

We start by recapitulating some elements of the complex-variable analysis to define the terminology. The coordinates of the plane are introduced as $x \equiv x_1$ and $y \equiv x_2$, while z has the following meaning:

$$z = x + iy \qquad \bar{z} = x - iy \qquad \text{where } i = \sqrt{-1} \tag{B.1}$$

A function of z, like

$$F(z) = \operatorname{Re}[F(z)] + i \operatorname{Im}[F(z)] \equiv \alpha(x, y) + i\beta(x, y) \tag{B.2}$$

has a single-valued derivative if the function is *analytic* in the domain; this implies that

$$F'(z) \equiv \frac{dF}{dz} = \frac{\partial \alpha}{\partial x} + i\frac{\partial \beta}{\partial x} = \frac{\partial \beta}{\partial y} - i\frac{\partial \alpha}{\partial y} \tag{B.3}$$

from which it follows that α and β are *conjugate functions*, satisfying the Cauchy-Riemann equalities

$$\frac{\partial \alpha}{\partial x} = \frac{\partial \beta}{\partial y} \qquad \frac{\partial \alpha}{\partial y} = -\frac{\partial \beta}{\partial x} \tag{B.4}$$

When these functions are further differentiated, it appears from (B.4) that they are both *harmonic* in the sense that

$$\nabla^2 \alpha = \nabla^2 \beta = 0 \qquad \nabla^2 \equiv \frac{\partial^2}{\partial x^2} + \frac{\partial^2}{\partial y^2} \tag{B.5}$$

It follows then that the derivative $F'(z)$ is also an analytic function.

The plane-elasticity problem under consideration is to find solutions of the bipotential equation

$$\nabla^2(\nabla^2\chi) = 0 \tag{B.6}$$

according to (A.35) and (A.36), where χ is the Airy stress function. We first introduce the mean normal stress in the plane as a variable $\tilde{\sigma}$ [compare (A.65)]

$$2\tilde{\sigma} = \sigma_{11} + \sigma_{22} = \nabla^2\chi \tag{B.7}$$

which is a harmonic function in view of (A.34) and (B.6). An analytic function can then be formed

$$S(z) = \tilde{\sigma} + i \operatorname{Im} S \tag{B.8}$$

where Im S is the conjugate of $\tilde{\sigma}$ through (B.4). By integration another analytic function appears, as suggested by Muskhelishvili [17]

$$\phi(z) = \int S(z)\, dz = P(x, y) + iQ(x, y) \tag{B.9}$$

such that, with (B.3),

$$\phi'(z) = \frac{\partial P}{\partial x} + i\frac{\partial Q}{\partial x} = \frac{\partial Q}{\partial y} - i\frac{\partial P}{\partial y} = \tilde{\sigma} + i \operatorname{Im} S$$

implying that

$$\frac{\partial P}{\partial x} = \frac{\partial Q}{\partial y} = \tilde{\sigma} \tag{B.10}$$

Proceeding from this, we can form a real function

$$R = 2\chi - xP - yQ \tag{B.11}$$

which turns out to be harmonic,

$$\nabla^2 R = 4\tilde{\sigma} - \nabla^2(xP) - \nabla^2(yQ) = 0 \tag{B.12}$$

because

$$\nabla^2(xP) = x\,\nabla^2 P + 2\frac{\partial P}{\partial x} = 2\tilde{\sigma} \qquad \nabla^2(yQ) = y\,\nabla^2 Q + 2\frac{\partial Q}{\partial y} = 2\tilde{\sigma}$$

where $\nabla^2 P = \nabla^2 Q = 0$ and (B.10) are observed. Therefore, the following representation of Airy's stress function must be valid

$$2\chi = xP + yQ + R \equiv \operatorname{Re}[\bar{z}\phi(z) + \psi(z)] \tag{B.13}$$

where the function R has been written as the real part of an analytic function $\psi(z)$. Equation (B.13) is the general solution of the bipotential equation, due to Goursat (see Ref. 17), in which P, Q, and R are harmonic functions.

When we differentiate (B.13), observing that $\partial \operatorname{Re}(\cdot)/\partial x = \operatorname{Re}[\partial(\cdot)/\partial x]$, etc., we derive in a straightforward way

$$2\frac{\partial\chi}{\partial x} = \operatorname{Re}(\phi + \bar{z}\phi' + \psi') \qquad 2\frac{\partial\chi}{\partial y} = \operatorname{Im}(\phi - \bar{z}\phi' - \psi') \tag{B.14}$$

and

$$\frac{\partial^2\chi}{\partial x^2} = \mathrm{Re}\,(\phi' + \tfrac{1}{2}\bar{z}\phi'' + \tfrac{1}{2}\psi'') = \sigma_{22}$$

$$\frac{\partial^2\chi}{\partial y^2} = \mathrm{Re}\,(\phi' - \tfrac{1}{2}\bar{z}\phi'' - \tfrac{1}{2}\psi'') = \sigma_{11} \tag{B.15}$$

$$-\frac{\partial^2\chi}{\partial x\,\partial y} = \tfrac{1}{2}\,\mathrm{Im}\,(\bar{z}\phi'' + \psi'') = \sigma_{12}$$

Equations (A.34) defining the stresses. To arrive at displacement formulas. we proceed from the kinematic relations (A.20) and Hooke's law, (A.32) or (A.33). In *plane strain*, for instance, (A.33) is written as

$$2G\varepsilon_{11} = (1-\nu)(\sigma_{11}+\sigma_{22}) - \sigma_{22} = 2G\frac{\partial u_1}{\partial x} = 2(1-\nu)\tilde{\sigma} - \frac{\partial^2\chi}{\partial x^2}$$

$$2G\varepsilon_{22} = 2G\frac{\partial u_2}{\partial y} = 2(1-\nu)\tilde{\sigma} - \frac{\partial^2\chi}{\partial y^2} \tag{B.16}$$

or, by integration,

$$2Gu_1 = 2(1-\nu)P - \frac{\partial\chi}{\partial x} + f_1(y) \qquad 2Gu_2 = 2(1-\nu)Q - \frac{\partial\chi}{\partial y} + f_2(x) \tag{B.17}$$

where (B.10) has been observed. The substitution of (B.17) into (A.20), (A.33), and (A.34), that is, in

$$2\varepsilon_{12} = \frac{\partial u_1}{\partial y} + \frac{\partial u_2}{\partial x} = \frac{1}{G}\sigma_{12} = -\frac{1}{G}\frac{\partial^2\chi}{\partial x\,\partial y} \tag{B.18}$$

leads to the condition

$$\frac{df_1}{dy} = -\frac{df_2}{dx} \tag{B.19}$$

which can be satisfied only if both sides of the equation equal the same real constant A. Hence

$$f_1 = Ay + B_1 \qquad f_2 = -Ax + B_2 \tag{B.20}$$

where B_1 and B_2 are further real constants. In relation to (B.17) we see that these constants represent a translation of the body as a rigid system, just as A corresponds to a rigid-body rotation. Such motion being of no concern in the present discussion, f_1 and f_2 will be taken as zero. Then introducing from (B.9) and (B.14) into (B.17), we are left with the displacements

$$4Gu_1 = \mathrm{Re}\,(\kappa\phi - \bar{z}\phi' - \psi') \qquad 4Gu_2 = \mathrm{Im}\,(\kappa\phi + \bar{z}\phi' + \psi') \tag{B.21}$$

where $\kappa = 3 - 4\nu$. The presentation in (B.21) is also aimed at including the case of *plane stress* by the alternative interpretation $\kappa = (3-\nu)/(1+\nu)$. This last result comes from a deduction quite similar to the one above, (A.33) having been replaced by (A.32) in the form

$$2G\varepsilon_{11} = \frac{\sigma_{11} + \sigma_{22}}{1 + \nu} - \sigma_{22}$$

and so forth.

Among the infinity of possible functions ϕ and ψ the *boundary conditions* will pick those which specify the unique solution. As a first condition we assume the plane $y = 0$ to be one of *symmetry*, such that

$$\sigma_{12} = 0 \qquad \text{for } y = 0 \tag{B.22}$$

This can be achieved by following Westergaard [18] and taking

$$\psi'' = -z\phi'' \tag{B.23}$$

so that

$$\psi' = -z\phi' + \phi + \text{const} \tag{B.24}$$

leading to the stresses, by (B.15),

$$\begin{aligned} \sigma_{11} &= \text{Re}\,\phi' - y\,\text{Im}\,\phi'' \\ \sigma_{22} &= \text{Re}\,\phi' + y\,\text{Im}\,\phi'' \\ \sigma_{12} &= -y\,\text{Re}\,\phi'' \end{aligned} \tag{B.25}$$

and the displacements (excluding rigid-body terms), by (B.21),

$$\begin{aligned} 2Gu_1 &= \frac{\kappa - 1}{2}\,\text{Re}\,\phi - y\,\text{Im}\,\phi' \\ 2Gu_2 &= \frac{\kappa + 1}{2}\,\text{Im}\,\phi - y\,\text{Re}\,\phi' \end{aligned} \tag{B.26}$$

In a similar way we can specify a class of solution having $y = 0$ as a plane of *antisymmetry*

$$\sigma_{22} = 0 \qquad \text{for } y = 0 \tag{B.27}$$

by assuming

$$\psi'' = -2\phi' - z\phi'' \tag{B.28}$$

or

$$\psi' = -\phi - z\phi' + \text{const} \tag{B.29}$$

This provides the stresses

$$\begin{aligned} \sigma_{11} &= 2\,\text{Re}\,\phi' - y\,\text{Im}\,\phi'' \\ \sigma_{22} &= y\,\text{Im}\,\phi'' \\ \sigma_{12} &= \text{Im}\,\phi' - y\,\text{Re}\,\phi'' \end{aligned} \tag{B.30}$$

and the displacements

$$\begin{aligned} 2Gu_1 &= \frac{\kappa + 1}{2}\,\text{Re}\,\phi - y\,\text{Im}\,\phi' \\ 2Gu_2 &= \frac{\kappa - 1}{2}\,\text{Im}\,\phi - y\,\text{Re}\,\phi' \end{aligned} \tag{B.31}$$

Note that the above solutions, beginning with (B.23), are somewhat specialized in that the right-hand side of (B.23) or (B.28) might have included an arbitrary constant, real or imaginary, respectively. However, the formulas as they stand are sufficient for present purpose.

B.2 PLANE STATES: SOLUTIONS AT AN INTERNAL CRACK

With the preceding results as a basis we can construct complete solutions valid in the vicinity of an isolated internal crack normal to a plate of large extent. The evaluation will proceed in two steps.

Step 1 Stresses and displacements are computed for the plate *without* a crack under the given external conditions. Of particular interest in the present context are the stresses $\sigma_{22}(1)$ and $\sigma_{12}(1)$, which will act along the crack line $y = 0$, $|x| < a$, referring to the local geometry of Fig. B.1.

Step 2 To simulate a crack the line $y = 0$, $|x| < a$, is subjected to tractions opposite those derived in step 1, together with whatever loading p_i may be *given* to act on the crack sides. (The latter contributions are also assumed to be reactive by pairs, as in Fig. B.1.) Then we evaluate the effects of prescribed boundary stresses along $y = 0$, $|x| < a$,

$$\sigma_{22}(2) = -\sigma_{22}(1) - p_2 \equiv -g_2(x) \qquad \sigma_{12}(2) = -\sigma_{12}(1) - p_1 \equiv -g_1(x) \qquad \text{(B.32)}$$

on a plate which is otherwise continuous and (infinitely) large. The solution will be the sum of the contributions from steps 1 and 2.

It is expedient before taking step 1 to decompose the action into one case which is symmetric (mode I) and one which is antisymmetric (mode II) with respect to $y = 0$. Then we can limit the discussion to one half plane ($y \geq 0$) and superpose the effects at the end.

Step 1 calls for no further comment. To step 2, which is of present concern, Sedov [19] has pointed out the general solutions

$$\phi' \equiv \begin{cases} \phi'_{\text{I}} = \dfrac{1}{\pi\sqrt{z^2 - a^2}} \displaystyle\int_{-a}^{a} \frac{g_2(\xi)\sqrt{a^2 - \xi^2}}{z - \xi}\, d\xi & \text{mode I} \qquad \text{(B.33)} \\[2ex] \phi'_{\text{II}} = \dfrac{-i}{\pi\sqrt{z^2 - a^2}} \displaystyle\int_{-a}^{a} \frac{g_1(\xi)\sqrt{a^2 - \xi^2}}{z - \xi}\, d\xi & \text{mode II} \qquad \text{(B.34)} \end{cases}$$

both having the desired property of predicting zero displacements infinitely far from the crack. This must be a consequence of the fact that each crack loading is an equilibrium group.

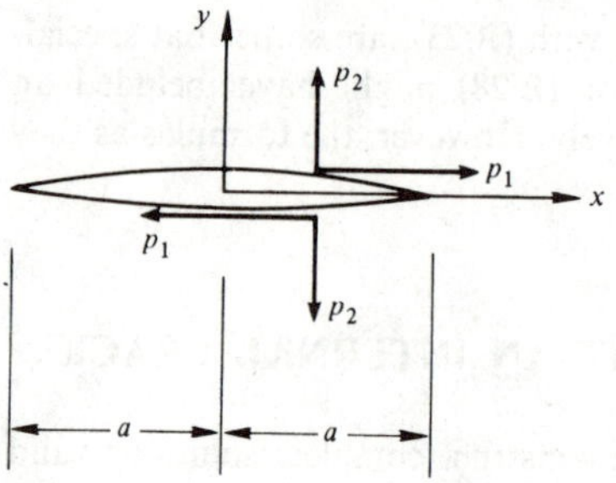

Figure B.1

To study the behavior close to the crack tip we further introduce

$$z + a \approx 2a \qquad z - \xi \approx a - \xi$$

to arrive at the asymptotic solutions at the right end, $z \to a$,

$$\phi'_{\mathrm{I}} = \frac{K_{\mathrm{I}}}{\sqrt{2\pi(z-a)}} \tag{B.35}$$

$$i\phi'_{\mathrm{II}} = \frac{K_{\mathrm{II}}}{\sqrt{2\pi(z-a)}} \tag{B.36}$$

where

$$K_{\mathrm{I}} = \frac{1}{\sqrt{\pi a}} \int_{-a}^{a} g_2(\xi) \sqrt{\frac{a+\xi}{a-\xi}}\, d\xi \tag{B.37}$$

and

$$K_{\mathrm{II}} = \frac{1}{\sqrt{\pi a}} \int_{-a}^{a} g_1(\xi) \sqrt{\frac{a+\xi}{a-\xi}}\, d\xi \tag{B.38}$$

Finally, the introduction of

$$z - a = re^{i\theta}$$

and the substitution from (B.35) in (B.25) and (B.26) and from (B.36) in (B.30) and (B.31) will lead to the local stresses and displacements for mode I

$$\begin{bmatrix} \sigma_{11} \\ \sigma_{22} \\ \sigma_{12} \end{bmatrix} = \frac{K_{\mathrm{I}}}{\sqrt{2\pi r}} \cos\frac{\theta}{2} \begin{bmatrix} 1 - \sin\frac{\theta}{2}\sin\frac{3\theta}{2} \\ 1 + \sin\frac{\theta}{2}\sin\frac{3\theta}{2} \\ \sin\frac{\theta}{2}\cos\frac{3\theta}{2} \end{bmatrix} \tag{B.39}$$

$$\begin{bmatrix} u_1 \\ u_2 \end{bmatrix} = \frac{K_I}{2G}\sqrt{\frac{r}{2\pi}}(\kappa - \cos\theta)\begin{bmatrix} \cos\frac{\theta}{2} \\ \sin\frac{\theta}{2} \end{bmatrix} \tag{B.40}$$

and for mode II

$$\begin{bmatrix} \sigma_{11} \\ \sigma_{22} \\ \sigma_{12} \end{bmatrix} = \frac{K_{II}}{\sqrt{2\pi r}}\begin{bmatrix} -\sin\frac{\theta}{2}\left(2+\cos\frac{\theta}{2}\cos\frac{3\theta}{2}\right) \\ \cos\frac{\theta}{2}\sin\frac{\theta}{2}\cos\frac{3\theta}{2} \\ \cos\frac{\theta}{2}\left(1-\sin\frac{\theta}{2}\sin\frac{3\theta}{2}\right) \end{bmatrix} \tag{B.41}$$

$$\begin{bmatrix} u_1 \\ u_2 \end{bmatrix} = \frac{K_{II}}{2G}\sqrt{\frac{r}{2\pi}}\begin{bmatrix} \sin\frac{\theta}{2}(2+\kappa+\cos\theta) \\ \cos\frac{\theta}{2}(2-\kappa-\cos\theta) \end{bmatrix} \tag{B.42}$$

Next these quantities can be transformed into polar components σ_r, σ_θ, $\tau_{r\theta}$, u_r, and u_θ, which turn out to be identical to the singular terms (for $n = 1$) derived in Sec. 2.1. This means that K_I and K_{II} are the stress intensity factors previously defined, implying that (B.37) and (B.38) are *general* formulas for these factors when the crack is small in a large plate. Although step 1 in itself has no singularities at the crack-tip locations, it will obviously contribute to the resulting singularities by determining the distribution functions $g_2(x)$ and $g_1(x)$.

A geometric interpretation of the displacement field (B.40) is shown in Fig. B.2, from which it also appears that $u_r = u_1$ and $u_\theta = -u_2$, giving alternatives to the more clumsy expressions (2.19).

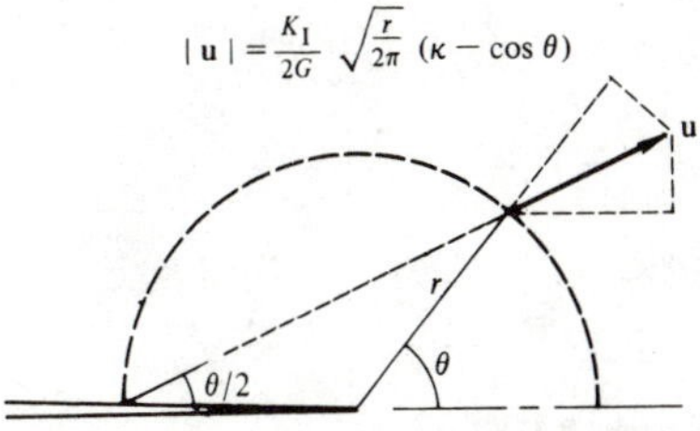

Figure B.2

Some examples of particular functions $g(x)$ are considered below.

Example B.1: A plate under uniaxial or biaxial tension orthogonal to the crack Referring to the figure, step 1 has the obvious solution $\sigma_{22} = \sigma_\infty$, $\sigma_{11} = \eta\sigma_\infty$, so that $g_2(x) = \sigma_\infty$ for all η. From (B.33) we can then obtain by integration

$$\phi_I' = \frac{\sigma_\infty z}{\sqrt{z^2 - a^2}} - \sigma_\infty$$

and so

$$\phi_I = \sigma_\infty \sqrt{z^2 - a^2} - \sigma_\infty z + \text{const}$$

Along $y = 0$ or $z = x$ the normal stress appears as

$$\sigma_{22}(x, 0) = \text{Re}\frac{\sigma_\infty x}{\sqrt{x^2 - a^2}} - \sigma_\infty + \sigma_\infty$$

the first two terms deriving from (B.25) and the last being the contribution from step 1. Thus

$$\sigma_{22} = \begin{cases} \dfrac{\sigma_\infty x}{\sqrt{x^2 - a^2}} & \text{for } |x| \geq a \\ 0 & \text{for } |x| < a \end{cases}$$

In the limit as $r = x - a \to 0$ we see from this that

$$\sigma_{22} = \frac{\sigma_\infty a}{\sqrt{2ar}} \equiv \frac{K_I}{\sqrt{2\pi r}} \qquad \text{with } K_I = \sigma_\infty \sqrt{\pi a}$$

Compare $a/W \to 0$ in (C.3), Appendix Table C.2. The last result would also follow directly from (B.37). The crack sides will have displacements as given by (B.26); thus

$$u_2(x, 0) = \frac{\kappa + 1}{4G} \sigma_\infty \,\text{Im} \sqrt{x^2 - a^2}$$

or, on both sides $y = 0^\pm$ of the crack,

$$u_2^\pm = \pm \frac{\kappa + 1}{4G} \sigma_\infty \sqrt{a^2 - x^2} \qquad \text{for } |x| \leq a$$

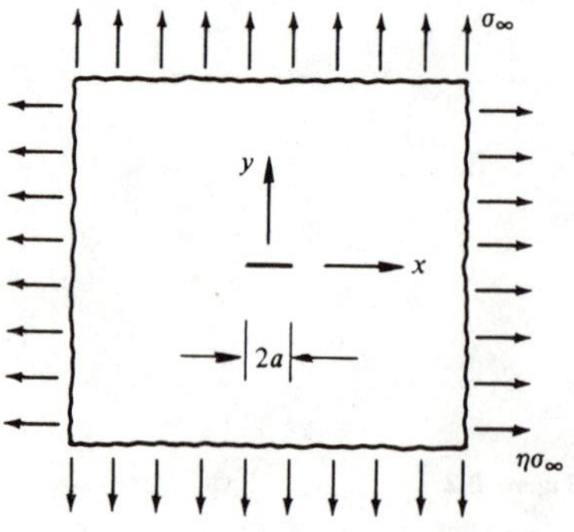

Example B.1

Likewise in step 2

$$u_{\text{I}}^{\pm} = 0 \qquad \text{for } |x| \le a$$

This means that the crack (within the present first-order frame of analysis) is deformed into an ellipse by the action considered.

Example B.2: Symmetrically and partially loaded crack For the loading case in part (*a*) of the figure we have from (B.33)

$$\phi_{\text{I}}' = \frac{p}{\pi\sqrt{z^2 - a^2}} \int_{-b}^{b} \frac{\sqrt{a^2 - \xi^2}}{z - \xi}\, d\xi \tag{a}$$

from (B.37)

$$K_{\text{I}} = \frac{p}{\sqrt{\pi a}} \int_{-b}^{b} \sqrt{\frac{a+\xi}{a-\xi}}\, d\xi = 2p\sqrt{\frac{a}{\pi}} \arcsin\frac{b}{a} \tag{b}$$

[compare (C.17)], and from (B.26)

$$u_2(x, 0) = \frac{\kappa + 1}{4G}\frac{p}{\pi} \operatorname{Im}\left[\int_a^x \int_{-b}^{b} \frac{\sqrt{a^2 - \xi^2}}{\sqrt{z^2 - a^2}(z - \xi)}\, d\xi\, dz\right] \tag{c}$$

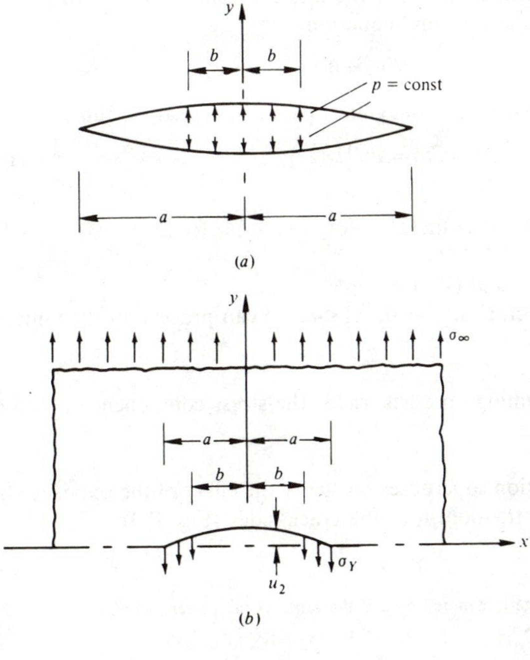

Example B.2

where the limit $z = a$ of integration has been included to satisfy $u_2 = 0$ in $z = x = a$.

If we now take $p = \sigma_Y$ in Eq. (c) and superpose the displacement due to the load case $g_2(x) = \sigma_\infty - \sigma_Y$ (compare Example B.1),

$$u_2(x, 0) = \frac{\kappa + 1}{4G}(\sigma_\infty - \sigma_Y)\,\mathrm{Im}\sqrt{x^2 - a^2}$$

we have the displacement u_2 along the crack for the situation in part (b) of the figure. Introducing finally

$$a - b \equiv c = \left[\left(\cos\frac{\pi\sigma_\infty}{2\sigma_Y}\right)^{-1} - 1\right]b$$

according to (2.44),† i.e., the length of the yield strip in a Dugdale model, we obtain

$$u_2(b, 0) \equiv \frac{\delta}{2} = \frac{4\sigma_Y b}{\pi E}\ln\left[\left(\cos\frac{\pi\sigma_\infty}{2\sigma_Y}\right)^{-1}\right]$$

with $\kappa = (3 - \nu)/(1 + \nu)$ as for plane stress. This is the result quoted in (2.46). Details of the integration of (c) have been omitted here.

B.3 ANTIPLANE STATES

In antiplane states, defined in Sec. 2.1, the displacement is normal to the x_1, x_2 plane and governed by the potential equation

$$\nabla^2 u_3 = 0$$

It can then be represented by the imaginary part of an analytic function Ω

$$Gu_3 = \mathrm{Im}\,[\Omega(z)] \tag{B.43}$$

such that

$$\sigma_{31} = G\frac{\partial u_3}{\partial x} = \mathrm{Im}\,\Omega' \qquad \sigma_{32} = G\frac{\partial u_3}{\partial y} = \mathrm{Re}\,\Omega' \tag{B.44}$$

in view of (A.31), (A.19), and (B.3).

For a small internal crack at $y = 0$, $|x| \le a$, we can proceed in the same way as in the plane problem:

Step 1 The analysis assuming no crack yields the stress component $\sigma_{32}(1)$ along $y = 0$, $|x| \le a$.

Step 2 The crack simulation superposes on step 1 the effect of the opposite stress $-\sigma_{32}(1)$ and some given traction p_3 at the crack sides (Fig. B.3).

†Note that the length formerly denoted by a is the same as the present length b.

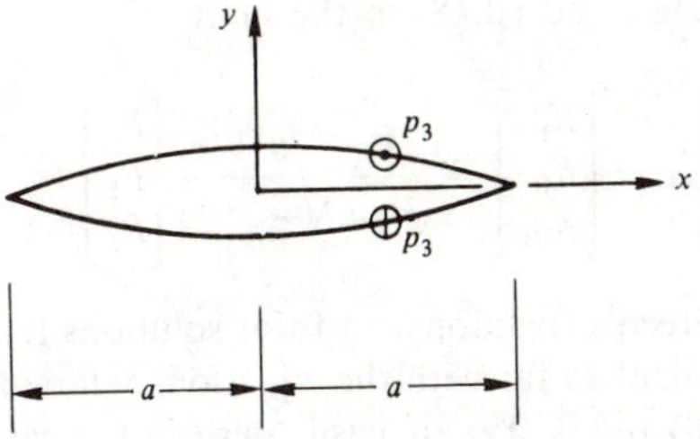

Figure B.3

Thus, we prescribe here the boundary stress

$$\sigma_{32}(2) = -\sigma_{32}(1) - p_3 \equiv -g_3(x) \tag{B.45}$$

along $y = 0$, $|x| \leq a$. The general solution to step 2 is

$$\Omega' = \frac{1}{\pi\sqrt{z^2 - a^2}} \int_{-a}^{a} \frac{g_3(\xi)\sqrt{a^2 - \xi^2}}{z - \xi}\, d\xi \tag{B.46}$$

analogous to (B.33). Close to the crack tip $z = a$ there will appear, in particular,

$$\Omega' = \frac{K_{\mathrm{III}}}{\sqrt{2\pi(z - a)}} \tag{B.47}$$

where

$$K_{\mathrm{III}} = \frac{1}{\sqrt{\pi a}} \int_{-a}^{a} g_3(\xi) \sqrt{\frac{a + \xi}{a - \xi}}\, d\xi \tag{B.48}$$

Compare (B.35) to (B.38). The introduction of $z - a = re^{i\theta}$ further provides, as $r \to 0$,

$$\begin{bmatrix} \sigma_{31} \\ \\ \sigma_{32} \end{bmatrix} = \frac{K_{\mathrm{III}}}{\sqrt{2\pi r}} \begin{bmatrix} -\sin\dfrac{\theta}{2} \\ \\ \cos\dfrac{\theta}{2} \end{bmatrix} \tag{B.49}$$

$$u_3 = \frac{K_{\mathrm{III}}}{G} \sqrt{\frac{2r}{\pi}} \sin\frac{\theta}{2} \tag{B.50}$$

identical to the singular solution ($n = 1$) in Sec. 2.1.

Example B.3: Concentrated loads on a crack Piling the load functions $g(x)$ toward the point $x = x_0$, with resultants as in the figure, we obtain from

(B.33), (B.34), (B.46), and (B.37), (B.38), and (B.48) in the limit

$$\begin{bmatrix} \phi_{\mathrm{I}}' \\ i\phi_{\mathrm{II}}' \\ \Omega' \end{bmatrix} = \frac{\sqrt{a^2 - x_0^2}}{\pi\sqrt{z^2 - a^2}(z - x_0)} \begin{bmatrix} P_2 \\ P_1 \\ P_3 \end{bmatrix} \qquad \begin{bmatrix} K_{\mathrm{I}} \\ K_{\mathrm{II}} \\ K_{\mathrm{III}} \end{bmatrix} = \frac{1}{\sqrt{\pi a}} \sqrt{\frac{a + x_0}{a - x_0}} \begin{bmatrix} P_2 \\ P_1 \\ P_3 \end{bmatrix}$$

These cases may be used as bricks (Green's functions) to form solutions for an arbitrary crack load and are equivalent so far with the equations referred to above. For instance, formulas (C.18) and (C.19) are easily derived but can also be anticipated through analogies with Example B.1.

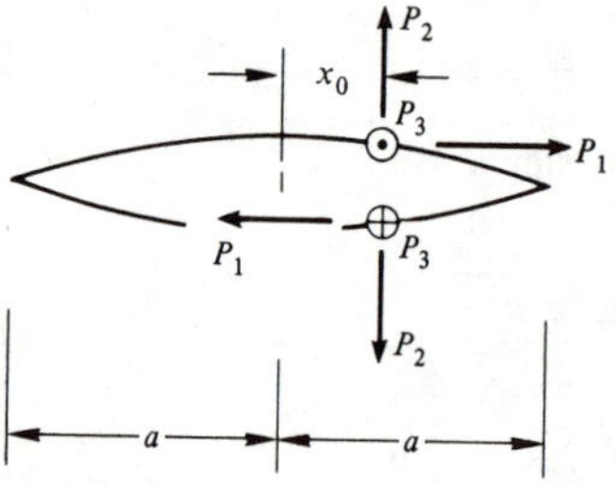

Example B.3

APPENDIX

C

STRESS INTENSITY FACTORS

Table C.1 Elliptical cracks (cases 1 and 2)

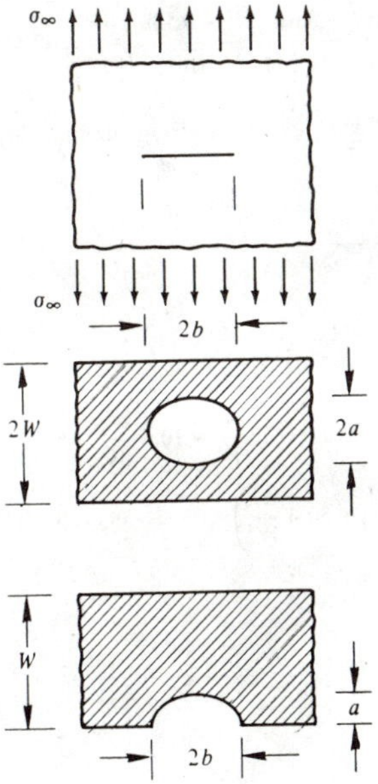

Elliptical or semielliptical (edge) crack under uniform normal remote stress, K_I referring to that point of the crack front where its value is greatest. For the ellipse (and for the semiellipse when $a/b < 0.8$) this is the symmetry point on the minor axis. Formula (C.2) is simply an approximate extrapolation from the limit $a/b = 0$. The formulas were derived for $a \ll W$ but may be used within some 10 percent error for all $a \leq 0.3W$. Further details will be found in Refs. 35, 65, and 66, among others.

By Refs. 5 and 10 to 12,

$$K_I \begin{cases} = \dfrac{1}{E_2}\sigma_\infty\sqrt{\pi a} & \text{elliptical}\dagger \quad \text{(C.1)} \\ \approx \dfrac{1.12}{E_2}\sigma_\infty\sqrt{\pi a} & \text{semielliptical} \quad \text{(C.2)} \end{cases}$$

E_2	1.0	1.016	1.051	1.097	1.151	1.211	1.277	1.345	1.418	1.493	$\pi/2$
a/b	0	0.1	0.2	0.3	0.4	0.5	0.6	0.7	0.8	0.9	1.0

† E_2 is the complete elliptic integral of the second kind with argument $\sqrt{1 - a^2/b^2}$. In the notation of Ref. 65, $E_2 \equiv \sqrt{Q}$.

Table C.2 Through-cracks in plates under remote normal stress (cases 3 to 6)

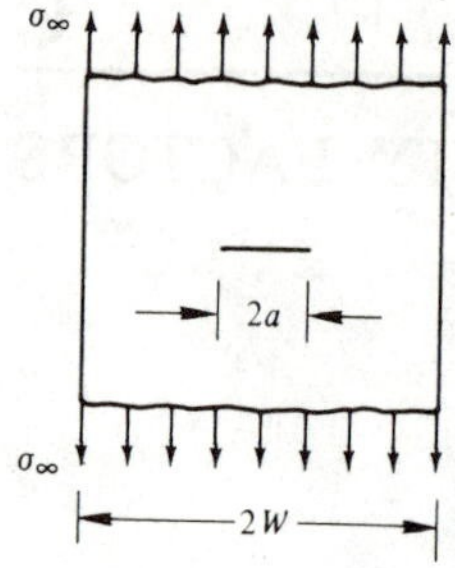

By Ref. 30, for all a/W under symmetry

$$K_\mathrm{I} = \sigma_\infty \sqrt{\pi a}\, \frac{1 - a/2W + 0.326\, a^2/W^2}{\sqrt{1 - a/W}} \tag{C.3}$$

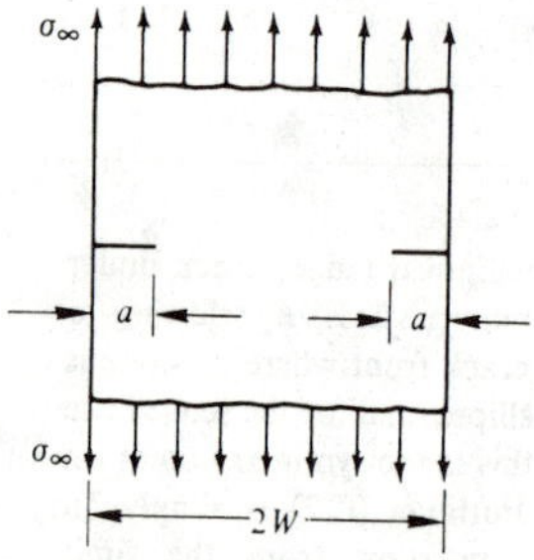

Based on Refs. 28 and 30; for all a/W

$$K_\mathrm{I} = \sigma_\infty \sqrt{\pi a}\, \frac{1.12 - 0.61a/W + 0.13a^3/W^3}{\sqrt{1 - a/W}} \tag{C.4}$$

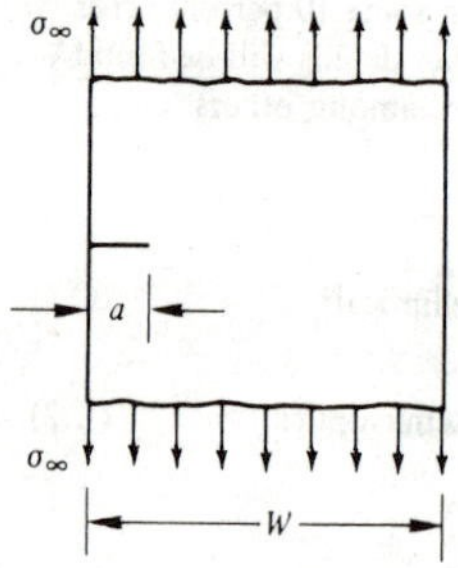

By Ref. 13, for $a/W < 0.7$

$$K_\mathrm{I} = \sigma_\infty \sqrt{\pi a}\left(1.12 - 0.23\frac{a}{W} + 10.6\frac{a^2}{W^2} - 21.7\frac{a^3}{W^3} + 30.4\frac{a^4}{W^4}\right) \tag{C.5}$$

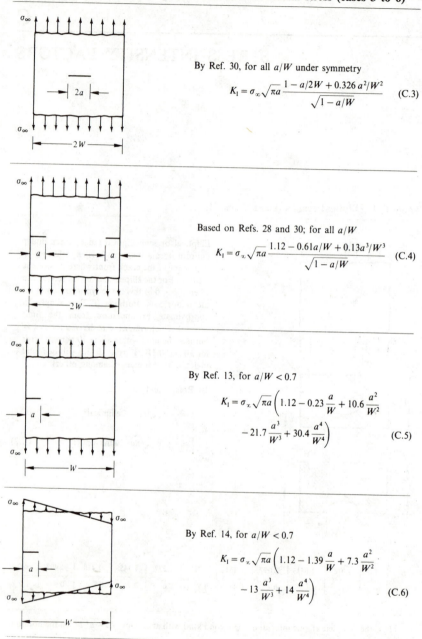

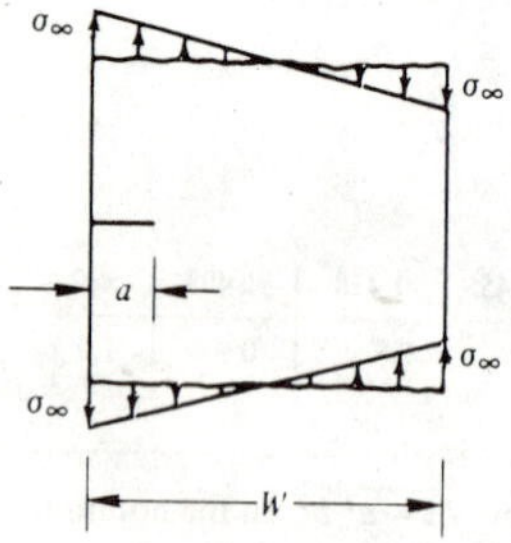

By Ref. 14, for $a/W < 0.7$

$$K_\mathrm{I} = \sigma_\infty \sqrt{\pi a}\left(1.12 - 1.39\frac{a}{W} + 7.3\frac{a^2}{W^2} - 13\frac{a^3}{W^3} + 14\frac{a^4}{W^4}\right) \tag{C.6}$$

Table C.3 Radial cracks around cylinders (cases 7 and 8)

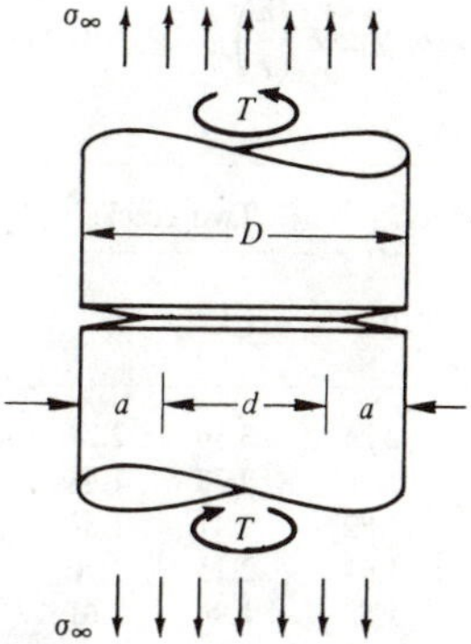

By Ref. 30, for all a/D

$$K_{\text{I}} = \sigma_{\infty}\sqrt{\pi a}\left(\frac{D}{d} + \frac{1}{2} + \frac{3}{8}\frac{d}{D} - 0.36\frac{d^2}{D^2} + 0.73\frac{d^3}{D^3}\right)\frac{1}{2}\sqrt{\frac{D}{d}} \tag{C.7}$$

$$K_{\text{III}} = \frac{16T}{\pi D^3}\sqrt{\pi a}\left(\frac{D^2}{d^2} + \frac{1}{2}\frac{D}{d} + \frac{3}{8} + \frac{5}{16}\frac{d}{D} + \frac{35}{128}\frac{d^2}{D^2} + 0.21\frac{d^3}{D^3}\right)\frac{3}{8}\sqrt{\frac{D}{d}} \tag{C.8}$$

Table C.4 Cracks in various plane geometries (cases 9 to 19)

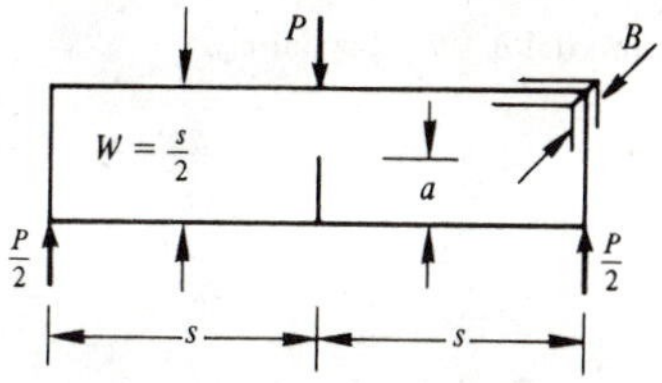

By Ref. 16, for $0.4 \le a/W \le 0.6$

$$K_{\text{I}} = 3.75\frac{PW}{B(W-a)^{3/2}} \tag{C.9}$$

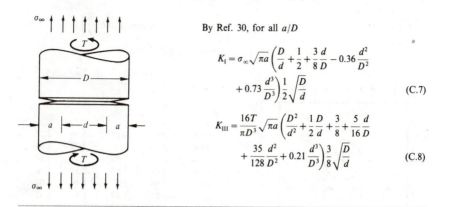

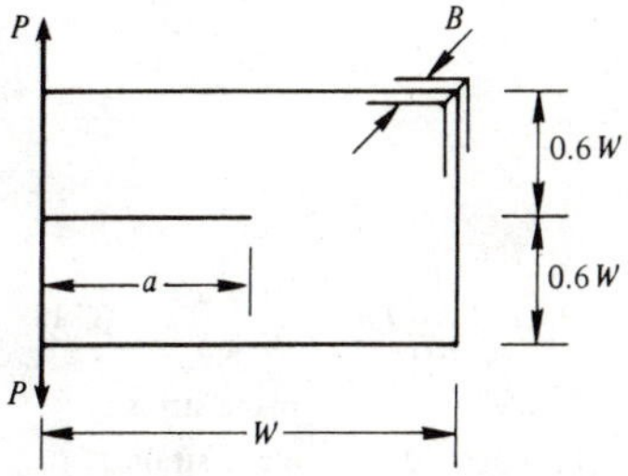

By Ref. 16, for $0.3 \le a/W \le 0.8$

$$K_{\text{I}} = 1.17\left(1.5 - \frac{a}{W} + 0.66\frac{a^2}{W^2}\right)\frac{P(2W+a)}{B(W-a)^{3/2}} \tag{C.10}$$

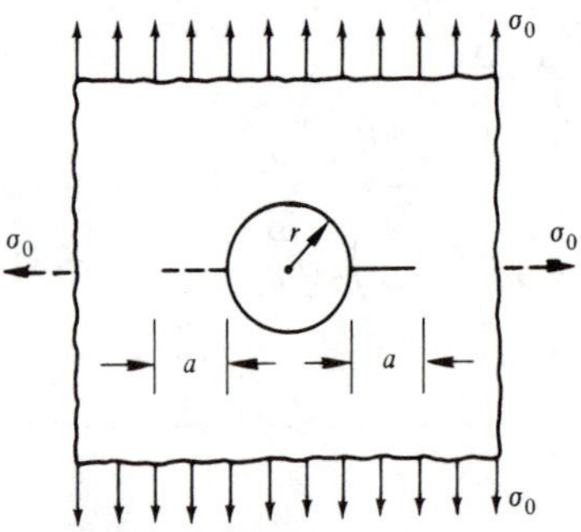

Crack(s) emanating from a hole

By Ref. 25,

$$K_1 = \sigma_0 \sqrt{\pi a}\, F\left(\frac{a}{r}\right) \qquad \text{(C.11)}$$

value of $F(a/r)$†

	One crack		Two cracks	
$\frac{a}{r}$	U	B	U	B
0.00	3.36	2.24	3.36	2.24
0.10	2.73	1.98	2.73	1.98
0.20	2.30	1.82	2.41	1.83
0.30	2.04	1.67	2.15	1.70
0.40	1.86	1.58	1.96	1.61
0.50	1.73	1.49	1.83	1.57
0.60	1.64	1.42	1.71	1.52
0.80	1.47	1.32	1.58	1.43
1.0	1.37	1.22	1.45	1.38
1.5	1.18	1.06	1.29	1.26
2.0	1.06	1.01	1.21	1.20
3.0	0.94	0.93	1.14	1.13
5.0	0.81	0.81	1.07	1.06
10.0	0.75	0.75	1.03	1.03
∞	0.707	0.707	1.00	1.00

†U = uniaxial σ_0, B = biaxial σ_0.

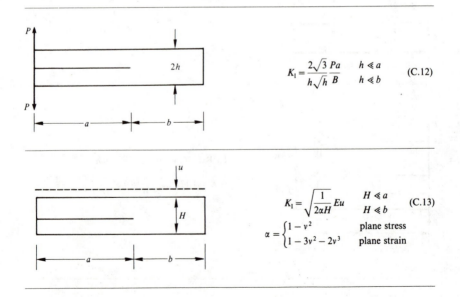

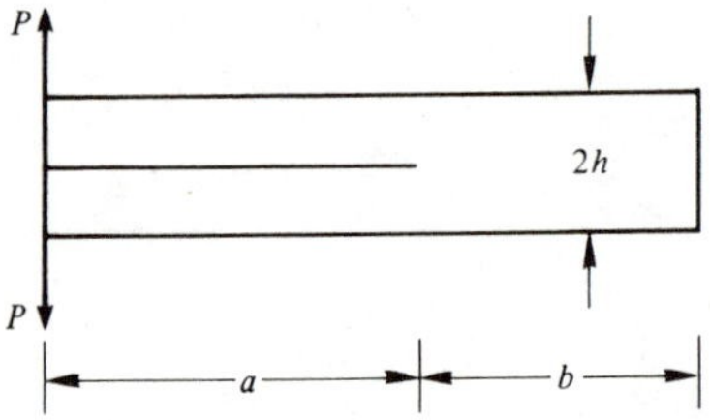

$$K_1 = \frac{2\sqrt{3}\,Pa}{h\sqrt{h}\,B} \qquad h \ll a \quad h \ll b \qquad \text{(C.12)}$$

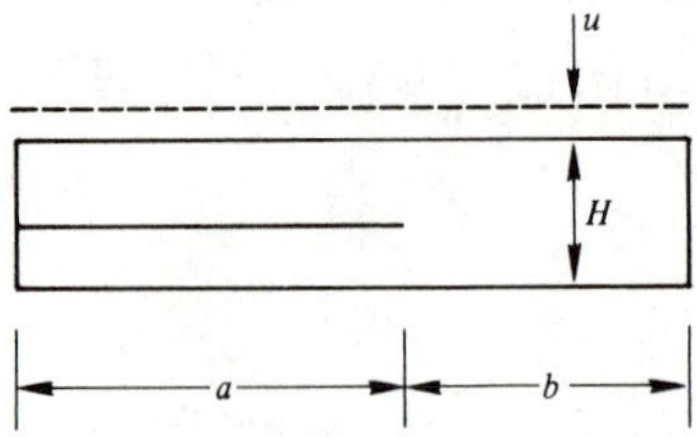

$$K_1 = \sqrt{\frac{1}{2\alpha H}}\,Eu \qquad H \ll a \quad H \ll b \qquad \text{(C.13)}$$

$$\alpha = \begin{cases} 1 - \nu^2 & \text{plane stress} \\ 1 - 3\nu^2 - 2\nu^3 & \text{plane strain} \end{cases}$$

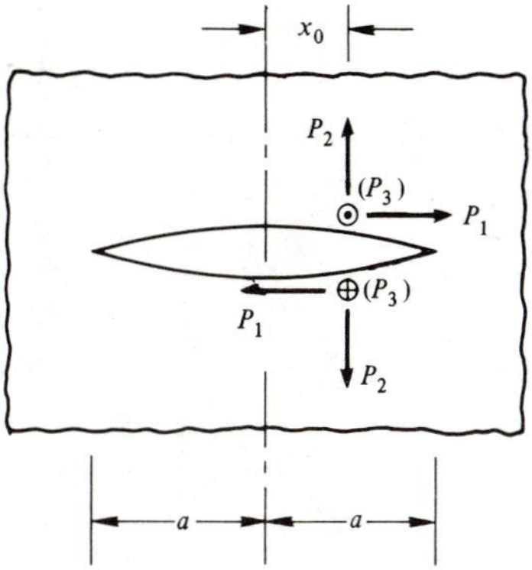

Small internal crack:

$$K_{\mathrm{I}} = \frac{P_2}{\sqrt{\pi a}} \sqrt{\frac{a + x_0}{a - x_0}} \tag{C.14}$$

$$K_{\mathrm{II}} = \frac{P_1}{\sqrt{\pi a}} \sqrt{\frac{a + x_0}{a - x_0}} \tag{C.15}$$

$$K_{\mathrm{III}} = \frac{P_3}{\sqrt{\pi a}} \sqrt{\frac{a + x_0}{a - x_0}} \tag{C.16}$$

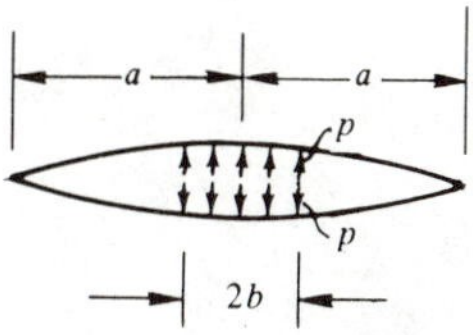

From, e.g., (C.14)

$$K_{\mathrm{I}} = \frac{2pb}{\sqrt{\pi a}} \frac{a}{b} \arcsin \frac{b}{a} \tag{C.17}$$

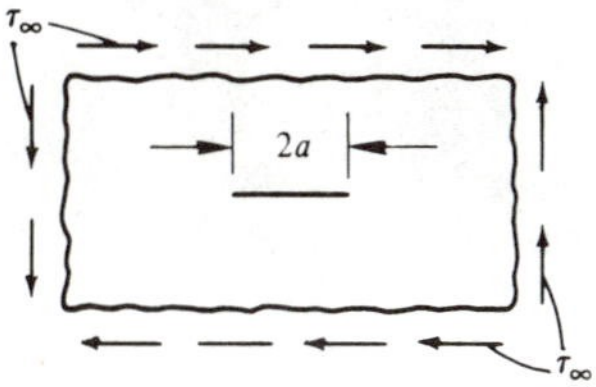

From, e.g., (C.15)

$$K_{\mathrm{II}} = \tau_\infty \sqrt{\pi a} \tag{C.18}$$

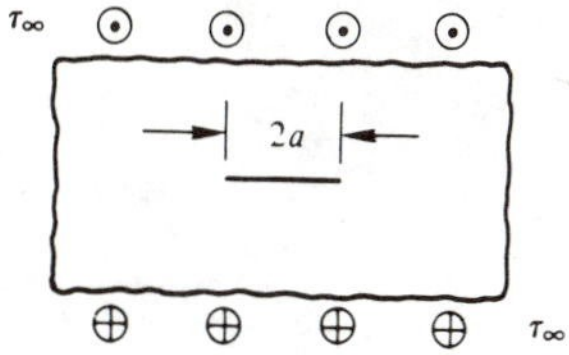

From, e.g., (C.16)

$$K_{\mathrm{III}} = \tau_\infty \sqrt{\pi a} \tag{C.19}$$

APPENDIX

D

INVARIANCE OF THE J INTEGRAL

It will be demonstrated that

$$\oint_{\Gamma} (w\,dx_2 - p_i u_{i,1}\,ds) = 0 \qquad f_{,j} \equiv \frac{\partial f}{\partial x_j}, \qquad f_{,3} = 0 \tag{D.1}$$

when the curve Γ with arc length s in the x_1, x_2 plane encloses a simply connected region Σ within which the work of deformation w is a single-valued function of the strains,

$$w = \int_0^{\varepsilon} \sigma_{ij}\,d\varepsilon_{ij} = \phi(\varepsilon_{ij}) \tag{D.2}$$

The state is assumed to be independent of x_3 and free of singularities within Σ. Body forces and inertia effects are excluded from consideration.

In the sequel use will be made of Gauss' divergence theorem in the plane

$$\oint_{\Gamma} n_j v_j\,ds = \int_{\Sigma} v_{j,j}\,d\Sigma \qquad d\Sigma \equiv dx_1\,dx_2 \tag{D.3}$$

v_j operating as vector components, where

$$\mathbf{n} = \{n_i\} = \{n_1, n_2, 0\} \tag{D.4}$$

is the outward unit normal, as well as

$$\oint_{\Gamma} \phi\,dx_2 = \int_{\Sigma} \phi_{,1}\,d\Sigma \tag{D.5}$$

where ϕ is any scalar function of x_1 and x_2.

We may envisage the curve Γ as the contour of a cylinder with unit axial extension along x_3. The second term in (D.1) can then be reformulated in view of boundary conditions (A.13)

$$p_i = n_j \sigma_{ji} \tag{D.6}$$

and (D.3) to become

$$\oint_{\Gamma} p_i u_{i,1}\, ds = \int_{\Sigma} (\sigma_{ji} u_{i,1})_{,j}\, d\Sigma \tag{D.7}$$

Further, using (D.5), we bring the total left-hand side (LHS) of (D.1) to the form

$$\text{LHS} = \int_{\Sigma} [w_{,1} - (\sigma_{ji} u_{i,1})_{,j}]\, d\Sigma \tag{D.8}$$

where

$$\begin{aligned} w_{,1} &= \frac{\partial w}{\partial \varepsilon_{ij}} \varepsilon_{ij,1} = \sigma_{ij}\varepsilon_{ij,1} = \tfrac{1}{2}\sigma_{ij}(u_{i,j} + u_{j,i})_{,1} \\ &= \tfrac{1}{2}\sigma_{ij}(u_{i,j} + u_{j,i} + u_{i,j} - u_{j,i})_{,1} = \sigma_{ji} u_{i,j1} \end{aligned} \tag{D.9}$$

In (D.9) the second equality derives from (D.2) and (A.28), the third equality from (A.19), and the fourth and the fifth from the fact that σ_{ij} is a symmetric tensor. Including the condition (A.7) for equilibrium, that is, $\sigma_{ji,j} = 0$ when $b_i = \rho \ddot{u}_i = 0$, we finally arrive at

$$w_{,1} = \sigma_{ji} u_{i,j1} + \sigma_{ji,j} u_{i,1} = (\sigma_{ji} u_{i,1})_{,j} \tag{D.10}$$

which is introduced in (D.8) to confirm (D.1).

The invariance is also implied if the material is inhomogeneous by stratification parallel to the crack (or x_1) direction. This will hold even if the transitions are discontinuous, as is the case with bimaterial interfaces along the x_1 direction.

APPENDIX

E

ANALYTICAL BASIS OF SECTION 3.2.1

This discussion presents a short review of the energy balance considering crack growth in an elastic body. First, the strain energy is viewed as depending on the crack area and controlled displacements

$$\Phi = \Phi[A, u_i(x_j)] \tag{E.1}$$

having the total differential

$$d\Phi = \left(\frac{\partial \Phi}{\partial A}\right)_u dA + (\delta\Phi)_A \tag{E.2}$$

where the last and the next-to-last term are the increments of Φ when A and $u_i(x_j)$ are fixed, respectively. The displacements can be controlled on the surface of the body but not in the interior, where body forces b_i might act. Therefore, in writing (E.2) the assumption $b_i = 0$ has tacitly been made. An alternative expression would follow from (3.14) and (3.15)

$$d\Phi = \begin{cases} -\mathcal{G}\,dA + \int_S p_i\,du_i\,dS & \text{distributed loads} \qquad \text{(E.3}a\text{)} \\ -\mathcal{G}\,dA + P_k\,du_k & \text{concentrated loads} \qquad \text{(E.3}b\text{)} \end{cases}$$

Equality of (E.2) and (E.3) for arbitrary dA and du leads to

$$\mathcal{G} = -\left(\frac{\partial \Phi}{\partial A}\right)_u \tag{E.4}$$

implying that the crack driving force can be derived as the partial derivative of the strain energy with respect to the crack area when displacements in the loaded surface are considered fixed. Further, the equality

$$(\delta\Phi)_A = \int_S p_i \, du_i \, dS \tag{E.5a}$$

or

$$(\delta\Phi)_A \equiv \left(\frac{\partial\Phi}{\partial u_k}\right)_A du_k = P_k \, du_k \tag{E.5b}$$

must hold, implying classical relations (Clapeyron's theorem, Castigliano's complementary theorem) for the intact elastic body.

If, on the other hand, the loads are viewed as controlled variables, body forces may also be included. An expedient transformation goes through introducing the complementary strain energy

$$\Omega = \int_S p_i u_i \, dS + \int_V b_i u_i \, dV - \Phi = \Omega(A, p_i, b_i) \tag{E.6a}$$

or

$$\Omega = P_k u_k - \Phi = \Omega(A, P_k) \tag{E.6b}$$

with the differential

$$d\Omega = \int_S p_i \, du_i \, dS + \int_S u_i \, dp_i \, dS + \int_V b_i \, du_i \, dV + \int_V u_i \, db_i \, dV - d\Phi$$

$$= \mathcal{G} \, dA + \int_S u_i \, dp_i \, dS + \int_V u_i \, db_i \, dV \tag{E.7a}$$

or

$$d\Omega = \mathcal{G} \, dA + u_k \, dP_k \tag{E.7b}$$

where $d\Phi$ has been introduced from (3.14) and (3.15). Equating this to the total differential

$$d\Omega = \left(\frac{\partial\Omega}{\partial A}\right)_P dA + (\delta\Omega)_A \tag{E.8}$$

for arbitrary increments dA and (dp_i, db_i) or dP_k leads to

$$\mathcal{G} = \left(\frac{\partial\Omega}{\partial A}\right)_P \tag{E.9}$$

where the index P indicates that the forces are to be considered fixed during differentiation. In addition, we must have

$$(\delta\Omega)_A = \int_S u_i \, dp_i \, dS + \int_V u_i \, db_i \, dV \tag{E.10a}$$

or

$$(\delta\Omega)_A \equiv \left(\frac{\partial\Omega}{\partial P_k}\right)_A dP_k = u_k \, dP_k \tag{E.10b}$$

again implying classical relations (Clapeyron's complementary theorem, Castigliano's theorem) for the intact elastic body.

Further, one might imagine the control to be of a mixed type, including both displacements and forces. The result is then

$$\mathscr{G} = -\frac{\partial \Pi}{\partial A} \tag{E.11}$$

with the controlled actions fixed and with Π given by

$$\Pi = \Phi - \int_{S_p} p_i u_i \, dS - \int_V b_i u_i \, dV \tag{E.12}$$

traction p_i being controlled on surface S_p. The last form generalizes those above, as $\Pi = \Phi$ for assumed displacement control and $\Pi = -\Omega$ for total load control. The proof of (E.11) follows the same lines as the proof of (E.9).

With *linear* elasticity it is a simple matter to express the energies Φ, Ω, and Π. As already implied, these scalars depend on the existing fields in a given geometry and not on how these fields have been produced. We can therefore chose the simplest path, namely that obtained when all forces and consequently all displacements are increased in the same ratio. Then, by (E.10) and (E.6) the results obviously become

$$\Omega = \frac{1}{2}\left(\int_S p_i u_i \, dS + \int_V b_i u_i \, dV\right) = \Phi \tag{E.13a}$$

or

$$\Omega = \tfrac{1}{2} P_k u_k = \Phi \tag{E.13b}$$

to be introduced in the respective relations (E.4) when $b_i = 0$, (E.9), and (E.11) and (E.12).

APPENDIX

F

ELEMENTS OF ELASTODYNAMICS

Dynamic fields and parameters are briefly discussed here in relation to the motion of a crack. Points of departure are the equations of motion (A.7), the kinematic constraint (A.19), and Hooke's law as expressed by the first of (A.31). The discussion follows essentially Refs. 21, 227, 243, and 248.

Cases of plane deformation when $u_3 \equiv u_z = 0$ are considered. A displacement field of the type $u_x = \partial\psi_1/\partial x$, $u_y = \partial\psi_1/\partial y$ is seen to be irrotational

$$|\text{curl}\,\mathbf{u}| = \left|\frac{\partial u_x}{\partial y} - \frac{\partial u_y}{\partial x}\right| = 0$$

whereas a field of the type $u_x = \partial\psi_2/\partial y$, $u_y = -\partial\psi_2/\partial x$ is isochoric or equivoluminal

$$\text{div}\,\mathbf{u} = \frac{\partial u_x}{\partial x} + \frac{\partial u_y}{\partial y} = 0$$

An arbitrary displacement field can conveniently be decomposed in terms of the potentials ψ_1 and ψ_2

$$u_x = \frac{\partial\psi_1}{\partial x} + \frac{\partial\psi_2}{\partial y} \qquad u_y = \frac{\partial\psi_1}{\partial y} - \frac{\partial\psi_2}{\partial x} \tag{F.1}$$

to yield by (A.31) the stresses

$$\begin{aligned}
\sigma_x &= 2G\left(\frac{\partial^2\psi_1}{\partial x^2} + \frac{\partial^2\psi_2}{\partial x\,\partial y} + \frac{\nu}{1-2\nu}\nabla^2\psi_1\right)\\
\sigma_y &= 2G\left(\frac{\partial^2\psi_1}{\partial y^2} - \frac{\partial^2\psi_2}{\partial x\,\partial y} + \frac{\nu}{1-2\nu}\nabla^2\psi_2\right)\\
\sigma_z &= \nu(\sigma_x + \sigma_y)\\
\tau_{xy} &= G\left(2\frac{\partial^2\psi_1}{\partial x\,\partial y} + \frac{\partial^2\psi_2}{\partial y^2} - \frac{\partial^2\psi_2}{\partial x^2}\right)
\end{aligned} \tag{F.2}$$

and by further substitution in (A.7) the equations of motion

$$c_1^2 \nabla^2 \psi_1 = \ddot{\psi}_1 \qquad c_2^2 \nabla^2 \psi_2 = \ddot{\psi}_2 \tag{F.3}$$

$$c_1 = \left(\frac{2G}{\rho}\frac{1-\nu}{1-2\nu}\right)^{1/2} = \left(\frac{E}{\rho}\frac{1-\nu}{1-\nu-2\nu^2}\right)^{1/2} \qquad c_2 = \left(\frac{G}{\rho}\right)^{1/2}$$

One can therefore conclude [compare the discussion of Eq. (5.4)] that an irrotational signal will propagate with a speed equal to c_1 while an equivoluminal signal has speed c_2. An example of the former type is a simple axial displacement: $\psi_1 = F_1(x)$, $\psi_2 = 0$, $u_x = F_1'(x)$, $u_y = 0$; and one of the latter type is a simple shear displacement: $\psi_1 = 0$, $\psi_2 = F_2(x)$, $u_x = 0$, $u_y = -F_2'(x)$, both signals traveling along the x axis.

We understand x and y above as axes fixed in space, so that within the small-displacement assumption the above acceleration terms $\ddot{u}$ and $\ddot{\psi}$ are time derivatives at a fixed location (x, y). Assuming the crack to extend along the x axis, we shall further find reason to employ local coordinates ξ and η_i ($i = 1$ or 2) having their origin at the moving crack tip. By definition

$$\xi = x - a \qquad \eta_i = \delta_i y \qquad i = 1, 2 \tag{F.4}$$

$$\delta_1 = \left[1 - \left(\frac{\dot{a}}{c_1}\right)^2\right]^{1/2} \qquad \delta_2 = \left[1 - \left(\frac{\dot{a}}{c_2}\right)^2\right]^{1/2}$$

in terms of the location $x = a$ and the speed $\dot{a}$ of the crack tip. The material time derivatives (at $x =$ const) of some physical variable $\beta[\xi(t), \eta_i, t]$ will then appear as

$$\dot{\beta} = \frac{\partial \beta}{\partial t} + \frac{\partial \beta}{\partial \xi}\frac{d\xi}{\delta t} = \frac{\partial \beta}{\partial t} - \dot{a}\frac{\partial \beta}{\partial \xi}$$

$$\ddot{\beta} = \frac{\partial \dot{\beta}}{\partial t} - \dot{a}\frac{\partial \dot{\beta}}{\partial \xi} = \frac{\partial^2 \beta}{\partial t^2} + \dot{a}^2\frac{\partial^2 \beta}{\partial \xi^2} - 2\dot{a}\frac{\partial^2 \beta}{\partial \xi\,\partial t} - \ddot{a}\frac{\partial \beta}{\partial \xi} \tag{F.5}$$

$\partial/\partial t$ denoting the local rate at $\xi =$ const. Polar coordinates may also be conveniently invoked; that is, $r_i = \sqrt{\xi^2 + \eta_i^2}$ and $\theta_i = \tan^{-1}(\eta_i/\xi)$, implying [with no sum over i to be understood in (F.6) to (F.9)]

$$\begin{aligned} \frac{\partial \beta}{\partial \xi} &= \cos\theta_i \frac{\partial \beta}{\partial r_i} - \frac{\sin\theta_i}{r_i}\frac{\partial \beta}{\partial \theta_i} = \frac{\partial \beta}{\partial x} \\ \frac{\partial \beta}{\partial \eta_i} &= \sin\theta_i \frac{\partial \beta}{\partial r_i} + \frac{\cos\theta_i}{r_i}\frac{\partial \beta}{\partial \theta_i} = \frac{1}{\delta_i}\frac{\partial \beta}{\partial y} \end{aligned} \tag{F.6}$$

Identifying β as ψ_i, we substitute from (F.5) in (F.3), to obtain the local formulation

$$\delta_i^2 c_i^2 \nabla_i^2 \psi_i = \frac{\partial^2 \psi_i}{\partial t^2} - 2\dot{a}\frac{\partial^2 \psi_i}{\partial \xi\,\partial t} + \ddot{a}\frac{\partial \psi_i}{\partial \xi} \tag{F.7}$$

where $\partial\psi_i/\partial\xi$ can be transformed to polar coordinates by means of (F.6), and

$$\nabla_i^2 \equiv \frac{\partial^2}{\partial\xi^2} + \frac{\partial^2}{\partial\eta_i^2} = \frac{\partial^2}{\partial r_i^2} + \frac{1}{r_i}\frac{\partial}{\partial r_i} + \frac{1}{r_i^2}\frac{\partial^2}{\partial\theta_i^2}$$

A solution to (F.7) is sought, in rather close analogy to the power-series approach of Sec. 2.1. As suggested by Nilsson [243],

$$\psi_i = r_i^{3/2} f_i(\theta_i, t) + g_i(r_i, \theta_i, t) \tag{F.8}$$

where a smallest permissible exponent (corresponding to $1 + \lambda$ in Sec. 2.1) equal to $\frac{3}{2}$ is anticipated. This insight has been inspired by the elastostatic results and can be verified in the sequel. $f_i(\theta_i, t)$ and $g_i(r_i, \theta_i, t)$ are assumed to be nonsingular to within the second derivatives, and the form (F.8) is then the most general one which leaves us with an $r^{-1/2}$ singularity in stresses. The substitution of (F.8) in (F.7), the polar form, leads next to

$$\begin{aligned} \tfrac{9}{4} f_i + \frac{\partial^2 f_i}{\partial\theta_i^2} = \frac{r_i}{\delta_i^2 c_i^2}\Bigg[& r_i \frac{\partial^2 f_i}{\partial t^2} - 3\dot{a}\cos\theta_i \frac{\partial f_i}{\partial t} + 2\dot{a}\sin\theta_i \frac{\partial^2 f_i}{\partial\theta_i\,\partial t} \\ & - \tfrac{3}{2}\ddot{a}\cos\theta_i f_i + \ddot{a}\sin\theta_i \frac{\partial f_i}{\partial\theta_i}\Bigg] + r_i^{1/2} h_i(r_i, \theta_i, t) \end{aligned} \tag{F.9}$$

where h_i includes the derivatives of g_i.

A state of *steady growth* may be achieved, in the sense that the crack moves at constant speed with time-independent fields sufficiently close to its tip; $\partial f_i/\partial t = 0$. In this case the right-hand bracket equals zero, by definition. However, in focusing on the region where the leading term of (F.8) dominates, $r_i \to 0$, we find that the total right-hand side of (F.9) is negligible under *all* circumstances. Accordingly, f_i must comprise the simple terms $\sin(3\theta_i/2)$ and $\cos(3\theta_i/2)$, with time dependence under general conditions entering only through their scaling factors.

A symmetry consideration pertaining to mode I growth leaves the solution

$$\psi_1 = B_1 r_1^{3/2} \cos\frac{3\theta_1}{2} \qquad \psi_2 = B_2 r_2^{3/2} \sin\frac{3\theta_2}{2} \tag{F.10}$$

where the factors $B_1(t)$ and $B_2(t)$ should be adapted to traction-free crack faces

$$\sigma_y(\theta = \theta_i = \pi) = 0 \qquad \tau_{xy}(\theta = \theta_i = \pi) = 0 \tag{F.11}$$

With some manipulation one gets from either of Eqs. (F.11) (since the eigenvalue $\frac{3}{2}$ was anticipated)

$$B_2 = -\frac{2B_1\delta_1}{1 + \delta_2^2} \tag{F.12}$$

and from (F.2) the corresponding near-tip stress field

$$\sigma_x = \tfrac{3}{4}GB_1\left[(1+2\delta_1^2-\delta_2^2)\frac{\cos(\theta_1/2)}{\sqrt{r_1}} - \frac{4\delta_1\delta_2}{1+\delta_2^2}\frac{\cos(\theta_2/2)}{\sqrt{r_2}}\right]$$
$$\sigma_y = \tfrac{3}{4}GB_1\left[-(1+\delta_2^2)\frac{\cos(\theta_1/2)}{\sqrt{r_1}} + \frac{4\delta_1\delta_2}{1+\delta_2^2}\frac{\cos(\theta_2/2)}{\sqrt{r_2}}\right] \tag{F.13}$$
$$\tau_{xy} = \tfrac{3}{2}GB_1\left[\frac{\sin(\theta_1/2)}{\sqrt{r_1}} - \frac{\sin(\theta_2/2)}{\sqrt{r_2}}\right]$$

Finally when we rename the factor B_1 as

$$B_1 = \frac{4(1+\delta_2^2)}{3G[4\delta_1\delta_2-(1+\delta_2^2)^2]}\frac{K_\mathrm{I}}{\sqrt{2\pi}} \tag{F.14}$$

we recover a familiar result ahead of the crack tip, $\sigma_y = K_\mathrm{I}/\sqrt{2\pi\xi}$ as in the static case, but with K_I having the meaning here of a dynamic, time-dependent stress intensity factor.

Since

$$\tan\theta_i = \delta_i \tan\theta \qquad \text{and} \qquad r_i = r[1-(1-\delta_i^2)\sin^2\theta]^{1/2}$$

the result (F.13) is obviously consistent with the declared form (5.32). It also approaches (B.39) as $\dot{a}\to 0$, which becomes apparent if some care is exercised in going to this limit. A ratio given by (F.13) and reproduced in Fig. 5.9 is

$$\frac{\sigma_y(\theta=0)}{\sigma_x(\theta=0)} = \frac{4\delta_1\delta_2-(1+\delta_2^2)^2}{(1+\delta_2^2)(1+2\delta_1^2-\delta_2^2)-4\delta_1\delta_2} \tag{F.15}$$

which goes to zero when the crack speed $\dot{a}$ approaches the speed of the Rayleigh wave, defined by the vanishing of the numerator. The stresses by (F.13) can also be transformed into polar components σ_r, $\tau_{r\theta}$, and σ_θ, the latter having interesting directional properties illustrated in Fig. 5.8. Particle velocities appear as material derivatives by (F.1); thus

$$\dot{u}_x = -\frac{3}{4}\frac{B_1}{\sqrt{r}}\dot{a}\left[\frac{\cos(\theta_1/2)}{\sqrt{r_1}} - \frac{2\delta_1\delta_2}{1+\delta_2^2}\frac{\cos(\theta_2/2)}{\sqrt{r_2}}\right]$$
$$\dot{u}_y = -\frac{3}{4}\frac{B_1}{\sqrt{r}}\dot{a}\delta_1\left[\frac{\sin(\theta_1/2)}{\sqrt{r_1}} - \frac{2}{1+\delta_2^2}\frac{\sin(\theta_2/2)}{\sqrt{r_2}}\right] \tag{F.16}$$

We finally turn to the problem of energy transfer associated with a moving crack tip. A contour Γ joining the crack faces as in Fig. 2.12 is defined, and the line integral in (F.17),

$$\mathcal{G} = \frac{1}{\dot{a}}\left\{\lim_{\Gamma\to 0}\int_\Gamma [p_i\dot{u}_i\,ds + \dot{a}(\phi + \tfrac{1}{2}\rho\dot{u}_i\dot{u}_i)\,dy]\right\} \tag{F.17}$$

with p_i according to (D.4) and (D.6), is considered. The integrated first term of the bracket is the rate of work exerted per unit thickness from the outside material on that inside Γ. The integrated second term can be interpreted as the total elastic and kinetic energy carried by particles through Γ, per unit time and thickness, if that contour is seen as stationary with respect to the moving crack tip. When Γ is shrunk toward the tip, the sum therefore becomes the energy extracted per unit time from the surrounding continuum for crack motion to be maintained. Again, this equals the crack speed times $\mathcal{G}$ (or C), the energy consumed per unit of crack advance.

Obviously, the above line integral must turn into the J integral in the limit as $\dot{a} \to 0$ (see also Secs. 2.6 and 3.2.3); however, in the dynamic case we cannot expand the path of integration to enclose any finite region, conserving the integral, since wavelike disturbances are not simultaneously sensed throughout a body. Only if (hypothetically) the fields enclosed by alternative paths are stationary with respect to the crack tip can such path independence be claimed.

The introduction of (F.13) and (F.16) in (F.17) results in (5.33) and (5.34). A case of plane stress will deviate but slightly from the plane strain considered here, mainly in that the plate velocity c_p replaces c_1.

APPENDIX

G

SOME CONVERSION FACTORS

For older forms used in the literature current SI forms and forms consistent with SI patterns are indicated in the footnotes.

Table G.1 Stress

	$MN\,m^{-2}$†	ksi‡	$N\,mm^{-2}$	$kp\,mm^{-2}$§
1 $MN\,m^{-2}$† equals	1	0.1450	1	0.1019
1 ksi‡	6.895	1	6.895	0.7031
1 $N\,mm^{-2}$	1	0.1450	1	0.1019
1 $kp\,mm^{-2}$§	9.807	1.4223	9.807	1

†1 $MN\,m^{-2} \equiv 1$ MPa.

‡Initial-letter abbreviations are gradually being replaced by forms more in line with SI forms; kips in^{-2} is now preferred to ksi.

§The kilopond (kp) was used in Europe as a force unit prior to SI forms; 1 kp is equal to 1 kilogram force.

Table G.2 Stress intensity factor, fracture toughness

	$MN\,m^{-3/2}$†	ksi $\sqrt{in}$‡	$N\,mm^{-3/2}$	$kp\,mm^{-3/2}$§
1 $MN\,m^{-3/2}$† equals	1	0.910	31.62	3.224
1 ksi $\sqrt{in}$‡	1.099	1	34.75	3.542
1 $N\,mm^{-3/2}$	0.03162	0.02878	1	0.1019
1 $kp\,mm^{-3/2}$§	0.3102	0.2823	9.807	1

†1 $MN\,m^{-3/2} \equiv 1$ MPa $m^{1/2}$.

‡In keeping with the second note to Table G.1, the form kips $in^{-3/2}$ is preferred.

§See the corresponding note to Table G.1.

APPENDIX
H
BIBLIOGRAPHY

Some of the relevant literature is listed and commented on briefly. Reference is made to the background material applied directly and to some additional work which may provide data and clues to further studies or consultations. No attempt has been made to present a total review, and the basis for selection is subjective and sometimes arbitrary. Typical of the subject in its present documented state is the large scatter into specialized articles, with reviews and textbooks in the minority. The reason may be the high priority given to pioneering research in several directions, occupying the time and complicating the task of would-be chroniclers.

The timing of this work has essentially excluded from consideration material published after 1981.

CHAPTER 1

1. Inglis, C. E.: Stresses in a Plate Due to the Presence of Cracks and Sharp Corners, *Trans. Inst. Nav. Archit.*, **55:**219–241 (1913).
2. Griffith, A. A.: The Phenomena of Rupture and Flow in Solids, *Phil. Trans. R. Soc. London*, **A221:**163–198 (1921).
3. Griffith, A. A.: The Theory of Rupture, pp. 55–63 in *Proc. 1st Int. Cong. Appl. Mech., Delft, 1925.*
4. Gordon, J. E.: "The New Science of Strong Materials," Penguin Books, Harmonsworth, England, 1968.
5. Irwin, G. R.: Fracture, pp. 551–590 in S. Flügge (ed.), "Encyclopedia of Physics," vol. 6, Springer, Berlin, 1958.
6. Knott, J. F.: "Fundamentals of Fracture Mechanics," Butterworths, London, 1973.
7. Broek, D.: "Elementary Engineering Fracture Mechanics," 3. ed., Nijhoff, The Hague, 1982.
8. Rivlin, R. S., and A. G. Thomas: *J. Polym. Sci.*, **10:**291–318 (1953).

Discussion

Reference 1 is a pioneering work on the analysis of stresses in notched elastic plates. It provided a necessary basis for the two papers of Griffith [2, 3] which are outstanding in the history of strength of materials, both for the idea they advanced and for the example they set in method and presentation. The subsequent total silence after this outburst is partly explained by a bibliographical note in Ref. 4. Indeed, so singly involved and so far ahead was Griffith that later progress in the field of fracture mechanics had to wait until the 1940s. Prominent in the new development was Irwin, whose article [5] is partly a review and systematization, and partly the introduction of new working concepts. References 6 and 7 are widely recognized textbooks, having their roots in the fields of materials science and experimental mechanics but differing considerably in emphasis.

Example 1.1, concerned with a crack along a narrow plate with kinematical constraints, is possibly the world's simplest example of a fracture-mechanical analysis. Our particular deduction seems to have been described first in Ref. 8, with a more detailed discussion of the energy release.† An alternative analysis due to Rice [101] is presented as Example 3.4.

CHAPTER 2

9. Williams, M. L.: *J. Appl. Mech.*, **24:**109–114 (1957).
* Irwin [5].
10. Irwin, G. R., *J. Appl. Mech.*, **29:**651–654 (1962).
11. Sneddon, J. N.: *Proc. Phys. Soc. London*, **187:**229–260 (1946).
12. Green, A. E., and J. N. Sneddon: *Cambridge Phil. Soc.*, **46:**159–163 (1950).
13. Gross, B., J. E. Srawley, and W. F. Brown, Jr.: *NASA Tech. Note* D-2395, 1964.
14. Gross, B., and J. E. Srawley: *NASA Tech. Note* D-2603, 1965.
15. Brown, W. F., Jr., and J. E. Srawley: Plane Strain Toughness Testing of High Strength Metallic Materials, *ASTM STP* 410, 1967.
16. Gross, B., and J. E. Srawley: *Eng. Fract. Mech.*, **4:**587–589 (1972).
17. Muskhelishvili, N. I.: "Some Basic Problems of the Mathematical Theory of Elasticity," Noordhoff, Groningen, 1963.
18. Westergaard, H. M.: *J. Appl. Mech.*, **61:**A-49 (1939).
19. Sedov, L. I.: "A Course in Continuum Mechanics," vol. 4, Volters-Noordhoff, Groningen, 1972.
20. Sih, G. C., and H. Liebowitz: pp. 67–190 in [20*a*].
20*a*. Liebowitz, H. (ed.): "Mathematical Fundamentals," vol. 2 of "Fracture," Academic, New York, 1968.
21. Rice, J. R.: pp. 191–311 in [20*a*].
22. Isida, M.: *J. Appl. Mech.*, **33:**674–675 (1966).
23. Isida, M.: *Eng. Fract. Mech.*, **2:**61–79 (1970).
24. Isida, M.: *Eng. Fract. Mech.*, **5:**647–665 (1973).
25. Bowie, O. L.: *J. Math. Phys.*, **35:**60–71 (1956).
26. Bowie, O. L., and D. M. Neal: *Eng. Fract. Mech.*, **2:**181–182 (1979).

†Disturbances to the homogeneous strain fields within the plate occur locally at its right edge and, more severely, in a transition region which includes the crack tip. They can be formally accounted for by adding a right-hand term $\Delta\Phi$ to (1.3), such that $\Delta\Phi/\Phi \to 0$ as $b/H \to \infty$. Insofar as the latter disturbance mainly translates with the crack tip, $\Delta\Phi$ will operate as a constant and (1.5) will be in accord with this sharper estimation. Indeed, it turns out that hardly more than $a > H$ and $b > H$ is required for (1.5) to apply if the crack moves centrally.

27. Bowie, O. L.: *J. Appl. Mech.*, **31:**208–212 (1964).
28. Keer, L. M., and J. M. Freedman: *Int. J. Eng. Sci.*, **11:**1265–1275 (1973).
29. Bueckner, H.: pp. 82–83 in Fracture Toughness Testing and Its Application, *ASTM STP* 381, 1965.
30. Koiter, W. T., and J. P. Benthem: pp. 131–178, in G. C. Sih (ed.), "Mechanics of Fracture," Noordhoff, Leyden, vol. 1, 1973.
31. Feddersen, C.: Discussion, p. 77, Plane Strain Crack Toughness Testing, *ASTM STP* 410, 1967.
32. Hellan, K.: Elementer av Analysen i Statisk Bruddmekanikk, *Rep. Inst. Mek.*, NTH, Trondheim, 1973.
33. Paris, P. C., and G. C. Sih: pp. 31–81 in Fracture Toughness Testing and Its Application, *ASTM STP* 381, 1965.
34. Sih, G. C.: "Handbook of Stress Intensity Factors," Lehigh University, Bethlehem, Pa., 1973.
35. Rooke, D. P., and D. J. Cartwright: "Compendium of Stress Intensity Factors," H.M. Stationery Office, London, 1976.
36. Tada, H., P. C. Paris, and G. R. Irwin: "The Stress Analysis of Cracks Handbook," Del Research, Hellertown, Pa., 1973.
37. Formelsamling i Hållfasthetslära, *Inst. Haallfasthetslaera, KTH, Stockholm Publ.* 104, 1978.
38. Carlsson, J.: "Brottmekanik," Ingenjörsförlaget, Stockholm, 1976.
39. Chan, S. K., I. S. Tuba, and W. K. Wilson: *Eng. Fract. Mech.*, **2:**1–17 (1970).
40. Mowbray, D. F.: *Eng. Fract. Mech.*, **2:**173–176 (1970).
41. Aamodt, B., P. G. Bergan, and H. F. Klem: pp. 911–921 in *Proc. 2d Int. Conf. Pressure Vessel Technol., San Antonio, 1973.*
42. Parks, D. M.: *Int. J. Fract.*, **10:**487–502 (1974).
43. Hellen, T. K.: *Int. J. Numer. Methods Eng.*, **9:**187–207 (1975).
44. Hayes, D. J.: *Int. J. Fract. Mech.*, **8:**157–165 (1972).
45. Tracey, D. M.: *Eng. Fract. Mech.*, **3:**255–265 (1971).
46. Benzley, S. E.: *Int. J. Numer. Methods Eng.*, **8:**537–545 (1974).
47. Henshell, R. D., and K. G. Shaw: *Int. J. Numer. Methods Eng.*, **9:**495–507 (1975).
48. Barsoum, R. S.: *Int. J. Numer. Methods Eng.*, **10:**551–564 (1976).
49. Bäcklund, J., F. Nilsson, and K. Markström: *Int. J. Fract.*, **13:**250–252 (1977).
50. Bueckner, H. F.: *Trans. ASME*, **80:**1225–1229 (1958).
51. Bueckner, H. F.: *Z. Angew. Math. Mech.*, **50:**529–540 (1970).
52. Rice, J. R.: *Int. J. Solids Struct.*, **8:**751–758 (1972).
53. Carlsson, A. J.: Use of the Reciprocity Relation for Determination of Crack Parameters, pts. I and II, *Inst. Haallfasthetslaera, KTH, Stockholm Rep.* 2, 3, 1975.
54. Labbens, R. C., J. Heliot, and A. Pellissier-Tanon, pp. 448–470 in Cracks and Fracture, *ASTM STP* 601, 1976.
55. Paris, P. C., R. M. McMeeking, and H. Tada: pp. 471–489 in Cracks and Fracture, *ASTM STP* 601, 1976.
56. Petroski, H. J., and J. D. Achenbach: *Eng. Fract. Mech.*, **10:**257–266 (1978).
57. Bowie, O. L., and P. G. Tracy: *Eng. Fract. Mech.*, **10:**249–256 (1978).
58. Petroski, H. J.: *Int. J. Fract.*, **15:**217–230 (1979).
59. Cruse, T. A.: *Int. J. Solids Struct.*, **5:**1259–1274 (1969).
60. Cruse, T. A., and W. Van Buren: *Int. J. Fract. Mech.*, **7:**1–15 (1971).
61. Brebbia, C. A.: "The Boundary Element Method for Engineers," Pentch, London, 1980.
62. Shah, R. C., and A. S. Kobayashi: *Int. J. Fract.* **9:**133–146 (1973).
63. Hartranft, R. J., and G. C. Sih: pp. 179–238, in G. C. Sih (ed.), "Mechanics of Fracture," Noordhoff, Leyden, vol. 1, 1973.
64. Smith, F. W., and D. R. Sorensen: *Int. J. Fract.*, **12:**47–57 (1976).
65. Raju, I. S., and J. C. Newman, Jr.: *Eng. Fract. Mech.*, **11:**817–829 (1979).
66. Hayashi, K., and H. Abé: *Int. J. Fract.*, **16:**275–285 (1980).
67. Tan, C. L., and R. T. Fenner: *Int. J. Fract.*, **16:**233–245 (1980).
68. Williams, M.L.: *J. Appl. Mech.*, **28:**78–82 (1961).
69. Sih, G. C., P. C. Paris, and F. Erdogan: *J. Appl. Mech.*, **29:**306–312 (1962).

70. Isida, M.: pp. 1–43, in G. C. Sih (ed.), "Mechanics of Fracture," Noordhoff, Leyden, vol. 3, 1977.
71. Heming, F. S., Jr.: *Int. J. Fract.*, **16**:289–304 (1980).
72. Folias, E. S.: *Int. J. Fract. Mech.*, **1**:20–46 (1965).
73. Folias, E. S.: *Int. J. Fract. Mech.*, **1**:104–113 (1965).
74. Folias, E. S.: *Int. J. Fract. Mech.*, **3**:1–11 (1967).
75. Erdogan, F., and J. J. Kibler: *Int. J. Fract. Mech.*, **5**:229–237 (1969).
76. Erdogan, F.: pp. 161–199, in G. C. Sih (ed.), "Mechanics of Fracture," Noordhoff, Leyden, vol. 3, 1977.
77. Barsoum, R. S., R. W. Loomis, and B. D. Stewart: *Int. J. Fract.*, **15**:259–280 (1979).
78. Bowie, O. L., and C. E. Freese: *Int. J. Fract. Mech.*, **8**:49–58 (1972).
79. Snyder, M. D., and T. A. Cruse: *Int. J. Fract.*, **11**:315–328 (1975).
80. Konish, H. J., Jr.: pp. 99–116, in Fracture Mechanics of Composites, *ASTM STP* 593, 1975.
81. Williams, M. L.: *Bull. Seismolog. Soc. Am.*, **49**:199–204 (1959).
82. Rice, J. R., and G. C. Sih: *J. Appl. Mech.*, **32**:418–423 (1965).
83. Malyshev, B. M., and R. L. Salagnik: *Int. J. Fract. Mech.*, **1**:119–127 (1965).
84. Erdogan, F.: *Eng. Fract. Mech.*, **4**:811–840 (1972).
85. Ashbaugh, N.: *Int. J. Fract.*, **11**:205–219 (1975).
86. Bogy, D. B.: *J. Appl. Mech.*, **38**:377–386 (1971).
87. Theocaris, P. S., and E. E. Gdoutos: *Int. J. Fract.*, **13**:763–773 (1977).
88. Atkinson, C.: *Int. J. Fract.*, **13**:807–820 (1977).
89. Cominou, M.: *J. Appl. Mech.*, **4**:631–636, 780–781 (1977).
90. Sanders, J. L.: *NASA Tech. Rep.* R13, 1959.
91. Poe, C. C.: pp. 79–97 in Damage Tolerance in Aircraft Structures, *ASTM STP* 486, 1971.
92. Vlieger, H.: *Eng. Fract. Mech.*, **5**:447–477 (1973).
93. Irwin, G. R.: pp. 63–78 in *Proc. 7th Sagamore Ordnance Mater. Res. Conf.*, vol. 4, Syracuse University Press, Syracuse, N.Y., 1961.
94. Dugdale, D. S.: *J. Mech. Phys. Solids*, **8**:100–104 (1960).
95. Barenblatt, G. I.: pp. 56–131, in H. L. Dryden and T. von Karman (eds.), "Advances in Applied Mechanics," Academic, New York, vol. 7, 1962.
96. Goodier, J. N.: Mathematical Theory of Equilibrium Cracks, pp. 1–66 in [20*a*].
97. Goodier, J. N. and F. A. Field: pp. 103–118 in D. C. Drucker and J. J. Gilman (eds.), "Fracture of Solids," Gordon and Breach, New York, 1963.
98. Bilby, B. A., A. H. Cottrell, and K. H. Swindon: *Proc. R. Soc.* **A272**:304–314 (1963).
99. Eshelby, J. D.: *Solid State Phys.*, **3**:79–144 (1956).
100. Sanders, J. L.: *J. Appl. Mech.*, **27**: 352–353 (1960).
101. Rice, J. R.: *J. Appl. Mech.*, **35**:379–386 (1968).
102. Cherepanov, G. P.: *Int. J. Solids Struct.*, **4**:811–831 (1968).
103. Hult, J., and F. A. McClintock: pp. 51–58 in *Proc. 9th Int. Cong. Appl. Mech., Brussels, 1956.*
104. Rice, J. R.: *Int. J. Fract. Mech.*, **2**:426–447 (1966).
105. Rice, J. R.: *J. Appl. Mech.*, **34**:287–298 (1967).
106. Rice, J. R.: *J. Mech. Phys. Solids*, **22**:17–26 (1974).
107. Levy, N., P. V. Marcal, W. J. Ostergren, and J. R. Rice: *Int. J. Fract. Mech.*, **7**:143–156 (1971).
108. Tracey, D. M.: *J. Eng. Mat. Technol.*, **98**:146–151 (1976).
109. Sorensen, E. P.: pp. 151–174 in Elastic-Plastic Fracture, *ASTM STP* 668, 1979.
110. Rice, J. R., and G. F. Rosengren: *J. Mech. Phys. Solids*, **16**:1–12 (1968).
111. Hutchinson, J. W.: *J. Mech. Phys. Solids*, **16**:13–31 (1968).
112. McClintock, F.A.: pp. 48–227 in [112*a*].
112*a*. Liebowitz, H. (ed.): "Engineering Fundamentals and Environmental Effects," vol. 3 of "Fracture," Academic, New York, 1969.
113. Hutchinson, J. W.: *J. Mech. Phys. Solids*, **16**:337–347 (1968).
114. Hilton, P. D., and J. W. Hutchinson: *Eng. Fract. Mech.*, **3**:435–451 (1971).
115. Goldman, N. L., and J. W. Hutchinson: *Int. J. Solids Struct.*, **11**:575–591 (1975).
116. Amazigo, J. C.: *Int. J. Solids Struct.*, **10**:1003–1015 (1974).

117. McMeeking, R. M.: *J. Mech. Phys. Solids*, **25**:357–381 (1977).
118. Chitaley, A. D., and F. A. McClintock: *J. Mech. Phys. Solids*, **19**:147–163 (1971).
119. Amazigo, J. C., and J. W. Hutchinson: *J. Mech. Phys. Solids*, **25**:81–97 (1977).
120. Rice, J. R., and E. P. Sorensen: *J. Mech. Phys. Solids*, **26**:163–186 (1978).
121. Rice, J. R., W. J. Drugan, and T-L. Sham: pp. 189–221 in Fracture Mechanics, *ASTM STP* 700, 1980.
122. Johnson, W., and P. B. Mellor: "Engineering Plasticity," Van Nostrand Reinhold, London, 1973.
123. Hill, R.: *J. Mech. Phys. Solids*, **1**:19–30 (1952).
124. Hill, R., and J. W. Hutchinson: *J. Mech. Phys. Solids*, **23**:239–264 (1975).
125. Needleman, A.: *J. Mech. Phys. Solids*, **20**:111–127 (1972).
126. Storåkers, B.: *Arch. Mech.*, **27**:821–839 (1975).
127. Hill, R.: pp. 1–75, in C. C. Yih (ed.), "Advances in Applied Mechanics," Academic, New York, vol. 18, 1978.
128. Miles, J. P.: *Arch. Mech.*, **32**:909–931 (1980).
129. Marciniak, Z., and K. Kuczynski: *Int. J. Mech. Sci.*, **9**:609–620 (1967).
130. Støren, S., and J. R. Rice: *J. Mech. Phys. Solids*, **23**:421–441 (1975).
131. Rice, J. R.: pp. 207–220 in *Proc. 14th Int. Cong. Theoret. Appl. Mech.*, North-Holland, Delft, 1976.
132. Hutchinson, J. W., and V. Tvergaard: *Int. J. Solids Struct.*, **17**:451–470 (1980).
133. Tvergaard, V., A. Needleman, and K. K. Lo: *J. Mech. Phys. Solids*, **29**:115–142 (1981).
134. McClintock, F. A., and G. R. Irwin: pp. 84–113 in Fracture Toughness Testing and Its Applications, *ASTM STP* 381, 1965.
135. Ewing, D. J. F., and R. Hill: *J. Mech. Phys. Solids*, **15**:115–124 (1967).
136. Ewing, D. J. F., and C. E. Richards: *J. Mech. Phys. Solids*, **22**:27–36 (1974).
137. Green, A. P.: *Q. J Mech. Appl. Math.*, **6**:223–239 (1953).
138. Green, A. P., and B. B. Hundy: *J. Mech. Phys. Solids*, **4**:128–145 (1956).

Discussion

In Ref. 9 the elastic-stress field at a crack tip was derived by a power-series assumption. This has been reproduced in Sec. 2.1 following a presentation in Ref. 32. Irwin [5] introduced the concept of a stress intensity factor for each of the respective modes.† Values were derived for elliptical and semielliptical [10] cracks on the basis of Refs. 11 and 12, two papers which solved the problem of an embedded elliptical crack in an infinite body.

References 13 to 16 provided relevant solutions for cracks through plates, applying Williams' [9] power-series and pointwise satisfaction (collocation) of the remote boundary conditions. Complex-variable methods introduced in Refs. 17 to 22 and practiced further in Refs. 22 to 29 provided sharpened and extended estimates for a variety of plane geometries. Reference 30 contains some new results but is mainly a superstructure built upon known special solutions by the use of systematic interpolation. Formulas (C.1) to (C.11) of Appendix C list results and estimates from Refs. 10 to 16 and 25 to 30. References 33 to 38 provide compilations of stress intensity factors far more extensive (e.g., Refs. 34 and 35) than the present short survey.

†In the original and several later versions the stress intensity factors were defined to be $1/\sqrt{\pi}$ times as large as those considered here.

The approximate formula (C.3) [30] is in good accord with the exact results of Isida [22–24]. Just as accurate and even simpler, but somewhat lacking in the rational basis, is a secant formula suggested [31].

References 39 to 49 offer a representative variety of applications of the finite-element method to cracked elastic bodies. An extrapolation toward the limit $r = 0$ is practiced in Ref. 39, while Ref. 40 suggests the compliance technique based on potential energy. Direct [41] and further developed [42–44] applications of the last principle are powerful methods for determining the stress intensity factor. The use of various singular elements is advocated [45–49, 77], this being a field which has attracted the attention of several numerical analysts.

One particular group of energy-based methods [44, 50–58] uses Betti's reciprocity theorem by introducing so-called weight functions. Related to these methods, in a sense, are the boundary-integral methods [59–61, 66, 67, 79, 80], in which integral equations in surface variables are discretized into algebraic equations.

For the solution of particular elasticity problems we note further results for elliptical and semielliptical cracks [62–67, 34, 35], for plates in bending when cracked through the thickness† [68–71, 33–35], for cracks through curved shells [72–77, 34, 35], for cracks in anisotropic plates [78–80, 20, 34], for cracks at the boundary between dissimilar elastic media [81–89, 23, 34], and for plates with stiffeners [90–92, 34, 35, 23, 24].

Irwin [93] presents his evaluation of the yield length, and Dugdale [94] introduced the more stringent model for plane stress. Reference 96 is a comprehensive review of the Dugdale model and its formal relation to the Barenblatt treatment [95] of a crack with cohesive end zones in an elastic body. References 97 and 98 derived the COD [Eqs. (2.46) and (2.47)] for the Dugdale model. Further solutions by that model are available [57, 58].

References 99 to 102 represent the main contributions to the introduction of the J integral; Eshelby [99] was first, and Rice [101] had several ideas about the possibilities of the tool even for nonelastic states.

The first rigorous elastoplastic analysis around a crack [103] was directed at

†By established bending theories the plate can be assigned a linear stress distribution through layers of lamina, each deforming essentially in a plane-stress condition. It is then reasonable to obtain the maximum stress intensity factor, considering mode I, as that of the crack-opening face of the plate when this extreme layer with its far-field stresses is treated as being independent of the others. With provision for the particular nature of the free-edge boundary conditions within their Kirchhoff frame of description, this is the level of approach used in Refs. 68 to 70. However, contrary to the assumptions of the above theories, initial normals to the plate surface may deform appreciably in the vicinity of the crack and local membrane forces be produced, due to some closure (as opposed to interpenetration) of the crack sides over the thickness. Reference 71 represents a first step toward taking such closure into consideration. The author is not aware of any rigorous three-dimensional analysis of this problem yet undertaken.

A series of papers by Knowles, Wang, Hartranft, Sih, and others is referred to in Ref. 71, working within the more accurate Reissner or Mindlin frame of analysis. Since closure is ignored, such solutions may represent significant improvements over Kirchhoff results if sufficiently large in-plane forces are being superposed to offset the interpenetration.

mode III. Since then the somewhat simpler problem of Sec. 2.7 has gone through the hands of Rice [21] for review. He proceeded more generally [104] by including boundary effects [Eqs. (2.77) and (2.78)] and by allowing [105] for strain hardening [Eq. (2.80)].

Elastoplastic analyses were carried out for a mode I crack in plane strain in Ref. 21, where we also find a first estimation of the COD, later supplemented by the simple deduction of Eq. (2.94) as presented in Ref. 106. Several numerical evaluations by the finite-element method were reported [107–109, 117] with no totally reliable value of the multiplier having yet been settled upon.†

The extension to include strain hardening (Sec. 2.8.2) was a dead heat between Refs. 110 and 111, the first of which anticipated the nature of the singularity at the outset, through invariance of J, while limiting attention to plain strain. A further discussion appears in Ref. 112. Reference 111 has a continuation [113], and Refs. 114 to 116 are related papers, the last being concerned with mode III. Reference 117 is a detailed study of crack-tip fields using second-order (large-displacement) theory.

Problems concerning instabilities and bifurcations have been extensively discussed [122–133], with recent emphasis on the localization of ultimate deformation into narrow bands [123, 124, 129–133]. The limit-load solutions in Sec. 2.9 can be found as follows: case 1 [134] (see also Example A.3, Appendix A); case 2*a* [135], case 2*b* [123] (see also example A.1); case 3 [136]; case 4 [137]; and case 5 [138].

CHAPTER 3

* Griffith [2].

139. Irwin, G. R.: *J. Appl. Mech.*, **24:**361–364 (1957).

* Bueckner [50].

140. Irwin, G. R.: pp. 557–592 in Fracture Mechanics, *Proc. 1st Symp. Nav. Struct. Mech., Stanford University, 1958*, Pergamon, New York, 1960.
141. Rice, J. R.: pp. 309–340 in T. Yokobori et al. (eds.), *Proc. 1st Int. Conf. Fract., Sendai, 1965*, vol. 1.

* Goodier [96].
* Sih and Liebowitz [20].
* Rice [21].
* Sorensen [109].
* Chitaley and McClintock [118].
* Amazigo and Hutchinson [119].
* Rice and Sorensen [120].
* Rice et al. [121].

142. Gurney, C., and J. Hunt: *Proc. R. Soc. London*, **A299:**508–524 (1967).
143. Gurney, C., and K. M. Ngan: *Proc. R. Soc. London*, **A325:**207–222 (1971).
144. Strifors, H.: pp. 63–66 in D. M. R. Taplin (ed.), Fracture 1977, *Proc. 4th Int. Conf. Fract., Waterloo, Ontario, 1977*, vol. 3.

†With recent evaluations [222] a value close to 0.64 seems to have been established.

145. Nguyen, Q. S.: pp. 315–330 in S. Nemat-Nasser (ed.), Three-Dimensional Constitutive Relations and Ductile Fracture, *Proc. IUTAM Symp. Dourdan, France, 1980,* North-Holland, Amsterdam, 1981.
146. Begley, J. A., and J. D. Landes: pp. 1–20 in Fracture Toughness, *ASTM STP* 514, 1972.
147. Bucci, R. J., P. C. Paris, J. D. Landes, and J. R. Rice: pp. 40–69 in Fracture Toughness, *ASTM STP* 514, 1972.
148. Rice, J. R., P. Paris, and J. Merkle: pp. 231–245 in Progress in Flaw Growth and Fracture Toughness Testing, *ASTM STP* 536, 1973.
149. Paris, P. C.: pp. 3–27 in Flaw Growth and Fracture, *ASTM STP* 631, 1977.
150. Sumpter, J. D. G., and C. E. Turner: pp. 312–329 in Cracks and Fracture, *ASTM STP* 601, 1976.
151. Turner, C. E.: pp. 23–210 in D. G. H. Latzko (ed.), "Post-Yield Fracture Mechanics," Applied Science, London, 1979.
152. Hickerson, J. P.: pp. 62–71 in Flaw Growth and Fracture, *ASTM STP* 631, 1977.
153. Merkle, J. G., and H. T. Corten: *J. Pressure Vessel Technol.*, **96:**286–292 (1974).
154. Landes, J. D., H. Walker, and G. A. Clarke: pp. 266–287 in Elastic-Plastic Fracture, *ASTM STP* 668, 1979.
155. Srawley, J.E.: *Int. J. Fract.*, **12:**470–474 (1976).
156. Paris, P. C., H. Ernst, and C. E. Turner: pp. 338–351 in Fracture Mechanics, *ASTM STP* 700, 1980.
157. Ernst, H., P. C. Paris, M. Rossow, and J. W. Hutchinson: pp. 581–589 in Fracture Mechanics, *ASTM STP* 677, 1979.
158. Andersson, H.: *J. Mech. Phys. Solids*, **21:**337–356 (1973).
* Hellan [32].
159. Hellan, K.: Energy Considerations in Static Fracture Mechanics, *Inst. Mek. N.T.H. Trondheim, Publ.* 73:6, 1973.
160. Andersson, H.: *J. Mech. Phys. Solids*, **22:**285–308 1974.
161. Hellan, K.: *Eng. Fract. Mech.*, **8:**501–506 (1976).
162. Miller, K. J., and A. P. Kfouri: *Proc. Inst. Mech. Eng.*, **190:**48/76, 571–584 (1976).
163. Kfouri, A. P., and J. R. Rice: pp. 43–59 in D. M. R. Taplin (ed.), Fracture 1977, *Proc. 4th Int. Conf. Fract., Waterloo, Ontario, 1977,* vol. 1.
164. Hellan, K., and I. Lotsberg: *Int. J. Fract.*, **13:**539–543 (1977).
165. Atluri, S. N., et al.: pp. 56–66 in A. R. Luxmoore and D. R. I. Owen (eds.), Numerical Methods in Fracture Mechanics, *Proc. Int. Conf., Swansea, 1978.*
166. Lotsberg, I.: pp. 496–507 in A. R. Luxmoore and D. R. I. Owen (eds.), Numerical Methods in Fracture Mechanics, *Proc. Int. Conf., Swansea, 1978.*
167. Kanninen, M. F., et al.: pp. 121–150 in Elastic-Plastic Fracture, *ASTM STP* 668, 1979.
168. Knowles, J. K., and E. Sternberg: *Arch. Rational Mech. Anal.*, **44:**187–211 (1972).
169. Budiansky, B., and J. R. Rice: *J. Appl. Mech.*, **40:**201–203 (1973).
170. Bui, H. D.: *Eng. Fract. Mech.*, **6:**287–296 (1974).
171. Strifors, H.: *Int. J. Solids Struct.*, **10:**1389–1404 (1974).
172. Carlsson, J.: pp. 139–160 in G. C. Sih et al. (eds.), "Prospects of Fracture Mechanics," Noordhoff, Leyden, 1974.
173. Rice, J. R., and N. Levy: pp. 277–293 in A. S. Argon (ed.), "Physics of Strength and Plasticity," MIT Press, Cambridge, Mass., 1969.
174. Bui, H. D., A. Ehrlacher, and Q. S. Nguyen: *J. Méc.*, **19:**697–723 (1980).
175. Cherepanov, G. P.: *J. Appl. Math. Mech.*, **31:**503–512 (1967).
176. Gurtin, M. E.: *Int. J. Solids Struct.*, **15:**553–560 (1979).
177. Strifors, H.: Thermomechanical Theory of Fracture, *Inst. Haallfasthetslaera, KTH, Stockholm, Rep.* 27, 1980.
178. Strifors, H.: Thermomechanical Theory of Fracture Based upon the Linear Strain Tensor, *Inst. Haallfasthetslaera, KTH, Stockholm, Rep.* 28, 1980.
179. Gurtin, M. E.: *Z. Angew. Math. Phys.*, **30:**991–1003 (1979).
* Barenblatt [95].
180. Hillerborg, A., M. Modéer, and P. E. Petersson: *Cem. Concr. Res.*, **6:**773–781 (1976).

181. Modéer, M.: A Fracture Mechanics Approach to Failure Analyses of Concrete Materials, *Univ. Lund, Div. Build. Mater. Rep.* TVBM-1001, 1979.
182. Hillerborg, A.: *Int. J. Cem. Compos.* **4**:177–184 (1980).
183. Bäcklund, J.: *Comput. Struct.*, **13**:145–154 (1981).
184. Janson, J.: *Eng. Fract. Mech.*, **9**:891–899 1977.
185. Janson, J., and J. Hult: *J. Méc. Appl.*, **1**:69–84 (1977).

Discussion

Griffith [2] is the classical introduction of the energy balance related to crack growth, with important supplements [139, 50, 140]; all are directed at essentially elastic response. The crack driving force has been expressed in terms of the stress intensity factors [139, 140] and derived by the compliance method [140]. Bueckner [50] considers boundary conditions and the possibility of generating solutions by superposition involving the reciprocity law. Subsequent work related to essential elasticity includes energy considerations for anisotropic media [20] and a comprehensive analytical and geometrical discussion of crack growth in cases of isotropic, nonlinear response [142, 143]. (Compare Sec. 3.2.1 and Appendix E).

The energy balance has been given a rational basis even for elastoplastic materials in a paper [141] that influenced the exposition in Secs. 3.1 and 3.3. A thorough discussion of crack-tip behavior within a first-order frame of analysis [96] shows that the total loss of potential energy goes into continuum plastic dissipation for the Dugdale model (Sec. 3.3.2). References 141 and 96 indicate the need for singular stresses at the crack tip if the separation work is to be nonzero. It is further suggested [144, 145, 178] that this requirement be sharpened into linear elastic local behavior, implying a stress singularity r^s as hard as $s = -\frac{1}{2}$.

With the slow motion of a crack tip in an elastoplastic solid a wake of unloading will be produced, together with singularities of stress and strain which are not of the same type as for a stationary crack. This complex case has been analyzed [118] for mode III, partially [21] for mode I, and extensively [119] for modes III and I. A numerical investigation has also been undertaken [109], and further discussions on ideal-plastic fields during steady crack growth have been presented [120, 121]. The above singularities are explored in Reference 119.

The equality of the J integral and the crack driving force is rigorously proved [21] for nonlinear elasticity, and this result is reinterpreted [146, 147] to have a meaning for elastoplastic materials before crack growth (Sec. 3.3.3). The derivation of a simple one-sample expression to approximate J, like Eqs. (3.43) and (3.44), has been dealt with extensively [147–157]. Contributions [168–172] extend the class of path-independent integrals so that the J integral can be regarded as the forward component of a vector which defines the loss of potential energy by translation of the crack in an arbitrary direction.

References 158, 32, and 159 suggested a numerical evaluation of the separation work (outlined in Sec. 3.3.4), and steady, slow crack motion [160] and initial cracking [161] in linear-hardening bodies were then analyzed. References 162 and 163 modeled crack growth by this same technique using a large-displacement-gradient formulation. Reference 165 includes an evaluation of the same case as Ref. 162, though possibly by first-order theory. The separation work under linear-hardening conditions was evaluated [160, 164] using excessively small elements and finding similar trends in the results $C > 0$ for plane stress and plane strain, respectively. The results in Ref. 161 were based on converging *ratios* of C values, and in both Refs. 161 and 164 the singularities in stress and strain would be as $r^{-1/2}$ at incipient virtual growth. Reference 166 considered stable crack growth by that same relaxation technique, whereas Ref. 167 suggested an explicit modeling of a finite process zone by special elements traveling with the crack. The nodal-release method is also applied [256, 261, 264, 265] to simulate dynamic crack growth as governed by critical separation work. An indirect method by modified J integration is suggested in Ref. 265.

In further discussions of the thermodynamics of fracture [173–179] Refs. 173 and 174 consider near-tip thermal fields specifically. References 145, 178, and 179 include the presence of cohesive zones as well. More detailed consideration of such zones [95, 180–185] covers applications to concrete and composites [180–183] and discusses the idea of continuous damage [184, 185]. In a somewhat special context the idea of cohesive zones has also been successfully applied to dynamic modeling [267–269, 278]. The reports in Refs. 177 and 178 represent a critical work of almost encyclopedic scope and volume.

CHAPTER 4

* Griffith [2], [3].
* Rice [21], [141].
* Goodier [96].
* Hellan [159].

186. Esterling, D. M.: *J. Appl. Phys.*, **47**:486–493 (1976).
187. Rice, J. R.: *J. Mech. Phys. Solids*, **26**:61–78 (1978).
* Irwin [5].

188. Orowan, E.: *Weld. J. Res. Suppl.*, **34**:157–160 (1955).
189. Irwin, G. R., and J. A. Kies: *Weld. J. Res. Suppl.*, **19**:193–198 (1954).
190. McClintock, F. A.: *J. Appl. Mech.*, **25**:581–588 (1958).
191. Gurney, C., and Y. W. Mai: *Eng. Fract. Mech.*, **4**:853–863 (1972); see also [142] and [143].
192. Clausing, D. P.: *Int. J. Fract. Mech.*, **5**:211–227 (1969).
193. Krafft, J. M., A. M. Sullivan, and R. W. Boyle: pp. 8–28 in *Proc. Crack Propag. Symp., Cranfield, England, 1961*, vol. 1.
194. Srawley, J. E., and W. F. Brown, Jr.: pp. 133–198 in Fracture Toughness Testing and Its Application, *ASTM STP* 381, 1965.
195. Wnuk, M. P.: *J. Appl. Mech.*, **41**:234–242 (1974).
196. McCartney, L. N.: *Int. J Fract.*, **14**:429–438 (1978).
* Broek [7].
* Cherepanov [102].
* Rice and Sorensen [120].
* Rice et al. [121].

197. Tetelman, A. S., and A. J. McEvily: "Fracture of Structural Materials," Wiley, New York, 1967.
* Knott [6].

198. Bluhm, J. I.: *ASTM Proc.*, **61**:1324–1331 1961.

199. ASTM Standard E 399-81, Standard Test Methods for Plane-Strain Fracture Toughness of Metallic Materials, 1981.
* Irwin [93].
200. Cottrell, A. H.: *Iron Steel Inst. Spec. Rep.* 69, 1961, p. 281.
201. Wells, A. A.: pp. 210–230 in *Proc. Crack Propag. Symp., Cranfield, England, 1961*, vol. 1.
202. Hahn, G. T., A. R. Rosenfield, and M. Sarrate: pp. 673–690 in M. F. Kanninen et al. (eds.), "Inelastic Behavior of Solids," McGraw-Hill, New York, 1970.
203. Otsuka, A., T. Miyata, S. Nishimura, and Y. Kashiwagi: *Eng. Fract. Mech.*, **7**:419–428 (1975).
204. Berry, G., and R. Brook: *Int. J. Fract.*, **11**:933–938 (1975).
205. Shih, C. F., H. G. deLorenzi, and W. R. Andrews: pp. 65–120 in Elastic-Plastic Fracture, *ASTM STP* 668, 1979.
206. Rice, J. R., and M. A. Johnson: pp. 641–672 in M. F. Kanninen et al. (eds.), "Inelastic Behavior of Solids," McGraw-Hill, New York, 1970.
207. Broberg, K. B.: *J. Mech. Phys. Solids*, **19**:407–418 (1971).
* Begley and Landes [146].
208. Landes, J. D., and J. A. Begley: pp. 24–39 in Fracture Toughness, *ASTM STP* 514, 1972.
209. Begley, J. A., and J. D. Landes: *Int. J. Fract.*, **12**:764–766 (1976).
210. Robinson, J.N.: *Int. J. Fract.*, **12**:723–737 (1976).
211. Chipperfield, C. G.: *Int. J. Fract.*, **12**:873–886 (1976).
212. Underwood, J. H.: pp. 312–329 in Cracks and Fracture, *ASTM STP* 601, 1976.
213. Markstrøm, K.: *Eng. Fract. Mech.*, **9**:637–647 (1977).
214. McMeeking, R. M., and D. M. Parks: pp. 175–194 in Elastic-Plastic Fracture, *ASTM STP* 668, 1979.
215. Shih, G. F., and M. D. German: *Int. J. Fract.*, **17**:27–44 (1981).
216. Witt, F. J.: *Eng. Fract. Mech.*, **14**:171–178 (1981).
217. Begley, J. A., and J. D. Landes: pp. 246–263 in Progress in Flaw Growth and Fracture Toughness Testing, *ASTM STP* 536, 1973.
218. Paris, P. C., H. Tada, A. Zahoor, and H. Ernst: pp. 5–36 in Elastic-Plastic Fracture, *ASTM STP* 668, 1979.
219. Hutchinson, J. W., and P. C. Paris: pp. 37–64 in Elastic-Plastic Fracture, *ASTM STP* 668, 1979.
* Ernst et al. [157].
* Kanninen et al. [167].
220. Kanninen, M. F., C. H. Popelar, and D. Broek: *Nucl. Eng. Design*, **67**:27–55 (1981).
221. Hellan, K.: SINTEF *Rep.* STF16 A79094, Trondheim, 1979.
222. Shih, C. F.: *J. Mech. Phys. Solids*, **29**:305–326 (1981).

Discussion

References 2 and 3 present Griffith's energy postulate on cracking. They discuss also failure by skew cracks in tension or compression fields and the assessment of high strength realized by reducing the cross section, e.g., fibers or whiskers. Discussions of the Griffith postulate are available in simpler terms [21, 141, 96, 159] and on a fully thermodynamic basis [186, 187]. The ideas leading to the use of LFM for locally yielded bodies were advanced in Refs. 5 and 188, whereas Ref. 141 extends the Griffith concept to the cracking of bodies under large-scale yielding conditions (Sec. 4.2).

Many papers are concerned with the fracture mechanics of stably growing cracks [141, 189–196, 7, 102, 120, 121]. An important contribution was the introduction of the *R* curve, as first suggested [189] and later well established, e.g., [193]. Reference 194 discusses various consequences of postulating an *R* curve, and a good overview [7, chap. 8] exists. In these cases localized yielding was assumed to occur, so that the linear theory (relating R to a stress intensity factor) could be invoked, with no need to consider details of crack-tip behavior. More detailed theories [195, 196, 102, 120, 121] apply as well to large-scale yielding, thus aiming at an extended field of prediction with considerable practical interest.

Effects of plate thickness, toughness, and yield strength on the development of fracture are treated in Refs. 193, 194, 198, and 6, chap. 5. The dimensional requirements (4.13) are set down in Ref. 199, and the yield-length "correction," $a + c/2$ replacing a, is presented in Ref. 93.

References 200 and 201 suggested the COD as a critical quantity; related observations of the crack-tip geometry can be found, e.g., [202–205]. A relationship between the critical COD and typical data applying to the microstructure of void nuclei has been derived [206] (see also Chap. 8). Reference 6, chap. 6 also reviews micro and macro aspects of the COD criterion.

Broberg [207] suggested the J integral as a critical quantity for crack initiation, for which strong experimental support was immediately provided [146, 208]. Some error in both measurement and evaluation entered the last two papers but in such a way that the two effects roughly offset each other [209]. With this fateful interference some general interest had been stirred, and confirming results followed, e.g., [210–213, 205]. Simplified ways of assessing the J integral were explored [147–157]. Critical discussions of the J integral as a single parameter capable of characterizing initial cracking [214, 215] are worthy of note. Occasionally the energy-based method of Witt [216, 217] may prove equivalent to the J criterion.

Methods and results of K_{Ic} testing are described in Chap. 7 [199, 378, 379, 401], where testing related to the COD and the J integral is commented upon [380–382, 151, 387, 388, 393]. R curve testing within localized yielding conditions has been reported and discussed [478–488].

Attempts to correlate stable crack growth and the J integral have been recounted [218, 219, 157, 167, 205]. In particular, Refs. 219 and 157 discuss the evaluation of J for growing (as opposed to stationary) cracks. References 205 and 167 also discuss the merits of other parameters, such as CTOA, in describing stable crack growth. There is a good survey of the present state of the art in such predictions [220]. Reference 221 reviews various implications for the J integral of test recordings. A comprehensive discussion of the relationship between the COD and the J integral [222] covers both stationary and growing cracks in power-law-hardening materials.

CHAPTER 5

223. Mott, N. F.: *Engineering*, **165:**16–18 (1948).
224. Roberts, D. K., and A. A. Wells: *Engineering*, **178:**820–821 (1954).
225. Berry, J. P.: *J. Mech. Phys. Solids*, **8:**194–216 (1960).
226. Eshelby, J. D.: *Proc. Phys. Soc.*, **A62:**307–314 (1949).
227. Yoffe, E. H.: *Phil. Mag.*, **12:**739–750 (1951).
228. Broberg, K. B.: *Arkiv Fys.*, **18:**159–192 (1960).

229. Broberg, K. B.: *J. Appl. Mech.*, **31**:546–547 (1964).
230. Craggs, J. W.: *J. Mech. Phys. Solids*, **8**:66–75 (1960).
231. Craggs, J. W.: pp. 51–63 in D. C. Drucker and J. J. Gilman (eds.), "Fracture in Solids," Wiley, New York, 1963.
232. Baker, B. R.: *J. Appl. Mech.*, **29**:449–458 (1962).
233. Carlsson, A. J.: *Trans. R. Inst. Technol., Stockholm*, 205, 1963.
234. Atkinson, C., and J. D. Eshelby: *Int. J. Fract. Mech.*, **4**:3–8 (1968).
* Rice [21].
235. Erdogan, F.: pp. 498–586 in [20*a*].
* Cherepanov [102].
236. Sih, G. C.: pp. 607–639 in M. F. Kanninen et al. (eds.), "Inelastic Behavior of Solids," McGraw-Hill, New York, 1970.
237. Eshelby, J. D.: *J. Mech. Phys. Solids*, **17**:177–199 (1969).
238. Freund, L. B.: *J. Mech. Phys. Solids*, **20**:129–152 (1972).
239. Freund, L. B.: *J. Elasticity*, **2**:341–349 (1972).
240. Nilsson, F.: *Int. J. Fract. Mech.*, **8**:403–411 (1972).
241. Achenbach, J. D.: pp. 1–57 in S. Nemat-Nasser (ed.), "Mechanics Today," Pergamon, New York, vol. 1, 1972.
242. Willis, J. R.: *Phil. Trans. R. Soc.*, **274**:435–491 (1973).
243. Nilsson, F.: *J. Elasticity*, **4**:73–75 (1974).
244. Kostrov, B. V.: *Int. J. Fract.*, **11**: 47–56 (1975).
245. Nilsson, F.: *Int. J. Solids Struct.*, **13**:542–548, 1133–1139 (1977).
246. Atkinson, C.: *Int. J. Eng. Sci.*, **12**:491–506 (1974).
247. Rose, L.R.F.: *Int. J. Fract.*, **12**:799–813 (1976).
248. Freund, L. B.: pp. 55–91 in S. Nemat-Nasser (ed.), "Mechanics Today," Pergamon, New York, vol. 3, 1976.
249. Shmuely, M., and Z. S. Alterman: *J. Appl. Mech.*, **40**:902–908 (1973).
250. Shmuely, M., and D. Peretz: *Int. J. Solids Struct.*, **12**:67–79 (1976).
251. Shmuely, M., *Int. J. Fract.*, **13**:443–454 (1977).
252. Popelar, C. H., and P. C. Gehlen: *Int. J. Fract.*, **15**:159–177 (1979).
253. Perl, M.: pp. 509–523 in D. R. J. Owen and A. R. Luxmoore (eds.), Numerical Methods in Fracture Mechanics, *Proc. 2d Int. Conf. Swansea, 1980*, Pineridge, Swansea, 1980.
254. Owen, D. R. J., and D. Shantaram; *Int. J. Fract.*, **13**:821–837 (1977).
255. Caldis, E. S., and D. R. J. Owen: pp. 553–568 in D. R. J. Owen and A. R. Luxmoore (eds.), Numerical Methods in Fracture Mechanics, *Proc. 2d Int. Conf. Swansea, 1980*, Pineridge, Swansea, 1980.
256. Keegstra, P. N. R., J. L. Head, and C. E. Turner: pp. 634–647 in A. R. Luxmoore and D. R. J. Owen (eds.), Numerical Methods in Fracture Mechanics, *Proc. Int. Conf. Swansea, 1978.*
257. Malluck, J. F., and W. W. King: pp. 648–659 in A. R. Luxmoore and D. R. J. Owen (eds.), Numerical Methods in Fracture Mechanics, *Proc. Int. Conf., Swansea, 1978.*
258. Rydholm, G., B. Fredriksson, and F. Nilsson: pp. 660–672 in A. R. Luxmoore and D. R. J. Owen (eds.), Numerical Methods in Fracture Mechanics, *Proc. Int. Conf., Swansea, 1978.*
259. Kobayashi, A. S., S. Mall, Y. Urabe, and A. F. Emery: pp. 673–684 in A. R. Luxmoore and D. R. J. Owen (eds.), Numerical Methods in Fracture Mechanics, *Proc. Int. Conf., Swansea, 1978.*
260. Brickstad, B., and F. Nilsson: *Int. J. Fract.*, **16**:71–84 (1980).
261. Brickstad, B., and F. Nilsson: pp. 478–487 in D. R. J. Owen and A. R. Luxmoore (eds.), Numerical Methods in Fracture Mechanics, *Proc. 2d Int. Conf., Swansea, 1980*, Pineridge, Swansea, 1980.
262. Nishioka, T., and S. N. Atluri: *J. Appl. Mech.*, **47**:570–585 (1980).
263. Nishioka, T., R. B. Stonesifer, and S. N. Atluri: pp. 489–507 in Numerical Methods in Fracture Mechanics, *Proc. 2d. Int. Conf., Swansea, 1980.*
264. Dahlberg, L., F. Nilsson, and B. Brickstad: pp. 89–108 in Crack Arrest Methodology and Application, *ASTM STP* 711, 1980.

265. Kishimoto, K., S. Aoki, and M. Sakata: *Eng. Fract. Mech.*, **13**:387–394 (1980).
266. Kanninen, M. F.: pp. 612–633 in D. R. J. Owen and A. R. Luxmoore (eds.), Numerical Methods in Fracture Mechanics, *Proc. 2d Int. Conf., Swansea, 1980*, Pineridge, Swansea, 1980.
267. Kanninen, M. F.: *Int. J. Fract.*, **10**:415–430 (1974).
268. Malluck, J. F., and W. W. King: *Int. J. Fract.*, **13**:655–665 (1977).
269. Gehlen, P. C., C. H. Popelar, and M. F. Kanninen: *Int. J. Fract.*, **15**:281–294 (1979).
270. Hellan, K.: *Int. J. Fract.*, **17**:311–320 (1981).
271. Hellan, K.: pp. 457–471 in D. R. J. Owen and A. R. Luxmoore (eds.), Numerical Methods in Fracture Mechanics, *Proc. 2d Int. Conf., Swansea, 1980*, Pineridge, Swansea, 1980.
272. Bilek, Z. J., and S. J. Burns: *J. Mech. Phys. Solids*, **22**: 85–95 (1974).
273. Burns, S. J., and Z. J. Bilek: *Metall. Trans.*, **4**:975–984 (1973).
274. Bilek, Z. J., and S. J. Burns: pp. 371–386 in G. C. Sih (ed.), "Dynamic Crack Propagation," Noordhoff, Leyden, 1973.
275. Freund, L. B.: *J. Mech. Phys. Solids*, **25**:69–79 (1977).
276. Hellan, K.: *Int. J. Fract.*, **14**:91–100, 173–184 (1978).
277. Popelar, C. H.: pp. 24–37 in Crack Arrest Methodology and Application, *ASTM STP* 711, 1980.
278. Sokolowski, M.: *Polska Akad. Nauk, Eng. Trans.*, **25**:369–393 (1977).
* Goodier and Field [97].
279. Atkinson, C.: *Arkiv Fys.*, **35**:469–476 (1968).
280. Kanninen, M. F.: *J. Mech. Phys. Solids*, **16**:215–228 (1968).
281. Glennie, E. B.: *J. Mech. and Phys. Solids*, **19**:255–272, 329–338 (1971).
282. Achenbach, J. D., and M. F. Kanninen: pp. 649–670 in N. Perrone et al. (eds.), Fracture Mechanics, *Proc. 10th Symp. Nav. Struct. Mech., Washington, 1978*, University Press of Virginia, Charlottesville, Va., 1978.
283. Achenbach, J. D., M. F. Kanninen, and C. H. Popelar: *J. Mech. Phys. Solids*, **29**:211–225 (1981).
284. Achenbach, J. D., and V. Dunayevsky: *J. Mech. Phys. Solids*, **29**:283–303 (1981).
285. Slepyan, L. J.: *Izv. Akad. Nauk. SSSR, Mekh. Tverd. Tela*, **9**:57–67 (1974).
286. Hahn, G. T., M. Sarrate, and A. R. Rosenfield: *Int. J. Fract. Mech.*, **5**:187–210 (1969).
287. Hahn, G. T., M. Sarrate, M. F. Kanninen, and A. R. Rosenfield: *Int. J. Fract.*, **9**:209–222 (1973).
288. Erdogan, F., F. Delale, and J. A. Owzarek, *J. Pressure Vessel Technol.*, **99**:90–99 (1977).
289. Emery, A. F., W. J. Love, and A. S. Kobayashi: *J. Pressure Vessel Technol.*, **99**:122–136 (1977).
290. Parks, D. M. and L. B. Freund: *J. Pressure Vessel Technol.*, **100**:13–17 (1978).
291. Parks, D. M., and L. B. Freund: pp. 359–378 in Crack Arrest Methodology and Application, *ASTM STP* 711, 1980.
292. McGuire, P. A., S. G. Sampath, C. H. Popelar, and M. F. Kanninen: pp. 341–358 in Crack Arrest Methodology and Application, *ASTM STP* 711, 1980.
293. Achenbach, J. D., and R. Nuismer: *Int. J. Fracture Mech.*, **7**:77–88 (1971).
294. Nilsson, F.: *Int. J. Solids Struct.*, **9**:1107–1115 (1973).
295. Kelley, J. W., and C. T. Sun: *Eng. Fract. Mech.*, **12**:13–22 (1979).
296. Mall, S., A. S. Kobayashi, and Y. Urabe: pp. 498–510 in Fracture Mechanics, *ASTM STP* 677, 1979.
297. Mall, S., A. S. Kobayashi, and F. J. Loss: pp. 70–85 in Crack Arrest Methodology and Applications, *ASTM STP* 711, 1980.
298. Mall, S.: pp. 539–551 in D. R. J. Owen and A. R. Luxmoore (eds.), Numerical Methods in Fracture Mechanics, *Proc. 2d Int. Conf., Swansea, 1980*, Pineridge, Swansea, 1980.
299. Hahn, G. T., R. G. Hoagland, and A. R. Rosenfield: pp. 1333–1338 in D. M. R. Taplin (ed.), Fracture 1977, vol. 2, *Proc. 4th Int. Conf. Fract., Waterloo, Ontario, 1977.*
300. Atkinson, C.: *Appl. Mech. Rev.*, **32**:123–135 (1979).

Discussion

Since the presentation in Chap. 5 covers only a few of the simpler or obviously significant aspects of fracture dynamics, we look here into some more material pertinent to the general development and special applications.

References 223 to 225 may be viewed as the first papers in which the Griffith energy balance was supplemented by kinetic terms. They are not completely dynamic analyses because the necessary field considerations were replaced by more simplified gross estimations.

Modern fracture dynamics can be viewed as having started with Ref. 226, a precursor of Yoffe's paper [227] on the motion of a running crack, which is theoretically important while somewhat academic in the sense that the crack was modeled to close at the trailing edge in such a way that its length remained constant. Far more realistic is the Broberg model [228, 229], which considers an internal crack to grow from zero length with constant opposite velocities of the two ends. Even if a somewhat artificial restraint is implied, in that the resistance must be proportional to the crack length in order to comply with the motion, the solution has had considerable impact on later research in the field. Craggs [230] and Baker [232] investigated semi-infinite cracks (as penetrating from the edge far into a body) under partly unrealistic loading conditions, but Ref. 232 is the first paper which allowed for the transient conditions when an existing crack starts moving from rest. This approach set an example for later work in the field, discussed below. References 227, 230, and 232 derived basic relationships of the type presented in Eq. (5.32), 227 discussing also crack branching in relation to the derived θ dependence of planes carrying the largest σ_θ. Reference 233 is experimentally directed at the same phenomenon.

Energy flow to the crack tip by a dynamic version of the J integral discussed in Ref. 234 confirmed Broberg's solution, in contrast to one given in Ref. 231. Similar considerations can be found in papers [102, 236] where the crack driving force is related to the dynamic stress intensity factor under assumed steady conditions. This relation [Eqs. (5.33) and (5.34)] was later shown [239, 243] to apply under quite general conditions.

The consideration of nonsteady or transient conditions started by Baker [232] continued through a series of later contributions, e.g., [237–239, 243–245]. Typical mathematical tools in such elastic analyses are integral transformation (Laplace or Fourier), the use of Wiener-Hopf technique for the typical boundary conditions, judicious application of the superposition principle, and the construction of similarity solutions in certain instances. The last cases contain no characteristic length, an example being Broberg's crack growing from zero initial length, addressed from this alternative viewpoint [231]. Similarity has been exploited further, e.g., [241, 242].

References 246 and (in part) 242 deal with crack growth along the junction between elastically dissimilar models. Cracking along striplike plates is considered in Ref. 240, and many good reviews of the current states of knowledge are available [21, 235, 241, 247, 248, 266, 300].

In addition to the analytical treatment, the development of which has been outlined above, numerical methods have found increased application. General interest is focused on the use of finite-difference methods [249–253] and finite-element methods [254–265]. Special uniaxial models have been developed to describe the double-cantilever beam with a crack running in the symmetry plane (the experimental part is commented on in Sec. 7.3). This type of specimen during fast cracking until arrest has been considered [267–271], one final version [269] responding to a criticism [268] and another [270, 271] being somewhat simpler in mathematical formulation. Slower, controlled propagation of the crack has also been analyzed [272–274].

It is remarkable that simple uniaxial models in fracture dynamics were so late in being addressed. The double-cantilever beam was a beginning, and there were further suggestions [275–278]. Reference 275 refers to the double cantilever but considers only the shear waves and arrives at greatly simplified (if inaccurate) conclusions. Reference 276 deals with the debonding of an elastic strip from a rigid substrate, and in the case of central action this problem has the same mathematical basis as Ref. 275. Section 5.2 relates to this work. A so-called double-torsion beam was considered [277], again using the same mathematics, and Ref. 278 also investigated central debond, here to include an elastic contact with the substrate, in analogy with Kanninen's beam model.

With few exceptions analyses of dynamic crack growth have been formally based on linear elasticity. Yielding of the material has been incorporated [97, 279–281] but then within the restrictions of the thin-plate Dugdale model, to be confined to a yield strip leading the crack. Near-tip analyses have been undertaken [282–284] for bilinear (linear-strain-hardening) materials. These are extensions to the dynamic case of the Amazigo-Hutchinson solution [119] for slow growth. Mode III is treated in Ref. 282 and mode I in Ref. 283. Ideal (perfect) plasticity is considered in Ref. 284, and Ref. 285 assumes a total-strain (nonlinear-elasticity) response in the loaded region, in contrast to the flow-theory analyses mentioned above.

Reference 255 also includes plastic response in the analysis of pipeline fractures. Further investigations of dynamic pipeline fracture are available [254, 286–292].

Attention above has been focused mainly on running cracks. With judicious interpretation one result [232] also has a bearing on dynamic loading applied to a stationary crack (compare Ref. 293), while separate analytical approaches to this last problem have been made, e.g., [248, 294]. Even in this case recent activity has been centered on the application of numerical methods, e.g., [295–298].

CHAPTER 6

301. McClintock, F. A., and A. S. Argon: "Mechanical Behavior of Materials," Addison-Wesley, Reading, Mass., 1966.
* Tetelman and McEvily [197].
302. Dieter, G. E.: "Mechanical Metallurgy," 2d ed., McGraw-Hill, New York, 1976.
303. Coffin, L. F., Jr., *Metall. Eng. Q.*, **3**:15–24 (1963).
304. Manson, S. S.: "Thermal Stress and Low-Cycle Fatigue," McGraw-Hill, New York, 1965.
305. Hertzberg, R. W.: "Deformation and Fracture Mechanics of Engineering Materials," Wiley, New York, 1976.
* Broek, [7].
306. Paris, P. C., M. P. Gomez, and W. E. Anderson: *The Trend in Eng.*, **13**:9–14 (1961).
307. Paris, P. C., and F. Erdogan: *J. Basic Eng.*, **85**:528–534 (1963).

308. Liu, H. W.: *J. Basic Eng.*, **83**:23–31 (1961).
309. Liu, H. W.: *J. Basic Eng.*, **85**:116–122 (1963).
310. McClintock, F. A.: pp. 170–174 in Fatigue Crack Propagation, *ASTM STP* 415, 1967.
311. Rice, J. R.: pp. 247–309 in Fatigue Crack Propagation, *ASTM STP* 415, 1967.
312. Donahue, R. J., H. McI. Clark, P. Atamo, R. Kumble, and A. J. McEvily: *Int. J. Fract. Mech.*, **8**:209–219 (1972).
313. Klesnil, M., and P. Lukas: *Eng. Fract. Mech.*, **4**:77–92 (1972).
314. Richards, C. E., and T. C. Lindley: *Eng. Fract. Mech.*, **4**:951–978 (1972).
315. Forman, R. G., V. E. Kearney, and R. M. Engle: *J. Basic Eng.*, **89**:459–464 (1967).
316. Hoeppner, D. W., and W. E. Krupp: *Eng. Fract. Mech.*, **6**:47–70 (1974).
317. Hult, J.: *J. Mech. Phys. Solids*, **6**:47–52 (1957).
318. McClintock, F. A.: pp. 65–102 in D. C. Drucker and J. J. Gilman, (eds.), "Fracture of Solids," Wiley, New York, 1963.
319. Wnuk, M. P.: *J. Appl. Mech.*, **41**:234–242 (1974).
320. Janson, J.: *Eng. Fract. Mech.*, **10**:651–657 (1978).
321. Führing, H., and T. Seeger: *Eng. Fract. Mech.*, **11**:99–122 (1979).
322. Elber, W.: *Eng. Fract. Mech.*, **2**:37–45 (1970).
323. Elber, W.: pp. 230–242 in Damage Tolerance in Aircraft Structures, *ASTM STP* 486, 1971.
324. Newman, J. C., Jr.: pp. 281–301 in Mechanics of Crack Growth, *ASTM STP* 590, 1976.
325. Nakagaki, M., and S. Atluri: *Fatigue Eng. Mater. Struct.*, **1**:421–429 (1979).
326. Williams, J. F., and D. C. Stouffer: *Eng. Fract. Mech.*, **11**:547–557 (1979).
327. Matsuoka, S., and K. Tanaka: *Eng. Fract. Mech.*, **11**:703–715 (1979).
328. Marci, G., and P. F. Packman: *Int. J. Fract.*, **16**:133–153 (1980).
329. Pelloux, R. M., M. Faral, and W. M. McGee: pp. 35–48 in Fracture Mechanics, *ASTM STP* 700, 1980.
330. Weiss, V., and J. Mautz: pp. 154–168 in Cracks and Fracture, *ASTM STP* 601, 1976.
331. Chu, H. P.: pp. 245–263 in Fracture Toughness and Slow-Stable Cracking, *ASTM STP* 559, 1973.
332. Vosikovsky, O.: *Eng. Fract. Mech.*, **11**:595–602 (1979).
333. Wästberg, S.: *Inst. Haallfasthetslaera, KTH, Stockholm Publ.* 209, 1980.
334. Schijve, J.: *Aeronaut. J.*, **75**:517–532 (1970).
335. Schijve, J.: *AGARD Conf. Proc. no.* 118, *Symp. Random Load Fatigue, 1972*, pp. 3-3 to 3-119.
336. Gardner, F. H., and R. I. Stephens: pp. 225–244 in Fracture Toughness and Slow-Stable Cracking, *ASTM STP* 559, 1973.
337. Haibach, E., *Darmstadt Lab. Betriebfestigk. T.M.* 50/70, *1970.*
338. Schijve, J.: pp. 3–34 in Fracture Mechanics, *ASTM STP* 700, 1980.
339. McCartney, L. N.: *Int. J. Fract.*, **14**:213–232 (1978).
340. Palmgren, A.: *Z. Vereines deutscher Ing.*, **68**:339–341 (1924).
341. Miner, M. A.: *J. Appl. Mech.*, **A12**:159–164 (1945).
342. Schijve, J.: *Eng. Fract. Mech.*, **11**:167–221 (1979).
343. Donaldsen, D. R., and W. E. Anderson: pp. 375–441 in *Proc. Crack Propag. Symp., Cranfield, England, 1960*, vol. 2.
344. Broek, D., and J. Schijve: *Aircr. Eng.*, **38**(11):31–33 (1966).
345. Griffiths, J. R., and C. E. Richards: *Mater. Sci. Eng.*, **11**:305–310 (1973).
346. Dowling, N. E., and J. A. Begley: pp. 82–103 in Mechanics of Crack Growth, *ASTM STP* 590, 1976.
* Paris [149].
347. Leis, N. E., and A. Zahoor: pp. 65–96 in Fracture Mechanics, *ASTM STP* 700, 1980.
348. Sadananda, K., and P. Shahinian: pp. 152–163 in Fracture Mechanics, *ASTM STP* 700, 1980.
349. McEvily, A. J. and R. W. Staehle (eds.): "Corrosion Fatigue," National Association of Corrosion Engineering, Houston, 1972.
350. Gasc, C., J. Petit, P. Bouchet, and J. DeFouquet: pp. 867–872 in D. M. R. Taplin (ed.), Fracture 1977, *Proc. 4th Int. Conf. Fract., Waterloo, 1977*, vol. 2.
351. Haagensen, P. J.: pp. 905–910 in D. M. R. Taplin (ed.), Fracture 1977, *Proc. 4th Int. Conf. Fract., Waterloo, Ontario, 1977*, vol. 2.

352. Craig, H. L., T. W. Crooker, and D. W. Hoeppner, (eds.), Corrosion-Fatigue Technology, *ASTM STP* 642, 1978.
353. Salivar, G. C., D. L. Creighton, and D. W. Hoeppner: *Eng. Fract. Mech.*, **14**:337–352 (1981).
354. Wei, R. P., and G. W. Simmons: *Int. J. Fract.*, **17**:235–247 (1981).
355. Coffin, L. F., Jr.: *Metall. Trans.*, **2**:3105–3113 (1971).
356. Schaefer, A. O., and R. M. Curran (eds.): Symposium on Creep-Fatigue Interaction, *ASME MPC*-3, 1976.
357. Radon, J. C., C. M. Branco, and L. E. Culver: *Int. J. Fract.*, **13**: 595–610 (1977).
358. Plumtree, A., and N. G. Persson: pp. 821–830 in D. M. R. Taplin (ed.), Fracture 1977, *Proc. 4th Int. Conf. Fract., Waterloo, 1977*, vol. 2.
359. Scarlin, R. B.: *Metall. Trans.*, **8A**:1941–1948 (1977).
360. Scarlin, R. B.: pp. 849–857 in D. M. R. Taplin (ed.), Fracture 1977, *Proc. 4th Int. Conf. Fract., Waterloo, Ontario, 1977*, vol. 2.
361. Logan, H. L.: "The Stress Corrosion of Metals," Wiley, New York, 1966.
362. Brown, F. B.: *Metall. Rev.*, **13**:171–183 (1968).
363. Clark, W. G., Jr.: pp. 138–153 in Cracks and Fracture, *ASTM STP* 601, 1976.
364. Clark, W. G., Jr.: pp. 97–110 in Fracture Mechanics, *ASTM STP* 700, 1980.
365. Broberg, H.: Creep Damage and Rupture, dissertation, Chalmers Tekniska Högskola, Gøteborg, 1975.
366. Nikbin, K. M., G. A. Webster, and C. E. Turner: pp. 627–634 in D. M. R. Taplin (ed.), Fracture 1977, *Proc. 4th Int. Conf. Fract., Waterloo, Ontario, 1977*, vol. 2.
367. Coleman, M. C., A. T. Price, and J. A. Williams: pp. 649–662 in D. M. R. Taplin (ed.), Fracture 1977, *Proc. 4th Int. Conf. Fract., Waterloo, Ontario, 1977*, vol. 2.
368. Dimelfi, R. J. and W. D. Nix: *Int. J. Fract.*, **13**:341–348 (1977).
369. Ewing, D. J. F.: *Int. J. Fract.*, **13**:101–117 (1978).
370. Janson, J.: *Eng. Fract., Mech.*, **10**:795–806 (1978).
371. Landes, J. D., and J. A. Begley: pp. 128–148 in Mechanics of Crack Growth, *ASTM STP* 590, 1976.
372. Yokobori, T., and H. Sakata: *Eng. Fract. Mech.*, **13**:509–522 (1979).
373. Yokobori, T., H. Sakata, and T. Yokobori, Jr.: *Eng. Fract. Mech.*, **13**:522–532 (1979).
374. Fu, L. S.: *Eng. Fract. Mech.*, **13**:307–330 (1980).
375. Riedel, H., and J. R. Rice: pp. 112–130 in Fracture Mechanics, *ASTM STP* 700, 1980.
376. Sadananda, K., and P. Shahinian: *Eng. Fract. Mech.*, **15**:327–342 (1981).
377. Dahlberg, L.: Exempelsamling i Brottmekanik, Institutionen för Hållfasthetslära, KTH, Stockholm, 1976.

Discussion

General introductions [301, 197, 302, 305] include empirical gross estimations by S-N or Wöhler diagrams. Empirical relations for low-cycle-fatigue lives (when plasticity is extensive, so that ΔK does not govern the behavior) have been considered [301, 197, 302] with reference to the original work [303, 304]. In particular, comprehensive discussions of loading-induced fatigue [305, chaps. 12, 13; 7, chap. 10] and environmentally induced fatigue [305, chap. 11] are available.

Among the first rational macroscopic examinations of the rate of crack growth, Refs. 308 and 309 used dimensional and energetic arguments to predict quadratic dependence on ΔK. These views were further elaborated [311] in the direction of Eq. (6.3), while deductive and empirical approaches were systematically reviewed. Here the field of local yielding upon load reversal is also first evaluated, the result being a smaller zone of reversal than of monotonic yielding, as recorded in the present text. References 310 and 312, among others, assume a linear dependence of the rate of growth on the COD, concluding again with a quadratic dependence on ΔK. In particular, the discussion of Eq. (6.7) originates from Ref. 312. On the other hand, arguments and data were reviewed [306, 307] in support of a relation like (6.10) with an exponent n larger than 2 in general.

The threshold value ΔK_t in relation to Eqs. (6.7) and (6.11) was respectively considered in Refs. 312 and 313. Reference 314 has arguments in support of the second parametric version of Eq. (6.13) and offers clarifying views on the relationship between physical mechanisms and macroscopic laws. The first, often applied version of Eq. (6.13) was suggested in Ref. 315, the whole series [312–315] referring to a large number of experimental results. Some of the empirical relations suggested for crack growth rate are reviewed in Ref. 316.

A detailed evaluation of the rate of growth has been attempted [317–321], characteristic internal variables affecting the development being partly assumed. The possibility of crack closure even when the stress intensity factor is positive, due to plastic deformation in the wake of the crack tip, has been noted [322, 323]; this important aspect is prominent in later research, e.g., [321, 324–329, 338]. The influence of the mean stress level has also been discussed [330–333], as well as the threshold effect and environmental response [330]. References 334 to 338 and 327 present experience related to varying load spectra and including such effects as delayed growth upon overload. Causes of the delay are discussed, e.g., [327, 338, 339]. A conclusion of such research is that the simple, linear damage theory [340, 341] has a limited but important range of application. A good review of the prediction capabilities [342] includes varying amplitudes and closure effects. Thickness effects are considered in Refs. 343 to 345.

The possibility that the J integral governs fatigue crack growth under large-scale yielding conditions has been discussed [346, 149, 347, 348].

Environmental effects covered [347–373] include corrosion fatigue [349–354], high-temperature fatigue [347, 355–360], stress corrosion [361–364], and creep fracture [365–376]. The last two topics are not fatigue behavior in relation to cyclic load but are included for completeness.

CHAPTER 7

- * ASTM [199].

378. Brown, W. F., Jr., and J. E. Srawley: Plane Strain Crack Toughness Testing of High Strength Metallic Materials, *ASTM STP* 410, 1967,
379. Brown, W. F., Jr. (ed.): Review of Developments in Plane Strain Fracture Toughness Testing, *ASTM STP* 463, 1970.
380. Burdekin, F. M., and D. E. W. Stone: *J. Strain Anal.*, **1**:145–153 (1966).
381. Ingham, T., G. R. Egan, and D. Elliot: pp. 200–216 in *Inst. Mech. Eng.*, *Conf. Pract. Appl. Fract. Mech. to Pressure Vessel Technol., London, 1971.*
382. British Standards Institution: Methods for Crack Opening Displacement Testing, BS 5762-1979, 1979.

- * Turner [151].
- * Robinson [210].
- * Chipperfield [211].

383. Paranjpe, S. A. and S. B. Banerjee: *Eng. Fract. Mech.*, **11**:43–53 (1979).
384. de Castro, P. M. S. T., J. Spurrier, and P. Hancock: *Int. J. Fract.*, **17**:83–95 (1981).
* Shih [222].
* Underwood [212].
385. Hickerson, J. P.: *Eng. Fract. Mech.*, **9**:75–85 (1977).
* Markstrøm [213].
* Kanninen et al. [167].
* Shih et al. [205].
386. Vassilaros, M. G., J. A. Joyce, and J. P. Gudas: pp. 251–270 in Fracture Mechanics, *ASTM STP* 700, 1980.
* Landes et al. [154].
* Paris et al. [156].
387. ASTM Standard E813-81: Standard Test for J_{Ic}, A Measure of Fracture Toughness, 1981.
388. Landes, J. D., and J. A. Begley: pp. 57–81 in Developments in Fracture Mechanics Test Method Standardization, *ASTM STP* 632, 1977.
389. Landes, J. D., and J. A. Begley: pp. 170–186 in Fracture Analaysis, *ASTM STP* 560, 1974.
* Paris [149].
390. Keller, H. P., and D. Munz: pp. 217–231 in Flaw Growth and Fracture, *ASTM STP* 631, 1977.
391. Munz, D.: pp. 406–425 in Elastic-Plastic Fracture, *ASTM STP* 668, 1979.
392. Pisarski, H. G.: *Int. J. Fract.*, **17**:427–440 (1981).
393. Dawes, M. G.: pp. 307–333 in Elastic-Plastic Fracture, *ASTM STP* 668, 1979.
394. Shih, C. F., H. G. de Lorenzi, and W. R. Andrews: *Int. J. Fract.*, **13**:544–548 (1977).
395. Ritchie, R. O., W. L. Server, and R. A. Wullaert: *Int. J. Fract.*, **14**:329–334 (1978).
396. Willoughby, A. A.: *Int. J. Fract.*, **15**:R125–126 (1979).
397. Ritchie, R. O., W. L. Server, and R. A. Wullaert: *Int. J. Fract.*, **15**:R139–141 (1979).
398. Carlsson, A. J.: *Trans. R. Inst. Technol. Stockholm*, 189, 1962.
399. Anctil, A. A., E. B. Kula, and E. diCesare: *ASTM Proc.*, **63**:799–810 (1963).
400. Gilbey, D. M., and S. Pearson: *R. Aircr. Establ., Farnborough Tech. Rep.* 66402, 1966.
401. Srawley, J. E., and W. F. Brown, Jr.: pp. 133–198 in Fracture Toughness Testing and Its Application, *ASTM STP* 381, 1965.
402. Davies, J., D. F. Cannon, and R. J. Allen: *Nature*, **225**: 1240–1242 (1970).
403. Pisarski, H. G., and S. J. Garwood: *Weld. Inst. Res. Bull.*, **19**:97–101 (1978).
404. Okumura, N., T. V. Venkatasubramanian, B. A. Unvala, and T. J. Baker: *Eng. Fract. Mech.*, **14**:617–625 (1981).
405. McCartney, L. N., P. E. Irvin, G. T. Symm, P. M. Cooper, and A. Kurzfeld: *Nat. Phys. Lab. Tech. Rep.* DMA(B)3, 1977.
406. Baudin, G., and H. Policella: pp. 1957–1964 in D. Francois (ed.), Advances in Fracture Research, *Proc. 5th Int. Conf. Fract. Cannes, 1981.*
407. Bardal, E., T. Berge, M. Grövlen, P. J. Haagensen, and B. M. Förre: *SINTEF Rep.* STF16 A81057, Trondheim, 1981.
408. Jilkén, L.: Dissertation, Linköpings Tekniska Högskola, Linköping, 1978.
409. Jilkén, L., and J. Bäcklund: *Stiftelsen Svensk Skeppsforskning (SSF) Rap.* 5610:9, Göteborg, 1977.
410. Gerberich, W. W., M. Stout, K. Jatavallabhula, and D. Atteridge: *Int. J. Fract.*, **15**:491–514 (1979).
411. Thaulow, C.: *Mater. Sci. Eng.*, **40**:133–135 (1980).
412. Kerkhof, F.: pp. 498–552 in K. Vollzath and G. Thomer (eds.), "Kurtzzeitphysik," Springer, Vienna and New York, 1967.
413. Greenwood, J. H.: *Int. J. Fract.*, **8**:183–193 (1972).
414. Beevers, C. J. (ed.): "The Measurement of Crack Length and Shape during Fracture and Fatigue," Engineering Materials Advisory Services Ltd., Warley, U.K., 1980.
415. Robinson, J. N., and A. S. Tetelman: pp. 139–158 in Fracture Toughness and Slow-Stable Cracking, *ASTM STP* 559, 1974.
416. Fields, B. A., and K. J. Miller: *Eng. Fract. Mech.*, **9**:137–146 (1977).

417. Lereim, J., and P. W. Lohne: *Int. J. Fract.*, **16:**R.223–R.228 (1980).
418. Clarke, G. A., W. R. Andrews, P. C. Paris, and D. W. Schmidt: pp. 27–42 in Mechanics of Crack Growth, *ASTM STP* 590, 1976.
419. Joyce, J. A., and J. P. Gudas: pp. 451–468 in Elastic-Plastic Fracture, *ASTM STP* 668, 1979.
420. Wilson, A. D.: pp. 469–492 in Elastic-Plastic Fracture, *ASTM STP* 668, 1979.
421. Bamford, W. H., and W. H. Bush: pp. 553–580 in Elastic-Plastic Fracture, *ASTM STP* 668, 1979.
422. Turner, C. E., pp. 93–114 in Impact Testing of Materials, *ASTM STP* 466, 1970.
423. Instrumented Impact Testing, *ASTM STP* 563, 1974.
424. Ireland, D. R.: Dynamic Fracture Toughness, *Weld. Inst.–ASTM Int. Conf., Cambridge, 1976*, pap. 5.
425. IIW Commission X, UK Briefing Group on Dynamic Testing, Dynamic Fracture Toughness, *Weld. Inst.–ASTM Int. Conf., Cambridge, 1976*, pap. 11.
426. Fearnehough, G. D.: Dynamic Fracture Toughness, *Weld. Inst.–ASTM Int. Conf., Cambridge, 1976*, pap. 29.
427. Chioclov, D. D.: Dynamic Fracture Toughness, *Weld. Inst.–ASTM Int. Conf. Cambridge, 1976*, pap. 14.
428. Gunleiksrud, A., J. Troset, and C. Thaulow: *SINTEF Rep.* STF16 A80089, Trondheim, 1980.
429. Hahn, G. T., R. G. Hoagland, and A. R. Rosenfield: pp. 267–280 in G. C. Sih et al. (eds.), "Prospects of Fracture Mechanics," Noordhoff, Leyden, 1974.
430. Hahn, G. T., R. G. Hoagland, M. F. Kanninen, and A. R. Rosenfield: *Eng. Fract. Mech.*, **7:**583–591 (1975).
431. Hoagland, R. C., A. R. Rosenfield, P. C. Gehlen, and G. T. Hahn: pp. 177–202 in Fast Fracture and Crack Arrest, *ASTM STP* 627, 1977.
432. Gehlen, P. C., R. G. Hoagland, and C. H. Popelar: *Int. J. Fract.*, **15:**69–84 (1979).
433. Hahn, G. T., et al.: *Battelle-Columbus Lab. Rep.* NUREG/CR-0825, BMI-2025, 1978, pp. 2.49–2.54.
434. Manogg, P.: Dissertation 4/64, University of Freiburg, 1964.
435. Theocaris, P. S.: *J. Mech. Phys. Solids*, **20:**265–279 (1972).
436. Theocaris, P. S., and N. I. Ioakimidis: *Eng. Fract. Mech.*, **12:**613–615 (1979).
437. Rosakis, A. J.: *Eng. Fract. Mech.*, **13:**331–347 (1980).
438. Theocaris, P. S., and G. A. Papadoupoulos: *Eng. Fract. Mech.*, **13:**683–698 (1980).
439. Kalthoff, J. F., J. Beinert, and S. Winkler: pp. 161–176 in Fast Fracture and Crack Arrest, *ASTM STP* 627, 1977.
440. Kalthoff, J. F., J. Beinert, S. Winkler, and W. Klemm: pp. 109–127 in Crack Arrest Methodology and Applications, *ASTM STP* 711, 1980.
441. Kalthoff, J. F., S. Winkler, and J. Beinert: *Int. J. Fract.*, **13:**528–531 (1977).
442. Theocaris, P. S., and F. Katsamanis: *Eng. Fract. Mech.*, **10:**197–210 (1978).
443. Theocaris, P. S., and J. Milios: *Int. J. Fract.*, **16:**31–51 (1980).
444. Theocaris, P. S., and J. Milios: *Int. J. Solids Struct.*, **17:**217–230 (1981).
445. Riley, W. F., and J. W. Dally: *Expert. Mech.*, **9:**27N–33N (1969).
446. Kobayashi, A. S., B. G. Wade, W. B. Bradley, and S. T. Chiu: *Eng. Fract. Mech.*, **6:**81–92 (1974).
447. Kobayashi, T., and J. W. Dally: pp. 257–273 in Fast Fracture and Crack Arrest, *ASTM STP* 627, 1977.
448. Kobayashi, A. S., and S. Mall: *Expert. Mech.*, **18:**11–18 (1978).
449. Kobayashi, T., and J. W. Dally: pp. 189–210 in Crack Arrest Methodology and Applications, *ASTM STP* 711, 1980.
- * Shmuely [251].
- * Mall et al. [296].
- * Burns and Bilek [273].

450. Burns, S. J., and C. L. Chow: pp. 228–240 in Fast Fracture and Crack Arrest, *ASTM STP* 627, 1977.
451. Carlsson, J., L. Dahlberg, and F. Nilsson: pp. 165–182 in G. C. Sih (ed.), Dynamic Crack Propagation, *Proc. Int. Conf. Lehigh University, 1972*, Noordhoff, Leyden, 1973.
- * Dahlberg et al. [265].

* Popelar et al. [252].
* Kanninen et al. [267].
452. Crosley, P. B., and E. J. Ripling: pp. 203–227, 372–391 in Fast Fracture and Crack Arrest, *ASTM STP* 627, 1977.
453. Theocaris, P. S., and E. Gdoutos: *J. Appl. Mech.*, **39**:91–97 (1972).
454. Theocaris, P. S., and A. Stassinakis: *Eng. Fract. Mech.*, **14**:363–372 (1981).
455. Barsom, J. M., and J. V. Pellegrino: *Eng. Fract. Mech.*, **5**:209–221 (1973).
456. Priest, A. H.: Dynamic Fracture Toughness, *Weld. Inst.–ASTM Int. Conf., Cambridge, 1976*, paper 10.
Carlsson [38].
457. Carlsson, J., S. G. Larsson, and K. Markstrøm: "Kompendium i brottmekanik," Institutionen för Hållfasthetslära, KTH, Stockholm, 1971.
458. Frost, N. E., L. P. Pook, and K. Denton: *Eng. Fract. Mech.*, **3**:109–126 (1971).
459. Pook, L. P., and A. F. Greenan: *Eng. Fract. Mech.*, **5**: 935–946 (1973).
460. Kaufman, J. G., R. L. Moore, and P. E. Schilling: *Eng. Fract. Mech.*, **2**:197–210 (1971).
461. Sullivan, A. M., and T. W. Crooker: *Eng. Fract. Mech.*, **9**:159–166 (1977).
462. Marandet, B., and G. Sanz: pp. 72–95 in Flaw Growth and Fracture, *ASTM STP* 631, 1977.
463. Pellini, W. S. et al.: *Naval Res. Lab. Rep.* 6300, Washington, 1965.
464. Hudson, C. M., and S. K. Seward: *Int. J. Fract.*, **14**:R151–R184 (1978).
* Hertzberg [305].
465. Hussain, M. A., S. L. Pu, and J. Underwood: pp. 2–28 in Fracture Analysis, *ASTM STP* 560, 1974.
466. Palaniswamy, K., and W. G. Knauss: pp. 87–148 in S. Nemat-Nasser (ed.), "Mechanics Today," Pergamon, New York, vol. 4, 1978.
467. Erdogan, F., and G. C. Sih: *J. Basic Eng.*, **85**:519–527 (1963).
468. Shah, R. C.: pp. 29–52 in Fracture Analysis, *ASTM STP* 560, 1974.
469. Pook, L. R.: *Eng. Fract. Mech.*, **3**:205–218 (1971).
470. Williams, J. G., and P. D. Ewing: *Int. J. Fract. Mech.*, **8**:441–446 (1972).
471. Finnie, I., and H. D. Weiss: *Int. J. Fract.*, **10**:136–137 (1974).
472. Wu, C. H.: *J. Appl. Mech.*, **45**:553–558 (1978).
473. Cotterell, B., and J. R. Rice: *Int. J. Fract.*, **16**:155–159 (1980).
474. Hayashi, K., and S. Nemath-Nasser: *J. Appl. Mech.*, **48**:521–524 (1981).
475. Sih, G. C.: *Int. J. Fract.*, **10**:305–321 (1974).
476. Sih, G. C. (ed.): "Mechanics of Fracture," Noordhoff, Leyden, vol. 3, 1977.
477. Badaliance, R.: *Eng. Fract. Mech.*, **13**:657–666 (1980).
478. Broek, D.: *Nat. Lucht- Ruimtevaartlab. Amsterdam*, NLR-TR M2152, 1965.
479. Brown, W. F., Jr. and H. M. Jones: pp. 63–101 in Review of Developments in Plane Strain Fracture Toughness Testing, *ASTM STP* 463, 1970.
480. Heyer, R. H., and D. E. McCabe: *Eng. Fract. Mech.*, **4**:393–412 (1972).
481. Heyer, R. H., and D. E. McCabe: *Eng. Fract. Mech.*, **4**:413–430 (1972).
482. Heyer, R. H.: pp. 3–16 in Fracture Toughness Testing by *R*-Curve Methods, *ASTM STP* 527, 1973.
483. Ripling, E. J., and E. Falkenstein: pp. 36–47 in Fracture Toughness Testing by *R*-Curve Methods, *ASTM STP* 527, 1973.
484. Gurney, C., Y. W. Mai, and R. C. Owen: *Proc. R. Soc. London*, **A340**:213–231 (1974).
485. Owen, R. C., Y. W. Mai, and C. L. Chow: *Int. J. Fract.*, **12**:3–17 (1976).
486. Mai, Y. W., A. G. Atkins, and R. M. Caddell: *Int. J. Fract.*, **12**:391–407 (1976).
487. Novak, S. R.: Resistance to Plane-Stress Fracture, *ASTM STP* 591, 1974.
488. ASTM Standard E561-81: Standard Practice for *R*-Curve Determination, 1981.

Discussion

Reference 199, the standard for K_{Ic} testing, evolved through extensive theoretical and practical studies [378, 379, 401]. Methods for COD testing have been developed and described [380–382, 151]. References 210, 211, 383, 384, and 222 discuss the relationship between COD and the J integral (Sec. 7.2), the basis being experimental (in the first four) or theoretical. Experimental determination of critical J values for initiation or continued stable crack growth has been extensively reported, e.g., [212, 385, 213, 167, 205, 386, 389, 417–421]. The relationship $J_{Ic} = (1 - \nu^2)K_{Ic}^2/E$ has been essentially confirmed when the load at first cracking is well recorded, e.g., [212, 385]. Deviations may appear, however, for certain transitional dependencies of the micro modes of fracture on the macro lengths [392, 393]. References 167 and 205 consider further variables, e.g., crack-opening angles, during stable growth. Ways to derive J values from load-displacement recordings, in the spirit of Eq. (7.2), have been discussed [147–157]. In particular, attention is called to guidelines [154, 156] for certain practical uses. For specifications and evaluations of the J test consult also Refs. 387 and 388. A review of testing procedures related to COD and J is available [393].

R-curve testing within localized yielding conditions is reported and specified in Refs. 478 to 488.

The necessary thicknesses or in-plane lengths for valid tests related to K_I, J, or COD have been discussed [378, 379, 388, 389, 149, 390–392]. Some doubt is cast on the sufficiency of the dimensional requirement (7.3b) by observed anomalies [392, 393]. Further geometry effects, such as those induced by applying side grooves, have been discussed, e.g., [386, 394–397].

References 398 to 404 describe two versions of the electric-resistance method for measuring crack length, with alternating [398, 403, 404] or direct current. An improved recent technique [405–407] uses low-frequency pulsed current, with promising results. Electromagnetic effects [408, 409], acoustic emission [401, 403, 410, 411], and ultrasonic detection [401, 403, 412, 413] are also being exploited. The problem of interpreting signals has been well illustrated [213]. Reference 414 gives a total review of various methods of measuring crack length.

The multispecimen technique of determining crack initiation by a backward intercept has been described in Ref. 389 in relation to the J integral. A possible alternative is the silicon-rubber-replica method [415, 416], which can be adapted to lower temperatures [417]; thereby only a single specimen need be tested. The same goal, in principle, is achieved by the method of small unloadings [418]. Both favorable [419, 420] and unfavorable [421] experiences have been reported.

Concerned with initial cracking under dynamic load are Refs. 422–428, 441, and 456. Some [422–425, 456] use the instrumented Charpy test to estimate K_{Id}. Ref. 426 and (advocating the COD) Refs. 427 and 428 suggest ways if yielding is extensive. K_{Id} is also measured optically [441] (see caustics below). Ref. 456 presents several test results.

Dynamic toughness K_{ID} for a running crack is dealt with in numerous papers [429–433, 439, 440, 442–451, 251, 296, 273, 265]. The methods of Refs. 252 and 267 are combined with experiments [429–431], the process having been systemetized [432] and critically evaluated [433]. The basis for the method of caustics (in transmission or reflection) has been provided [434–438], and Refs. 439, 440, 442, and 443 use the method in the context of running cracks. High-speed photography is carried out essentially by techniques described in Ref. 445, also in reported photoelastic investigations [446–449, 251, 296]. While blunting of the original crack tip is a means of accumulating energy for a subsequent fast jump, e.g., [429–433], driving the crack by a wedge may provoke a more controlled, slower motion [273, 450]. Further tests involving dynamic crack growth are reported, e.g., [451, 265, 452].

Several applications of the experimental and analytical method of caustics are also known in quasi-static cases, e.g., [453, 454].

Experimental results [455, 456, 38, 457–463] are referred to in the text. Reference 464 is an important compilation of sources of material data in metal fracture. Another compilation of such data [305] contains in particular some values of the threshold intensity ΔK_t.

There are contributions [465–471] both by theory [465–467] and experiment to the prediction of cracking under nonsymmetric action. Additional analyses have been performed, e.g., [472–474]. Reference 474 definitely established an analog of Eq. (4.12) with respect to the *kinked* end of the crack, and Ref. 472 is a comprehensive review of results. Formula (7.4) originates from Ref. 466 and is discussed further in Ref. 472. Reference 473 includes continuously curved cracks and predictions of crack path and its stability. The so-called strain-energy-density hypothesis in the estimation of skew cracking, introduced in Ref. 475, is further exploited in Refs. 476 and 477, for example.

CHAPTER 8

489. Beachem, C. D., and R. M. N. Pelloux: pp. 210–244 in Fracture Toughness Testing and Its Applications, *ASTM STP* 381, 1965.
490. Electron Fractography, *ASTM STP* 436, 1967.
491. Strauss, B. M., and W. H. Cullen (eds.): Fractography in Failure Analysis, *ASTM STP* 645, 1978.
* McClintock and Argon [301].
* Tetelman and McEvily [197].
* Dieter [302].
* Broek [7].
* Hertzberg [305].
492. Beachem, C. D.: pp. 243–349 in [492*a*].
492*a*. Liebowitz, H. (ed.): "Microscopic and Macroscopic Fundamentals," vol. 1 of "Fracture," Academic, New York, 1968.
493. American Society for Metals: "Metals Handbook," vol. 9, Metals Park, Ohio, 1974.
494. Curry, D. A.: Cleavage Mechanisms of Crack Extension in Steels, in Micromechanisms of Crack Extension, *Met. Soc.–Inst. Phys. Symp., Cambridge, March–April 1980.*
495. Knott, F.: Micromechanisms of Fibrous Crack Extension in Engineering Alloys, in Micromechanisms of Crack Extension, *Met. Soc.–Inst. Phys. Symp., Cambridge, March–April 1980.*
496. McClintock, F. A.: *J. Appl. Mech.*, **35**:363–371 (1968).
497. Rice, J. R., and D. M. Tracey: *J. Mech. Phys. Solids*, **17**:201–217 (1969).
* Rice and Johnson [206].
498. Tracey, D. M.: *Eng. Fract. Mech.*, **3**:301–315 (1971).
499. Hellan, K.: *Int. J. Mech. Sci.*, **17**:369–374 (1974).

500. McClintock, F.: pp. 307–326 in A. S. Argon (ed.), "Physics of Strength and Plasticity," The M.I.T. Press, Cambridge, Mass., 1969.
501. Needleman, A.: *J. Appl. Mech.*, **39**:964–970 (1972).
502. Nagpal, V., and F. A. McClintock: pp. 365–385 in A. Sawczuk (ed.), "Foundations of Plasticity," Noordhoff, Leyden, 1973.
503. Thomason, P. F.: *J. Inst. Met.*, **36**:360–365 (1968).
504. Thomason, P. F.: *Int. J. Fract. Mech.*, **7**:409–419 (1971).
505. Willoughby, A. A., T. J. Baker, and P. L. Pratt: The Influence of Void Shape and Orientation on Ductile Fracture, in Micromechanisms of Crack Extension, *Met. Soc.–Inst. Phys. Symp., Cambridge, March–April 1980.*
506. Tvergaard, V.: *Int. J. Fract.*, **17**:389–407 (1981).
507. Eriksson, K.: *Scand. J. Metall.*, **4**:131–139 (1975).
* Otsuka et al. [203].
508. Tanaka, J. P., C. A. Pampillo, and J. R. Low, Jr.: pp. 191–215 in Review of Developments in Plane Strain Fracture Toughness Testing, *ASTM STP* 463, 1970.
509. Cocks, A. C. F., and M. F. Ashby: Intergranular Fracture during Power-Law Creep under Multiaxial Stresses, *Camb. Univ. Eng. Dept.*, CUED/C/MATS/TR 58, 1979.
510. Argon, A. S., I. W. Chen, and C. W. Lau: Mechanics and Mechanisms of Intergranular Cavitation in Creeping Alloys, in Three-Dimensional Constitutive Relations and Ductile Fracture, *Proc. IUTAM Symp., Dourdan, June 1980.*
511. Rice, J. R.: Creep Cavitation at Grain Interfaces, in Three-Dimensional Constitutive Relations and Ductile Fracture, *Proc. IUTAM Symp., Dourdan, June 1980.*
512. Wood, W. A.: Some Basic Studies of Fatigue in Metals, in B. L. Averbach et al. (eds.), "Fracture," Wiley, New York, 1959.
513. Wood, W. A.: *Bull. Inst. Met.*, **3**:5–6 (1955).
514. Cottrell, A. H., and D. Hull: *Proc. R. Soc. London*, **242A**:211–217 (1957).
515. Avery, D. H., and W. A. Backofen: Nucleation and Growth of Fatigue Cracks, pp. 339–389 in D. C. Drucker and J. J. Gilman, (eds.), "Fracture of Solids," Wiley, New York, 1963.
516. Grosskreutz, J. C.: A Critical Review of Micromechanisms in Fatigue, in J. D. Burke, N. L. Reed, and V. Weiss (eds.), "Fatigue: An Interdisciplinary Approach," Syracuse University Press, Syracuse, N.Y., 1964.
517. Plumbridge, W. J., and D. A. Ryder: *Metall. Rev.*, **14**(136):119–142 (1969).
518. Laird, C.: pp. 131–168 in Fatigue Crack Propagation, *ASTM STP* 415, 1967.
519. Schijve, J., pp. 533–534 in Fatigue Crack Propagation, *ASTM STP* 415, 1967.
520. Pelloux, R. M. N.: *ASM Trans.*, **62**:281–285 (1969).
521. Bowles, C. Q., and D. Broek: *Inst. J. Fract. Mech.*, **8**:75–85 (1972).
* Richards and Lindley [314].
522. American Society for Metals: "Fatigue and Microstructure," Metals Park, Ohio, 1979.
523. Tomkins, B.: Micromechanisms of Fatigue Growth at High Stress, in Micromechanisms of Crack Extension, *Met. Soc.–Inst. Phys. Symp., Cambridge, March–April 1980.*
524. Ashby, M. F.: Micromechanisms of Fracture in Static and Cyclic Failure. *Camb. Univ. Eng. Dept.* CUED/C/MATS/TR.51, 1979.
525. Bradt, R. C., D. P. H. Hasselman, and F. F. Lange (eds.): "Fracture Mechanics of Ceramics," 2 vols., Plenum, New York, 1974.
526. Pratt, P. L.: Micromechanisms of Crack Extension in Ceramics, in Micromechanisms and Crack Extension, *Met. Soc.-Inst. Phys. Symp., Cambridge, March–April 1980.*
527. Gilbertson, L. N., and R. D. Zipp (eds.): Fractography and Materials Science, *ASTM STP* 733, 1981.
528. Kinloch, A. J.: Micromechanisms of Crack Extension in Polymers, in Micromechanisms and Crack Extension, *Met. Soc.–Inst. Phys. Symp., Cambridge, March–April 1980.*
529. Henning Kausch, H., J. A. Hassel, and R. I. Jaffee (eds.): "Deformation and Fracture of High Polymers," Plenum, New York, 1973.
530. Composite Materials: Testing and Design, *ASTM STP* 674, 1979.

531. Harris, B.: Micromechanisms of Crack Extension in Composites, in Micromechanisms and Crack Extension, *Met. Soc.–Inst. Phys. Symp., Cambridge, March–April 1980.*
532. Derucher, K. N., pp. 664–679 in [530].
533. Ouchterlony, F.: Review of Fracture Testing of Rock, *Solid Mech. Arch.*, **7**:131–211 (1982).
534. Fatigue of Composite Materials, *ASTM STP* 569, 1975.
535. Liu, H. W. (ed.): The Fatigue of Non-Metals, *Int. J. Fract.*, vol. 16 no. 6 (1980).
536. Budiansky, B., J. W. Hutchinson, and B. Slutsky: pp. 13–45 in H. G. Hopkins and M. J. Sewell (eds.), "The Mechanics of Solids," The Rodney Hill 60th Anniversary Volume, Pergamon, London, 1982.

Discussion

Reviews of fractographic techniques and results can be found in Refs. 489 to 491 and introductions to the micromechanisms of fracture in Refs. 301, 197, 302, 7, and 305. In particular, chaps. 2 and 11 of Ref. 7 should be noted as a comprehensive survey which has guided parts of the present exposition. Further discussions of microfracture are available [492–495].

Void growth in an ideally plastic matrix has been considered [496, 497, 206, 500, 502–504]. Also, strain hardening [498, 499, 501, 506] and creep conditions [499] are allowed for. In further treatments of cavitation in creep [509–511] the event considered is intergranular, as is generally the case, with diffusion appearing as a strong mechanism [510, 511]. Interaction of voids in a cracking or failure mechanism are also discussed [206, 500–506]. Experimental data for void effects are presented in some of these papers and mainly in Refs. 7, 507, 203, and 508. Reference 508 concerns a high-strength aluminum which has a low-energy (locally brittle) fracture process even though void growth is the mechanism.

References 512 to 514 consider the initiation of fatigue cracks, and Refs. 518 to 521 deal with mechanisms of stretch zones and striations. Further discussions on the micromechanisms of fatigue fracture are available [7, 515–517, 314, 522–524], the last of which [524] is more generally concerned with the design of *fracture-mechanism maps* showing the dominant mechanisms under given loading conditions as dependent on stress and temperature.

Nonmetallic microfracture has been extensively dealt with in the literature. Some insight into this heterogeneous field is provided for ceramic materials [525–527], for polymers [527–529], for composites [527, 530, 531], for concrete [532], and for rock [533]. Coverage of fracture in ceramics, composites, and concrete can also be found in Ref. 553. Fatigue as a special topic is considered in Refs. 534 and 535, for example.

CHAPTER 9

* Broek [7].
* Carlsson [38].
537. Rolfe, S. T., and J. M. Barsom: "Fracture Control in Structures," Prentice-Hall, Englewood Cliffs, N.J., 1977.

538. Gurney, T. R.: "Fatigue of Welded Structures," 2d ed., Cambridge University Press, London, 1979.
539. Masubuchi, K.: "Analysis of Welded Structures," Pergamon, London, 1980.
540. Munse, W. H.: pp. 371–448 in [540*a*].
540*a*. Liebowitz, H. (ed.): "Engineering Fracture Design," vol. 4 of "Fracture," Academic, New York, 1969.
541. Maddox, S. J.: *Int. J. Fract.*, **11**:221–243 (1975).
542. Berge, S., and H. Myhre: *Norwegian Marit. Res.* 1, 1977.
543. British Standard Institute: Guidance on Some Methods for the Derivation of Acceptance Levels for Defects in Fusion Welded Joints, *BSI Publ. Doc.*, 6493:80, 1980.
544. Duffy, A. R., G. M. McClure, R. J. Eiber, and W. A. Maxey: pp. 159–232 in [544*a*].
544*a*. Liebowitz, H. (ed.): "Fracture Design of Structures," vol. 5 of "Fracture," Academic, New York, 1970.
545. McGonnagle, W. J.: pp. 371–430 in [112*a*].
546. Weibull, W.: *Ingenioersvetenskapakad. Handl.* 151, 1939.
547. Weibull, W.: *Ingenioersvetenskapakad. Handl.* 153, 1939.
548. Weibull, W.: *J. Appl. Mech.*, **18**:293–297 (1951).
549. Batsdorf, S. B., and J. G. Crose: *J. Appl. Mech.*, **41**:459–464 (1974).
550. Stanley, P., A. D. Sivill, and H. Fessler: *Proc. 5th Int. Conf. Expert. Stress Anal., Udine, 1974*, pap. 22.
551. Matsuo, T.: *Eng. Fract. Mech.*, **14**:527–538 (1981).
552. Jayatilaka, A. de S., and K. Trustrum: *J. Mater. Sci.*, **12**:1426–1430 (1977).
553. Jayatilaka, A. de S.: "Fracture of Engineering Brittle Materials," Applied Science Publishers, London, 1979.
554. Hunt, R. A., and L. N. McCartney: *Int. J. Fract.*, **15**:365–375 (1979).
555. McCartney, L. N.: *Int. J. Fract.*, **15**:477–487 (1979).
556. Freudenthal, A. M.: pp. 592–619 in [20*a*].
557. Shih, T. T.: *Eng. Fract. Mech.*, **13**:257–271 (1980).
558. Carlsson, J.: pp. 417–428 in L. H. Larsson (ed.), "Advances in Elasto-Plastic Fracture Mechanics," Applied Science Publishers, London, 1980.
559. Nilsson, F.: *Inst. Haallfasthetslaera, KTH Stockholm Rapp.* 19, 1975.
* Bowie [25].
560. Hasebe, N., and J. Iida: *Eng. Fract. Mech.*, **10**:773–782 (1978).
561. Gross, B., and A. Mendelsohn: *Int. J. Fract. Mech.*, **8**:267–276 (1972).
562. Rooke, D. P., F. J. Baratta, and D. J. Cartwright: *Eng. Fract. Mech.*, **14**:397–426 (1981).
563. Bäcklund, J., and S. Sjöström: pp. 787–797 in A. R. Luxmoore and D. R. J. Owen (eds.), Numerical Methods in Fracture Mechanics, *Proc. Int. Conf., Swansea, 1978.*
564. Lotsberg, I., and P. Bergan: *Eng. Fract. Mech.*, **12**:33–47 (1979).
* Irwin [140].
565. Irwin, G. R.: pp. 204–229 in E. R. Parker (ed.), "Materials for Missiles and Spacecraft," McGraw-Hill, New York, 1963.
566. Darlaston, J. L., and R. P. Harrison: pp. 165–172 in P. Stanley (ed.), "Fracture Mechanics in Engineering Practice," Applied Sciences Publishers, London, 1976.
* Folias [72–74].
* Erdogan and Kilber [75].
* Erdogan [76].
567. Dowling, A. R., and C. H. A. Townley: *Int. J. Pressure Vessels Pip.*, **3**:77–108 (1975).
568. Harrison, R. P., and I. Milne: pp. 69–82 in P. Stanley (ed.), "Fracture Mechanics in Engineering Practice," Applied Sciences Publishers, London, 1976.
569. Harrison, R. P., K. Loosemore, I. Milne, and A. R. Dowling: *Cent. Elec. Generating Board Rep.* R/H/R6-Rev. 2, 1980.
570. Chell, G. G.: pp. 581–605 in Elastic-Plastic Fracture, *ASTM STP* 668, 1979.
571. Milne, I.: *Mater. Sci. Eng.*, **39**:65–79 (1979).

572. Burdekin, F. M., and M. G. Dawes: pp. 28–37 in *Proc. Inst. Mech. Eng. Conf. Pract. Appl. Fract. Mech. Pressure Vessel Technol. London, 1971.*
* Broek [478].
573. de Koning, A. U.: *Nat. Lucht- Ruimtevaartlab. Amsterdam* NLR MP 75035 U, 1975.
* Lotsberg [166].
* Kanninen et al. [167].
* Shih et al. [205].
* Rice et al. [120].
* Kanninen et al. [220].
574. Lereim, J.: *Eng. Fract. Mech.*, **18:**703–715 (1983).
575. Gurney, T. R.: *Weld. Inst. Res. Rep.* 88/1979, Cambridge, 1979.
576. Gurney, T. R.: The Influence of Thickness on the Fatigue Strength of Welded Joints, *2d Int. Conf. Behaviour Off-Shore Struct., London, 1979.*
577. Pook, L. P.: *Nat. Eng. Lab. Rep.* 561, Glasgow, 1974.
578. Berge, S., and K. Engesvik: *Proc. Int. Conf. Steel Marine Struct. Paris, 1981*, pap. 2.5.
* Berry [225].
579. Hasselman, D. P. H.: pp. 89–103 in W. W. Kriegel and H. Palmour (eds.), "Ceramics in Severe Environments," vol. 5 of "Materials Science Research," Plenum, New York, 1971.
580. Hasselman, D. P. H.: *J. American Ceramic Society*, **11:**600–604, (1969).
581. Rockley, D. P.: "An Introduction to Industrial Radiology," Butterworths, London, 1964.
582. Krautkrämer, J., and H. Krautkrämer: "Werkstoffprüfung mitt Ultraschall," Springer-Verlag, Berlin, Heidelberg, New York, 1980.
583. Birnbaum, G., and G. Free (eds.): Eddy-current Characterization of Materials and Structures, *ASTM STP* 722, 1981.
584. Williams, R. W.: "Acoustic Emission," Adam Hilger Ltd., Bristol, 1980.
585. American Society for Metals: "Case Histories in Failure Analysis," Metals Park, Ohio, 1979.

Discussion

The general role of fracture mechanics in the context of structural design and safety is illustrated in part II of Ref. 7, chap. 2 of Ref. 38, and Refs. 537 and 538. Practical problems related to weldments have been discussed [538–542], with emphasis on residual stresses [539]. Guidelines for estimating acceptance levels for defects have been proposed [543]. The design of pressure vessels and piping [544] is considered with later innovations in current issues of the *Journal of Pressure Vessel Technology* and the *International Journal of Pressure Vessels and Piping*, among others. Nondestructive test methods are reviewed in Refs. 7, 414, 545, and 581 to 584.

Systematic studies of fracture in statistical terms have been undertaken [546–548], with further contributions toward the inclusion of multiaxial stresses, crack parameters, and nonuniform fields [549–556]. References 556 to 558 are review articles, Refs. 556 and 557 also discuss the Weibull theory, and Ref. 553 is a simple introduction to that topic as well as to certain implementations of fracture-mechanical concepts. Well advanced in the latter respect is a report [559] concerned with failure probabilities under alternating loads.

The solutions discussed in Sec. 9.2 for cracks emanating from a circular hole and from a V notch were obtained in Refs. 25 and 560, respectively. The V notch without a crack is considered in Ref. 561. A comprehensive review of methods to determine stress intensity factors in general is available [562]. Numerical analyses of crack growth under cyclic load [563, 564] were presented in Sec. 9.3.

The leak-before-break criterion is discussed in several places, e.g., [7, chap. 15; 38, chap. 12; 544, 140, 565, 566]. Effects of the type represented by the footnote equation in Sec. 9.4 are considered in Refs. 72 to 76.

The failure-assessment diagram (Sec. 9.5) is introduced in Ref. 567 and discussed in Refs. 568 and 569. Extensions to include thermal and residual stresses [570] and some stable crack growth [571] in the assessment have been suggested. These are all semiempirical tools, loosely referred to the COD criterion, as is an earlier approach, the COD design curve [572].

Recordings [478, 573] and evaluations [573] of stable crack growth in thin sheets have been presented, and the process has been numerically simulated by nodal release [573, 166]. Arguments in favor of the CTOA as a material constant during established growth are implied in Refs. 167, 205, and 121. A two-parametric approach combines this description with a J-controlled initial growth [220].

The present Sec. 9.7 was inspired by an extensive discussion of thickness effects [574]. Investigations pertaining to this subject have otherwise been presented, e.g., [575–578]. Fatigue is dealt with specifically in Refs. 576–578. Emphasis is on numerical analysis [575, 577], experimental evidence [576], or both [578].

References 579 and 580 analyse cracking through severe thermal action making use of using the type of assessment advocated in Ref. 225. A discussion pertaining to the present problem of residual-stress-induced cracking will be found in chap. 10 of Ref. 539. That reference contains a wealth of material related to weldments and their fields of temperature, stress, and displacement.

Attention is finally called to a compilation of post-failure studies [585] which presents detailed metallographic evidence and discuss reasons for the total fracture.

INDEX

INDEX†

†Names are entered here only as related to particular topics of the text. Otherwise the Bibliography may be consulted.